T0184827

HANDBOOK OF THE
RUBI OF GREAT BRITAIN
AND IRELAND

Ac ne forte roges quo me duce, quo lare tuter,
Nullius addictus iurare in verba magistri.
HORACE: Ep. I. I.

HANDBOOK OF THE
RUBI OF GREAT BRITAIN
AND
IRELAND

BY THE LATE
W. C. R. WATSON

WITH 50 DRAWINGS BY
RUTH M. BALL & A. W. DARNELL

CAMBRIDGE
AT THE UNIVERSITY PRESS
1958

CAMBRIDGE UNIVERSITY PRESS
Cambridge, New York, Melbourne, Madrid, Cape Town,
Singapore, São Paulo, Delhi, Mexico City

Cambridge University Press
The Edinburgh Building, Cambridge CB2 8RU, UK

Published in the United States of America by Cambridge University Press, New York

www.cambridge.org
Information on this title: www.cambridge.org/9781107642294

© Cambridge University Press 1958

This publication is in copyright. Subject to statutory exception
and to the provisions of relevant collective licensing agreements,
no reproduction of any part may take place without the written
permission of Cambridge University Press.

First published 1958
First paperback edition 2013

A catalogue record for this publication is available from the British Library

ISBN 978-1-107-64229-4 Paperback

Cambridge University Press has no responsibility for the persistence or
accuracy of URLs for external or third-party internet websites referred to in
this publication, and does not guarantee that any content on such websites is,
or will remain, accurate or appropriate.

FOREWORD

WHEN the author died in 1954, he had nearly completed his work and only a few trivial emendations and additions have been made. All the descriptions were complete with the exception of the section *Appendiculati* which has been prepared from his account of the series and subseries. In the analytical key there was no method of arriving at the sections, subsections, series and subseries. As it seemed undesirable that the text should have to be consulted to arrive at the appropriate group, the group descriptions have been extracted from the text and incorporated into the key. Under the nomenclature only dates had been quoted, but it seemed desirable to have full references. This has been done with the exception of the group headings. No authority or reference for the group headings, many of which are new or have new status, had been given by the author, and it has only proved practicable to give basonym authorities. All new names and combinations have been published separately in a paper in *Watsonia*. All the type material from Watson's herbarium has been deposited at the British Museum (Nat. Hist.).

It is believed that the author would have wished to thank Miss Ruth M. Ball and Mr A. W. Darnell for drawing the illustrations; the authorities of the British Museum (Nat. Hist.) and the South London Botanical Institute for the use of specimens and books; and the many fellow botanists for help received during his life-long study of the genus *Rubus*, which has resulted in this, his most considerable work.

P. D. SELL

J. E. WOODHEAD

April 1957

v

PREFACE

Rubus is a remarkable genus in several respects. It has spread to all the continents; it shows signs of decay in its reproductive system whilst it is still in process of evolution, and so offers an opportunity of observing the origin of species. The best approach to a study of the genus is to try to identify the brambles met with, and in the endeavour, knowledge comes. The present work includes descriptions of all the forms found in Great Britain, Ireland and the Channel Isles, three-quarters of which are also found on the Continent.

On account of their relatively slow methods of distribution, their hardiness and the consequent preservation of old forms, the Rubi afford perhaps an unrivalled instrument for ascertaining the movement and succession of florulas and floras, if they are studied in connexion with the climatic and geological changes of the past.

If they may be comparatively unknown to most persons, they are nevertheless no strangers to any of us. From their wild habitats—woods and heaths, mountains and moors—they come to our hedges and garden fences, offering their flowers, like little roses, and then their fruit to our regard, after the cultivated bush fruits are all over. They readily accept the hospitality of a garden, and if taken in hand will be found, some of them, to yield delicious fruit if they are protected from the thrushes and allowed to ripen.

The principal object of the present work is to furnish the means of identifying any native species met with, and the means of understanding the causes and types of variation, and in some measure the probable natural relationships of the species. It has to be remembered, however, that the study of Rubi is still in an early stage of investigation, and there is room for much pioneering work in many directions.

A conspectus of species is prefixed to the descriptive part of the work, analysed to small ultimate groups of species, which are placed in order of frequency within their group. In addition the very common species are distinctively marked, and those other geographically restricted species are also suitably indicated. It is believed that these and other novel aids adopted in the main descriptions will speed identification and serve better than any other scheme.

Liberty has been taken to be very brief when brevity suffices; and the descriptions are never longer than is necessary. The normal style and order of description has every now and then been set aside, in the case of a common species, for a summary of the more striking characters. Interest languishes under a surfeit of routine analyses.

Experience will soon show that there is no 'hopeless wading through hundreds of descriptions', nor 'endless intermediates'; that one never comes across the fabled 'swelter of forms' or 'swarms of hybrids'; and that there is no need to use a microscope, nor to be an ecologist, a geneticist or a cytologist in order to come to the name and systematic position of any blackberry. This is not to say that there is not wide scope for the exercise of those useful modes of investigation.

The systematic arrangement of this work does not closely follow any previous author: it is partly on traditional lines, and partly an original arrangement adapted to suit the British species. In the matter of identification, although all the points of view have been considered that I could learn of, and all works consulted, general and local, British and western Continental, that I could manage to see, I have ended by following my own counsel, adopting nothing on the authority of anyone else.

In the account of the distribution of British and Irish species, because of the uncertainty of many of the older records I have relied almost entirely on my own determinations and my own fieldwork, carried out largely in vice-counties 1–5, 9, 11, 13–18, 20–22, 24, 27, 30, 36, 39, 40, 47, 49, 62 and H. 1 and 2, and abroad in south Belgium and west Germany around Aachen. Apart from this I have grown at least 300 of the British forms in my garden within the past forty years.

Since Rogers' Handbook was published in 1900 many species not before known for Great Britain and Ireland have been brought to light, and have increased the number from 200 to nearly 400. They are species that have mostly been already described on the Continent. Of the endemic species that have been added, some were collected and distributed in the last century, under wrong names; others were allowed to drop out owing to mistaken ideas about their systematic value, or because they had become extinct in their original station and no other was known. However, all of them are extant, many of them widely or very widely distributed, and they are important for the understanding of the evolution and history of the west European bramble flora. They have accordingly been incorporated in their place in the present Handbook.

It is pleasing to note that some of the species recently identified were already being collected by the fathers of British batology 100 and 140 years ago, contemporaneously with Libert and Lejeune, Weihe, Kaltenbach, Questier, Lefèvre and Mueller on the Continent. References in the distribution records commemorate some of these early British discoveries.

W. C. R. WATSON

BICKLEY, BROMLEY,
KENT
March 1954

CONTENTS

INTRODUCTION

> The analysis of polymorphic groups of forms must precede experimental in-
> vestigation, so as to give a definite direction to experiments....It is therefore
> my conviction that a painstaking investigation of the smaller groups of forms
> is an indispensable task that should precede all studies on the nature of organic
> species.   W. O. FOCKE (1877)

IT is so long since a work on *Rubus* appeared at home or abroad that
there is perhaps small wonder that with the march of knowledge considerable changes should have taken place in the outlook.

Until the end of the eighteenth century the RUBI of this country were
by all botanists brought under the six Linnaean species, *Rubus arcticus,
idaeus, Chamaemorus, saxatilis, caesius* and *fruticosus*, of which the last-
named included all European blackberries. Then when blackberries
began to be compared with one another differences were found, and
a few workers turned to analysing, describing and naming them.
Arrangements were drawn up, intended to enable the brambles to be
identified rather than to reflect their natural affinities and descent. In
about 1860 and onwards much thought was given to the latter aspect,
mainly from a morphological point of view. About the turn of the
century genetic experiments designed to throw light on the nature of
RUBI were begun in earnest, and cytological studies were commenced
about 1920. However, we are unfortunately as yet without a comprehensive fact-finding review from this angle that could be of any assist-
ance for systematics. Since Lidforss, there does not seem to have been
any geneticist or cytologist who has had more than the barest acquaintance with the group in nature: there has been in consequence no
planned series of investigation.★

Vaarama has, however, provided proof that both *R. caesius* and
R. saxatilis have unlike pairs of sets of chromosomes, which is to say
that they must be of hybrid origin: so those who accept these Linnaean
species—and who does not?—can no longer consistently reject the
blackberry tetraploids, which also are of hybrid origin. So much must
be obvious; but the meaning of what is obvious may sometimes be
missed: it is that the blackberry tetraploids, as well as *R. caesius* and
R. saxatilis, show the way of origin of the species; they are not merely

---

★ This must surely have been written before Watson had studied the work of
Gustafsson. (P.D.S. and J.E.W.)

hybrids, but hybrids plus something (as the late A. J. Wilmott wrote to me a few years ago concerning *R. saxatilis*), something that has gone into their production, whatever that something is, besides mere hybridization.

In many instances species were described and named when found growing in a single spot or a limited area; and some systematists, concerned at the mounting number of names, gave way to the temptation to reduce them arbitrarily to varieties and subspecies and severely to limit the number of their species. But varieties and even subspecies they are not, in the strict sense of these terms—they differ in form throughout. Other authors, with more reason, retained them as species, but created different grades of species to accommodate them and to express their views on their systematic rank, taking into account the extent of the area of distribution then known to them. A good many of these supposed local forms have since been found in other countries and not always in small quantity. *R. euryanthemus* is a case in point. At first mistaken for *R. fuscus*, then ranked as a variety of *R. pallidus*, it was in 1900 known only in south and mid England and Schleswig. Now it is known in the Hebrides and, under the name of *R. Schleicheri* microg. *chloroxylon* Sudre, from south to north France, Belgium, east to west Germany and Switzerland.

Very many British brambles have been grown from seed or rooted scions, and very many species have been studied upon a large number of bushes in nature. Minor variations have often been noticed of a recurring nature running through a proportion of the plants, but not such as to obscure their specific identity. Anything in the least resembling the alleged 'hybrid swarms' has never been observed in blackberries in nature, or in cultivation, other than in deliberate crosses in their second generation. But great and promiscuous is the mixture that occurs in nature in the case of secondary crosses of the Linnaean species *R. caesius*, the Dewberry, in damp places at very low altitudes especially on post-glacial soils. It is also perfectly true that the two diploid species *R. ulmifolius* and *R. tomentosus*, both of unassailable specific rank, are just those that offer the greatest variability, are the most prone to hybridizing, and are alleged by one cytologist to be self-sterile. On what an unstable footing then does the old conception of the genus rest!

By and large the tetraploid blackberries prove to have the same ability to reproduce their kind as truly at least as do the other sexually breeding species of the genus.

Introduction

It is common knowledge how confused and disputed the identification of RUBI in Britain has become in the last two or three decades. This has been due in part to the acceptance of identifications for a hundred years past from secondary and not from the original authorities for the species, and in part it must be added from insufficient acquaintance with the individual species in all their living aspects to enable sound determinations to be made, often from a dried specimen, and that improperly selected and dried. As a contributory cause also the peculiar belief or prejudgement has prevailed in some quarters that a species must be multiplex—a concept arising apparently from a conviction that a genus must not contain a large number of species.

Focke in his last work (1914) chooses seventy-one principal species and then, as he points out, using the freedom which a prodromus permits, assigns to this, that and the other many of the remaining European species, known or unknown to him, and omits the rest. Not that he assigns the non-principal species all as synonyms to the principal species, although he does so treat a certain number of them; he reverts to his early idea of instituting minor species of lower ranks in subordination to his principal species, calling them now 'forms', 'allied forms', 'prospecies' and 'subspecies'. Whatever may be the real value of this provisional, but his last, arrangement, it is at any rate certain that no one could trace the position in it of many of the species that he knows, except through the index.

Remembering that it is the duty of a scientist to expound nature as it is, and not to confound it by setting it forth as he would wish it to be, it seems right to keep separate what one has found by every test to be separate in nature; and by defining the species to the best of one's ability to furnish a secure basis for future work which shall have the aid of genetic and cytological investigations. That is in fact the purpose that I have kept before me: let a careful and complete analysis precede any attempt at a comprehensive synthesis. Already, however, the conception is emerging that the British bramble flora comprises not only a mosaic of distinct florulas, various in age and in composition, but also, systematically considered, comprising small disjunct groups of few related forms, and not few groups of many small forms which run one into another. To devise arrangements that bring these related species together, and at the same time serve the function of an easy 'key', has been the problem.

I. Environmental variations

There is no escape: since we cannot be the masters of Nature, nor create the
plants afresh after our own idea, we must submit to the laws of Nature, and by
intelligent application learn the characters that are inscribed upon plants.

LINNAEUS, *Genera Plantarum* (1764, 6th ed.)

One does not know much about brambles, until one has mastered the
range of changes a bramble plant undergoes due to age, season, climate,
exposure, soil, habitat and so on; neither is it possible for a given bush
or specimen to be named intelligently by comparison with another
specimen or figure unless one knows how to allow for these modifying
influences. One may instance particularly the deep shade of woods, an
excess or deficiency of soil moisture, atmospheric moisture, a spell of
hot weather, a run of cool rain, a shallow soil overlying rock or iron
pan, exposure to salty sea breezes, a frost in May, lopping in hedgerows.
The effects of these factors are discovered by observation of the
brambles growing under such conditions. That they are of practical
concern a few instances may prove.

R. carpinifolius sometimes grows in marshes and is then apt to make
lax panicles and somewhat laciniated leaflets. Such a specimen was
named *R. leucandrus* by Focke, and another *R. carpinifolius* var. *laxus* by
Sudre. *R. lacustris* Rogers found in Lakeland is a wet-soil form of
R. Lindebergii. *R. vestitus* growing in woods so far deceived Babington
that he published it as a new species, *R. Leightonianus*, and was corrected
by Leighton. Salter was only stopped by Borrer from describing woodland *R. vestitus* as a new species, *R. rotundifolius*. *R. ericetorum* var.
cuneatus Rogers & Ley is the same as *R. ericetorum* of Rogers. *R. ericetorum* ssp. *sertiflorus* var. *scoticus* Rogers & Ley is the same as *sertiflorus*
of Rogers. *R. botryeros* Focke, at first associated with *R. longithyrsiger*
by both Focke and Rogers, is the same as *R. oegocladus* of Rogers.
Rogers repeats *R. dumnoniensis* Bab. under the name of *R. cariensis*
Rogers (not Genev.). Sudre has this bramble under four names. The
bramble which Rogers has as *R. Borreri*, Sudre has figured under that
name and again under *R. retrodentatus*; he has described it also as
R. Schmidelyanus var. *breviglandulosus*.

These faults are due to remissness in observing the behaviour of the
species in different environments.

Turn now to the effect of temperature on flower colour. In this
country I have examined thousands and thousands of the flowers of
R. separinus. The petals are constantly pink or pale pink in the opening

4

bud; the styles are usually greenish but sometimes pink-based. I have found in Kent that the styles come rosy or red in new flowers opening after a run of hot dry weather. Genevier, who is the author of *R. separinus*, says that the petals are white, the styles rose, violaceous or red. Sudre, who collected in Genevier's station, says that he found the petals sometimes pinkish and the styles flesh-coloured. Well knowing the bleaching effect of strong sunlight I opened an advanced bud on one of Genevier's specimens and found the petals were *pink*. (Genevier says that the carpels are *glabrous*; on his specimens I found them *pilose*. Sudre says they are *glabrous*; on his specimens I found them *pilose*. Bouvet says they are *glabrous*. They are invariably *pilose* on British specimens.)

The foregoing illustration relates to the effect of settled, hot, dry weather. The following note by Focke on a sheet of *R. rhamnifolius* collected in his garden at Bremen in the summer of 1885 shows the opposite effect of rainstorms: 'Styli in solo arido et sub coelo sereno rubentes, post pluvias vero in eadem planta virentes.'

I have myself seen similar temperature and rain effects on *R. plintho-stylus* growing in my garden.

Where petals or stamens have been bleached after the flower has opened, the colour can often be seen remaining unaffected on the claw of the petals or the base of the filaments.

There also seems to be sometimes a colour change in the reverse direction. Petals and stamens which had every appearance of being pure white when growing, turn definitely light pink on being dried in the press.

II. Genetic intraspecific variations

Certain variable characters are found that are not known to be affected by any environmental condition and are presumably under genetic control. They are of a rather minor nature, and their presence or absence does not obscure the identity of the species. They are not due to hybridization; that is, they are not directly due to any fresh hybridization. Their appearance or disappearance in a new generation is apparently due to a recombination of genes that regularly takes place prior to the formation of gametes.

In some species one or other of these characters seems to have become fixed, and may be reckoned as one of the distinguishing characters of the species. In other species the same character may occur sometimes, not always. In yet other species it may not have been met with at all. For instance, pilose anthers occur always in *Balfourianus, danicus, Drejeri, gratus, mucronifer* and *sciocharis*; often in *affinis, plicatus, regillus, silvaticus, vestitus* and *Winteri*; never in *adscitus, cardiophyllus, foliosus, pallidus, rudis* and *Sprengelii*.

A survey is needed to learn in which species the following characters appear, and whether always.

1 *a*. Leaves 3-nate to 4-nate-pedate or 5-nate-pedate.
1 *b*. Leaves 3-nate to 5-nate-digitate or 5-nate-subdigitate, never 6-nate or 7-nate.
1 *c*. Leaves 3-nate to 5-nate-digitate and sometimes 6-nate or 7-nate.
2. Number of sepals and petals where the number anywhere exceeds 5.
3. Petals glabrous on the margin, or pilose.
4. Anthers pilose; always?
5. Carpels glabrous; pubescent; pilose; bearded.
6 *a*. Receptacle glabrous; or pilose.
6 *b*. Receptacle hirsute at the base with a brush of hairs protruding below the lowest carpels all round.

To observe the bush, the best way is to take a freshly opened flower and thumb down the nearer stamens outwards. This will give a clear view of the young carpels also. Receptacular hairs often grow out between the carpels and must not be mistaken for hairs on the carpels. To remove any doubt it is wise to take off a few of the upper carpels for examination; this will also expose the receptacular hairs clearly.

A good deal of information has been collected under the above heads, and some of it is included in the descriptions. Until it is much more complete it will be unsafe to begin to draw conclusions. In time it can be used to throw light on the method of seed production in particular instances, and perhaps on the relationships existing between the species concerned.

III. Chromosomes and genes

Wherever known, the chromosome number is shown in the description. In some cases counts have been made only on one bush; the chromosome number is then starred. There is a possibility that a different number will be found when a second or third bush is examined in these cases, as where two bushes have been examined two different numbers have several times been obtained.

Certain results are, however, already sufficiently clear, namely (1) there is only one basic number, 7; (2) the chromosomes are found always in exact multiples of the basic number, with exceedingly rare exceptions; (3) the prevailing number is 28 in body cells, 14 in germ cells.

Three dwarf seedlings which appeared in a family raised from *R. thyrsiger* self-pollinated each had 27 chromosomes: it was claimed that they represented the systematists' var. *parvifolius*. These dwarfs, however, proved to be comparatively infertile, whereas natural dwarfs are quite fertile. Moreover, several different natural dwarfs have now been studied and have been found to possess 28 chromosomes each. The implication, then, based on the *thyrsiger* seedlings refers only to that case; the cause of the dwarfing in the other cases remains unexplained. It is also not known whether dwarfing is a reversible condition.

Different opinions prevail as to whether the forms of chromosomes can be distinguished well enough to throw light on the origin and relations of species in *Rubus*. In *R. arcticus, idaeus, caesius, saxatilis,* and hybrids artificially created between some of those species, the chromosomes have been studied, distinguished, delineated, and conclusions have been drawn from their form; their pairing behaviour in meiosis has also been established; but there is a difference of opinion between cytologists as to whether allotetraploidy or autotetraploidy is indicated in the parent species concerned, *R. caesius* and *R. saxatilis*. No doubt in course of time the doubt will be resolved to the satisfaction of cytologists. From a morphological point of view it seems very probable that *R. caesius* and *R. saxatilis* are of hybrid origin, and that one parent in each case is *R. idaeus*.

As to the blackberries (MORIFERI), one cytologist urges that the morphology of the chromosomes is of outstanding importance and that every effort should be made to determine their forms. Against this it has been advanced more than once that the chromosomes are too small and too vaguely defined to make it likely that they can be of much use in this connexion. When one compares the different drawings that have been published of *R. caesius* chromosomes, the idea occurs that possibly the methods of fixing, staining and cutting may be different and may be the cause of the discrepant results, rather than that, as has been suggested, the chromosomes are really different in the different plants of *R. caesius* examined. One notes also that the unstainable centromere, which is probably differently located in the seven chromosomes and should help to differentiate them, is not drawn.

An additional method of recognizing allo- from autopolyploidy has been recommended by one cytologist, namely a morphological study of the bramble itself.

Improvements of technique and further consideration of the problem will perhaps bring enlightenment, and we may turn away from these aspects and see what relation is borne to the systematy of brambles by the number of chromosome sets found to be present. (Note that **x**=7 chromosomes.)

1. Diploids (**2x**). *R. ulmifolius, R. idaeus* and *R. arcticus* are the only British species concerned. *R. tomentosus* of the Continent is also **2x**. A **2x** seedling was obtained from *R. macrothyrsus*; the other seedlings of the same family were **4x**.

2. Triploids (**3x**). *R. nitidus*, belonging to SUBERECTI. *R. hylophilus*, belonging to CANDICANTES. *R. thyrsanthus* in the same group is usually **3x** on the Continent, but here **4x**, one bush. *R. Braeuckeri*, SPRENGELIANI. *R. rotundatus*, SYLVATICI, remarkable as being the only morifer known to have three satellite chromosomes.

3. Tetraploids (**4x**). The great majority of blackberries belong here.

4. Pentaploids (**5x**). Eight species of MORIFERI have been found to be **5x**, in four of which the number **4x** has also been obtained. Three varieties of *R. caesius*, and 11 out of 14 of its derivatives (TRIVIALES), were found to be **5x**, in two instances also producing a **4x** form.

5. Hexaploids (**6x**). Four species of MORIFERI have been found to be **6x**, in two of which the number **4x** has also been obtained. One of the TRIVIALES was found to be **6x**. The hexaploid *R. magnificus* included above also furnished a **3x** seedling. An odd **6x** seedling also appeared in a family of *R. pyramidalis* seedlings.

It remains to be seen what explanation the cytologists will wish to give of the causes and effects of these dual chromosome forms which occur in the same species and more than once in the same family of seedlings, and which do not disturb specific identification. It seems that they can be ruled out from having played any part in the origination of new species or even varieties; although they will probably be found to affect fertility and resistance, and therefore, in the long run, prejudice the survival of the chromosome form.

Perhaps it hardly needs to be mentioned that the phenomenon is widespread in other genera, as *Ranunculus Ficaria* **2x**, **3x**, **4x**, **5x**, **6x**, and *Caltha palustris* **4x**, **6x**, **7x**.

The form of the chromosome depends partly upon the number of clusters of genes arranged along the chromosome. These are stainable microstructures which are held to be associated with definite characters in the phenotype; they work together as crews with neighbouring crews, exerting their influences collectively. The genes are liable to change over in a block from one chromatid to another in synapsis, either as a one-sided move or as a mutual exchange, or they may undergo a reversal of order in a segment of their own chromatid. Such moves break the existing liaisons and reconstitute the co-operative, collective effects or products of their interactions, with corresponding results upon the future phenotype. Those results also seem to be compound, affecting more than one character at a time. It is rather singular that in *Rubus* more seems to be known, or can be predicated, of the genes, and of their operation—although the genes are ultra-microscopic—than seems to be known of the chromosomes themselves.

The effects of the genes are probably exerted through the cytoplasm, which is itself open to influences of the environment, judging from the changed effects upon the phenotype that result from changed external conditions. Causes of variation of different kinds are thus proceeding continually from the interaction of these two agencies; and the risks of coming to conclusions upon the observation of one agency without the other will be evident.

IV. Reproduction

R. Chamaemorus (**8x**) is dioecious. It has been observed that where the male plant is absent the female plant does not set seed. Where male and female plants are together, seed is set. Sexual reproduction is thus clearly indicated.

When the styles of a bramble, e.g. *R. procerus* or *R. caesius*, are cut off, fruit is not produced; but when flower buds of either of the same two species are enclosed in a muslin bag, in late summer and in the shade, the flowers open and set fruit, self-pollinated. Thus, for fruit-production pollination is necessary, which is not the case for clone formation.

Many species successfully cross together, whether fairly closely related as *R. ulmifolius* and *R. subinermoides* or remotely related (if at all) as *R. plicatus* and *R. Bellardii*. They may give a fertile hybrid intermediate between the parents, and may have pollen and fruit as good as or better than those do; or they may be infertile and propagate solely by rooting stem-tips. Such intermediates are found sometimes in nature growing in the presence of the parent species. Examples are given under R. *ulmifolius* and R. *propinquus*. All such plants, both parents and offspring, are clearly not clone-producing nor clone-produced.

Whoever observes brambles closely in nature, especially when they are in flower, will be aware that variations in the same species are frequent. Clones do not produce variations.

The claim has been made by cytologists and others that many species of *Rubus* are clones, and indeed that this is the usual method of reproduction except in diploids. The proof offered is that in R. *nitidoides* and R. *thyrsiger* the development of seed by apospory has been noted cytologically. A cell of the nucellus has budded and pushed into the embryo sac and, without the nucleus having undergone meiosis, has developed into a false egg.

But it has been observed in the same two species that two daughter cells (**2x**) in the normal embryo sac united again after having undergone meiosis. In this state (**4x**) they could, either with or without union with a sperm nucleus, grow into a true seed and pass on any variation acquired before meiosis. But no egg grows into a seed until the central nuclei in the embryo sac that form the endosperm have been fertilized by one of the two male generative nuclei proceeding from the pollen tube. Endosperm is apparently required for the growth of the egg. This is of course a sexual process, and it touches off the spontaneous development of the normal egg cell, the second male generative nucleus taking no part. Being thus only a unisexual process, it is termed pseudogamy. The egg so constituted (**4x**) carries the characters segregated in meiosis, and is not distinguishable from a bisexually formed egg (**4x**) in form or function. This agrees with the conclusion arrived at in the cytological investigation to which I have referred: 'We may regard apomictic processes which allow for segregation as subsexual, although the progeny are wholly maternal.' Another conclusion was that 'apomixis is an escape from sterility'.

It seems clear that a theory of sterility and a prevalent accompaniment of clone reproduction does not account for the state of things met with in *Rubus*: allotetraploidy, implying past and present hybridization which must be sexual intraspecific variation within limits; frequent high fertility; sterility in female plants where male plants are absent, and in bisexual plants when pollination is prevented. Neither need it be supposed that pseudogamy represents the normal method of reproduction in nature. The experiments demonstrating pseudogamy consisted, without exception, of hybridizations. Two alternatives were presented to the egg nuclei; cross-fertilization or spontaneous development without fertilization. In the former case a hybrid would result, in the latter case the character of the maternal type would be transmitted unchanged. It was not a case of an 'escape from sterility', but of an escape from a threat to the purity of the type. The natural method of reproduction in *Rubus* is by self-pollination or by pollination by another bush of the same species. This was how the three dwarf seedlings were raised from *R. thyrsiger*, as well as several other seedlings with imbricate leaves, variations which proved that neither clones nor apomixis had anything to do with the matter.

Allotetraploids or autotetraploids. The determination of these two states calls for clarification by cytologists. The case of *R. caesius* may be referred to in illustration. It seems to be agreed that the pollen is about 90 per cent perfect (90–100 per cent, Lidforss), and that mostly only bivalents are formed in meiosis, although one and rarely two quadrivalents may be formed. On morphological grounds alone it seems certain that *R. caesius* is of hybrid origin. The chromosome morphology points decidedly the same way; and high pollen fertility is not out of place for an autotetraploid. This species has been determined by one cytologist as an allotetraploid, and by another as an autotetraploid.

R. nitidoides in a **4x** state with 50 per cent good pollen is said to show five or six chromosome quadrivalents in meiosis, and this has been said to be autotetraploid. I think it is quite certain by morphological analysis that *R. nitidoides* is of hybrid origin from *R. gratus* crossed, probably, with *R. carpinifolius*.

A high degree of multivalency in meiosis in a tetraploid hybrid, *R. nitidoides* × *R. thyrsiger*, has been held to point to near affinity between the two parents. But *R. nitidoides* and *R. thyrsiger* lie systematically very far apart, and it is very doubtful whether they have anything at all in common. Some radical reconciliation of differences between the chromosome sets would seem to have been effected in the resultant hybrid. That it is partially and functionally an autotetraploid, as claimed, may be the case.

Hybrids: origin of species. When one turns from this baffling pairing behaviour to consider what kind of adjustments take place in the form of chromosomes when they meet in a hybrid combination, one is on slightly firmer ground, thanks to Vaarama's investigations on *R. idaeus*, *arcticus*, *saxatilis*, *caesius* and some of their hybrids. The pertinent facts may be summarized as follows:

	Satellite chromosomes	*A*-chromosomes
caesius 4**x**	2	2
saxatilis 4**x**	2	2
idaeus 2**x**	2	—
arcticus 2**x**	2	I
caesius × *saxatilis* 4**x**	2	I
caesius × *idaeus* 3**x**	I	I
saxatilis × *arcticus* 3**x**	I	—

In all **2x**, **3x** and **4x** species examined two satellite chromosomes were found (three in one **3x** species). How to explain the disappearance of some of the satellites in the formation of the **4x** species seems impossible, but, like the change of form of some of the *A*-chromosomes when introduced into a cross so that they can no longer be identified, it serves to bear out that a hybrid may not be merely a recombination of existing chromosomes or characters, but makes changes and achieves a new identity for itself. The homologies of the chromosomes are probably worked out and improved within the range of the material present. It seems very doubtful in the circumstances whether judgements about auto- or allotetraploidy based on the supposed homologies of pairing chromosomes can have any validity.

The usual assumption that diploids have good pollen and are very fertile, whilst tetraploids are the reverse, is unsound; but it seems to have been proved that triploids have nearly always very poor pollen.

V. The species in *Rubus*

Rubus: Inquirendum est in hybridam specierum originem!
L. REICHENBACH, *Fl. Germ. Excurs.* (1830–2), 599

It is the duty of the exploring botanist to expound, not to confound, nature: to bring to light what he meets with, not to impose his own ideas about what ought to be there. Consequently he should not shrink from accepting the fact that, despite heredity, variations occur and lead to the production of new types.

The criteria of a species in *Rubus* are that it possesses constant and inheritable characters running all through the plant and distinguishing it from all other species. Alongside this assemblage of characters, species are liable to vary from one generation to another in certain characters of a minor kind that are often found also in related species.

This faculty of variation does not invalidate the conception of a true breeding species. On the contrary, it is provided for and ensured in the mechanism of reproduction; it may be regarded as an asset, not a disability to the species. As in other groups the species of *Rubus* differ in size, that is in degree of distinctness; they may be different in mode of origin and in mode of reproduction. They may differ again in their behaviour; they may be able to cross with certain species but not with others; the cross may be sterile, or it may be fertile from the first, or, sterile at first, it may in time become fertile, or even very fertile.

Because of the fluctuating variations referred to, which are controlled to some extent by genes and are liable to be altered in any generation, the characters of the species may not all be present together in every individual; a few will be missing here or there, perhaps occasionally in forms occupying geographically separated areas. If, then, an ice-sheet or an invading sea blots out a part of the population of that species, the parts that remain may not be able to breed together as one unit and may develop along rather different lines. It is possible that a few geographical races may exist of some *Rubus* species in Wales and south Devon as compared with Sussex and Kent or the west of France.

VI. Ecesis and migration

If sown when ripe, seeds of brambles germinate readily in open soil from soon after mid-March until June. This has been my invariable experience over more than 40 years, and it agrees fairly well with that of Babington who says that more than forty kinds sown at Cambridge in the autumn usually appeared in the succeeding May or June. Genevier, however, at Angers in the west of France says that in his experience the seeds are difficult of culture and take two years to germinate.

Seedlings are to be found in nature near blackberry bushes wherever birds perch, at the foot of fences or hedges as well as on bare soil under the bushes themselves. They are recognized by the elliptical, ciliate seed-leaves. The seedlings will perish if they are too much exposed to the sunshine.

Species foreign to my neighbourhood which I have brought into my garden at Bickley I have subsequently met with near fences beside footpaths up to a distance of 200 yds. away, which gives an idea how far they may be transported by resident birds in autumn. The seeds are carried endozoically by birds that eat the berries, as the Turdidae,

Magpie, Crow, Golden Oriole, mountain birds as Grouse, Blackcock, Ptarmigan, Capercaillie, and by mammals as Man, Bear, Fox and Pine Marten. They may be carried also epizoically overland by birds, especially by Woodcock, which, with wet mud on its legs after a night's feeding in marshy ground, repairs at dawn to crouch and doze under bramble bushes and hollies during the daytime. Woods called 'Woodcock Wood' or 'Cockshoot Wood', or woods near them, especially on rather high ground, consequently often contain far-brought bramble species.

About 20 years ago a bush of *R. glanduliger* came up under a holly on Chislehurst Common at the highest point of the Tertiary plateau, where Woodcocks had been twice observed. Until then it was known only in west Sussex. As I used to pass the spot every week and had listed all the species, now 19, growing there on the green, I could be certain about its introduction.

Cultivated species of blackberry as *R. laciniatus, procerus* and loganberry often appear bird-sown from neighbouring gardens. The longest cultivated of these, *R. laciniatus*, was figured by Leonard Plukenet in *Phytographia* (1691) and was described by Philip Miller in the fourth edition of his *Gardeners' Dictionary* (1754); yet after 200–250 years it has not succeeded in spreading so far or increasing so much in any station that I have seen for anyone to mistake it for a wild bramble. *R. procerus* is spreading more aggressively. In New Zealand it has been troublesome from before 1895 (see Rogers' *Handbook* under *R. macrophyllus*, for which it was mistaken). In Australia *R. vulgaris* has been troublesome to farmers for 50 years past, and the occasion for Government action. It is probable that the rapid multiplication on cultivated soil has been by offsets formed at the tips of the main stem and its branches. By such means one sees some species in the absence of competition radiate as much as 30 ft. in a single year, and the young plants from the offsets come up in the following April as strongly as two- or three-year-old bushes and will flower and fruit in the second season.

R. retrodentatus is mainly a west of England bramble, but it appeared about 1930 on Tooting Bec Common, Surrey, beside the artificial lake, brought perhaps in mud on the leg of some aquatic bird.

It has often been said that birds on migration 'fly clean', that is to say they do not carry seeds internally or externally when they cross the seas and oceans to foreign lands. I feel inclined to doubt this.

A hard frost in April or May, after the new shoots of a bramble have well started to grow, kills them back in exposed positions but will not

kill the whole plant. The only species that I have lost in this way is
R. saxatilis. The effect produced on the new shoot of the year if it is
not killed too far back is to cause it to branch freely; on the shoot of
the previous year the effect is to cause the production of radical in-
florescences, which are excessively branched, leafy and floriferous, and
late to come into flower. Severe lopping, of course, has a similar effect.

Most British bramble species are decidedly forest plants, as the pre-
valence of the drip-tip type of leaflet shows; they nevertheless now
prefer the margins of woods, or glades within the wood. The presence
of holly and birch with some heather or bracken indicates that the soil
is very suitable for brambles. *R. Sprengelii, pyramidalis, gratus* and
carpinifolius are frequently met with growing among heather or gorse.
All RUBI die out where bracken gains the upper hand. Very few species
will be found upon chalk; on acid soils they will be abundant, especially
where the ground dries out near the surface in summer.

Species differ greatly in the degree to which they endure shade. One
notices that the same species, all low-growing, habitually penetrate
farthest into the depth of the wood, but do not often flower.

In the open or in the wood, by far the largest number of bushes will
be found at the roadside and along the paths.

I have often found northern slopes productive when the southern
slopes have been unproductive.

Clay, densely wooded or low-lying and wet, will as a rule furnish
few species other than TRIVIALES; but if it forms a ridge, or the clay is
mixed with gravel or 'head' at the surface, or contains seams of sand,
it makes good bramble ground. Thus clay-with-flints overlying chalk
will be profitable, although brambles will probably be absent from the
chalk itself.

Amongst those species often, although not exclusively, found
growing on clay are the following:

alterniflorus	*hostilis*	*pallidus*
apiculatus	*insectifolius*	*prionodontus*
Balfourianus	*Lindleianus*	*propinquus*
Bellardii	*macrophyllus*	*rudis*
caesius	*newbridgensis*	*subinermoides*
cuspidifer	*obcuneatus*	*ulmifolius*
granulatus	*oegocladus*	*vestitus*

The better clay soils have generally been brought under cultivation with
the exception of woods on the clay-with-flints; but one occasionally

comes upon a small piece of wet, clay heath, the drier parts of which support an interesting collection of brambles, or upon a wood preserved in a more or less natural state. It is my experience that brambles planted out on a clay soil thrive and maintain themselves quite as long as one wishes to keep them.

VII. Enemies, pests and diseases

Goats, certainly, and (?) sheep also, when they are tethered up, browse brambles, and I have seen bushes that have been entirely stripped of leaves by them. On the Continent some buprestid larvae are very destructive, and attempts have been made in New Zealand to control their weed RUBI by liberating imported buprestids upon them.

Attacks of the rust fungus, *Phragmidium* spp., occur on the leaves and stem of brambles, but are not serious. More damaging in Canada is the fungus *Hapalosphaeria deformans* which destroys the anthers of cultivated blackberries such as *R. laciniatus* and the Boysenberry and spoils the fruit. For some thirty years past it has been found on *R. latifolius* in Scotland. The anthers become first granulated and then powdered with the white spores.

A virus is probably the cause of arrested, tufted, dwarf growth, in which the deformed panicles do not flower and the plant dies off in a few years. I can find no fungus on such bushes.

The most prevalent and most troublesome disease, however, found upon brambles is the felt-disease, which is believed to be a kind of gall due to a mite-attack. An extremely dense coating of brassy coloured felt extends over branches, leaves, pedicels, calyces and fruit; it masks or replaces the true felt or hair covering and inhibits the production of glands and acicles. The disease seriously checks the growth of the bramble and gives it a very sickly look. Some parts of the plant may be found to be less attacked than others, but it is better not to collect from such bushes at all. Rooted tips carry the infection, but if a ripe or nearly ripe fruit can be found and the seeds are cleaned thoroughly before sowing, clean seedlings can be obtained.

VIII. Classification

The conception that an apparently almost innumerable multiplicity of forms exists is felt by many naturalists of the old school to be intolerable. The fact may be denied in the study: in the open field it will have to be accepted by everyone who looks around. It is certainly downright disquieting that in many genera (*Rubus, Rosa, Hieracium...*) there is such an extraordinary number of species and forms that really nature seems for once not to be created for the convenience of the systematist. FOCKE, *Syn. Rub. Germ.* (1877), 3

THE classification adopted in this work for MORIFERI has a twofold basis. Focke and Gremli collaborated for many years to discover the most suitable characters on which to base their arrangement of the forms of western Europe. They finally settled upon the comparative development in size and distribution of prickles, pricklets, acicles and stalked glands. Although no better master-plan has been discovered, it assumes that one will possess full knowledge of the climax of development of these structures in all species, and will not be misled by their imperfect evolution in young and shade-grown plants. On this basis, then, Focke proposed fifteen groups (including TOMENTOSI, which is not represented in Britain), and proceeded to distribute his species amongst them. Recognizing that some of them, whilst belonging technically to one group, showed natural affinities to species which he had assigned to another group, he often decided to place them where their natural affinity lay. This practice he further developed in his second and third, and most of all in his final work, whilst retaining all the time with one exception the original names for his groups. The result is that a beginner who wishes to identify an unknown bramble is at a loss to know where to look for it, although Focke has sometimes mentioned a species under two groups; but anyone who knows the name of the bramble in which he is interested, and simply wishes to see Focke's account of it and his views of its relationship to other species, will have no difficulty in finding what he wants through the index.

The classification adopted in this work is based in its main outlines upon the nature of the armature, and coincides more or less with Focke's main divisions. For the rest it may be sufficient to say that attention has been given to forming natural subsidiary groups fitting into those outlines. Where it departs from Focke's plan it will often be found to agree with the arrangement followed by Genevier and Sudre, or by one of them; this applies especially to SYLVATICI. The subdivisions of APICULATI, a very large group in Britain, are new.

From the mode of origin of most of the species it is natural to find that they often have affinities with more than one group, and this makes them resistant to perfect classification.

About 78 per cent of British species are found on the Continent; the remainder are endemic, and amongst them are to be found some that appear to be very old species. For instance, *R. Reichenbachii*, which is found around Hawkenbury, west Kent, appeared to Otto Kuntze to be so singular in having an eglandular stem with equal prickles, whilst the panicle was very glandular with unequal prickles, that he rejected Koehler's specimen as a mixture of pieces from bushes belonging to two different species. He would have understood the specimen better if he had known of the Madeiran *R. grandifolius*. In the same neighbourhood as *R. Reichenbachii* grows *R. longifrons* which is related to it.

Resemblances may be accompanied by more or less profound differences; this facilitates the identification of the species, but does not help the classification.

If the classification here adopted is found to be involved and hard to grasp, well and good! for then it reflects the situation that exists in nature.

IX. Collection and identification

The stem in many plants furnishes differences so essential that when it is wanting there is no certainty as to the species.

LINNAEUS, *Philosophia* (1780, 2nd ed.), 218

One *Rubus* referee used to say 'No stem-piece, no name'. In *Rubus* the stem is furnished with various types of deposit, as pruina or wax, and outgrowths such as hairs single, paired, or small and clustered, glands sessile, subsessile or stalked, acicles short or long, sometimes gland-tipped, pricklets and prickles. These are the most typically developed parts for purposes of identification, especially as to frequency, on the middle part of the non-flowering stem growing in the open and belonging to a mature and vigorous plant, and on a panicle growing on the middle part of the two-year-old stem. The character of the armature is, within limits, of more importance than the amount; it will be seen to vary even between the lower and the upper part of a single internode; it is always more advanced on the panicle than on the stem.

The colour of the unripe fruit is worthy of notice; for instance, very pale green in *R. dasyphyllus*, coral red in *R. Bakeranus*, blood red in *R. rufescens* and *R. obscurus*, fuscous in *R. phaeocarpus* and *R. euryanthemus*.

Very important is the direction taken by the sepals under the green fruit. It may be possible to judge this sometimes before the bush has passed entirely out of flower, but more often a second visit will have to be paid.

Before concluding that the stem and panicle are eglandular it is advisable to look at the central petiolule of a stem leaf, at the petiole of a leaf situated below the base of the panicle, and at the upper pedicels and the calyx, as it is at these points if anywhere that glands will be in evidence.

Before accepting that the stem is glabrous, the base and the growing point, or in a specimen the area close to a leaf-base, should be examined. In some species the hairs disappear with age. Felt should always be looked for with the aid of a strong lens, and it should be remembered that it may be present even where the leaf surface looks greenish. Pruina is most easily detected at the base of the stem, or sometimes at the base of the branches of the stem or of the flowering branch; it is always to be looked for on the side turned away from the sun, and it may sometimes be late in putting in an appearance.

It is a good plan to examine the top simple leaves of a panicle, and the small leaves belonging to axillary shoots pushing out from the main stem. If they are not felted beneath, felt will not be found anywhere on the plant.

Note should be taken of any point required for identification that will not show in the dried specimen, such as the colour of the *unclosing* petals (*observed through a lens*), filaments and styles, and the relative lengths of the two latter organs. A few petals should be pressed flat separately to show the shape and the margin clearly. The petals should not be too forcibly removed, otherwise the claw will be left in the flower. It is convenient to record whether the carpels and receptacle have been found glabrous or pilose; this has to be done on newly opened flowers, as the hairs soon fall off the carpels. The hairs on anthers are best seen on unburst anthers, as here also they soon fall off.

If the species is determined at the bush, the specimen can more easily be selected to show those characters that determine the identification. The specimen need only consist of one or, if thought necessary, two stem-pieces about 3 in. long and with one leaf attached, taken from towards the middle of the stem, but not within 3 ft. of the base, and a panicle showing buds and open flowers but also going into fruit, or accompanied by a piece of another panicle in green fruit showing the direction of the sepals. The panicle should be not less than 9–10 in.

long; it should be complete and should show the mode of divisions of the middle branches. The tip of the growing stem is not wanted, nor a panicle in ripe fruit.

A stem-piece comprising more than one leaf will prove a nuisance in pressing and mounting; it will have to lie towards the centre of the paper, where it will foul the panicle; and it will make piling and pressing and subsequent storage of mounted specimens difficult.

It is a good practice to turn the terminal leaflet over when putting the specimen in the press, so as to make certain that the underside of the leaf will be shown when it is mounted. At the first change of the drying papers—preferably after 12 hr.—the panicle should be laid out and any leaves covering the panicle should be turned away. If the panicle is too dense to give a good view of the rachis and pedicels, a forefront branch may be removed, and if necessary one that goes to the back as well.

In the past brambles have been collected in this country much too late, with disastrous results for the naming. As a rule in south-east England representative specimens can be obtained between midsummer and mid-July, although this will be too late for early-flowering species, whilst late July would not be too late for those that flower late.

In the descriptions of species that follow, figures and characters placed in brackets are of less frequency than those left unbracketed. The characters of the group and those of the subheading under which the species falls are not repeated as a rule in the description of the species. Where not otherwise stated the rachis of the panicle is to be understood to be armed like the stem. The direction of the sepals, whether reflexed patent or erect, is the direction of the sepals on the green fruit, not that on the flower or on the black fruit. When it is stated that the petals are pilose, or that they are glabrous, this should be understood to apply primarily to the apex of the petal.

X. Characteristics of the British-Irish bramble flora

The characteristics of the British-Irish bramble flora are brought out when a comparison is made of them statistically group by group with the brambles of a near and a far part of western Europe. North France with Belgium, and Bavaria with Pfalz and Alsace, have been chosen, as their bramble flora is well known and they suit the requirements for a just comparison with Britain. North France comprises the departments of *Nord, Pas-de-Calais, Somme, Ardennes, Aisne* and *Oise: Nord* is low and flat, *Ardennes* is upland to moderately mountainous, the other departments are hilly. Belgium is low in the north, high and mountainous in the south. Both north France and Belgium are entirely periglacial. Alsace consists of mountain (Vosges) and plain (Rhine); Pfalz consists of mountain (Hardt Mts.) and plain (Rhine); Bavaria consists of high Alps, high plateau, forests and the wide Danube plain.

	North France	Belgium	North France and Belgium	Britain and Ireland	Bavaria with Pfalz and Alsace
Sylvatici	42 (1)	33	59 (1)	94 (29)	23 (5)
Discolores	20	13	26	14	19 (2)
Sprengeliani	1	8	8	7 (2)	2
Vestiti	21 (1)	10	26 (1)	30 (6)	19 (2)
Subtomentosi	1	1	1	—	1
Tomentosi	—	—	—	—	2
Mucronati	6	7 (1)	12 (1)	16 (7)	7
Dispares	—	—	—	7 (4)	1
Radulae				18 (2)	
Rudes	62	57	103	—	86 (1)
Apiculati				87 (23)	
Grandifolii	4	4	5	20 (11)	3
Hystrices	21 (3)	21	36 (3)	40 (6)	34 (1)
Euglandulosi	25	54	69	20 (1)	78
Moriferi	203 (5)	208 (1)	345 (6)	353 (91)	275 (11)

(i) The bracketed figures stand for endemic species.

(ii) The unbracketed figures standing beside bracketed figures include endemic and non-endemic species.

(iii) Where unbracketed species stand alone they represent non-endemic species.

The greater frequency of EUGLANDULOSI in mountainous country (Belgium and the Bavaria group) will be noticed. The high proportion of endemics in Britain and Ireland in all groups except DISCOLORES and EUGLANDULOSI is also remarkable, especially in SYLVATICI, MUCRONATI, DISPARES and GRANDIFOLII, all of which groups possess characters shown in east Atlantic, Macaronese species; that is to say, short or slightly sigmoid prickles, large flowers, scarcity or absence of glands on the barren stem, but an abundance on the panicle, together with, in some instances, highly developed, intricately branched, pyramidal panicles. A high degree of endemism in species incorporating such characters must be regarded as marking an ancient flora.

On the other hand it is shown that a low proportion of endemics is general on the Continent. In Switzerland, too, there is apparently only one endemic species. Dr R Keller claimed three in 1922, but two of these have been found in other countries, R. *Barbeyi* Favrat & Gremli in France, Loire-et-Cher, and R. *Fischer-Oosteri* Sudre in Belgium, Vierset-Barse.

On the Continent species of HYSTRICES number about one-half of the species of EUGLANDULOSI, but here the proportion is reversed, the species of HYSTRICES being twice as numerous as those of EUGLANDULOSI. Our DISCOLORES number less than one-half of the VESTITI, but in north France, Belgium and the Bavaria group they number the same as the VESTITI. The EUGLANDULOSI have a wide distribution in the east and are numerously represented there. The DISCOLORES are more a Mediter-ranean, north African and Macaronese group.

We possess 13 (2) SUBERECTI, but 29 (13) TRIVIALES. There can hardly be any doubt, I think, that the SUBERECTI are far the older group of the two, as the same species are found over an extremely wide area of Europe. The species of TRIVIALES, on the other hand, are very local indeed, each country of Europe, especially those in glaciated regions, having a large proportion of them endemic, which in this instance is not indicative of an ancient pedigree, but rather of a facile origin. R. *caesius* is found crossing abundantly with MORIFERI and TRIVIALES, and TRIVIALES with TRIVIALES.

XI. Cultivating native blackberries for fruit

> I would address one general admonition to all; that they consider what are the true ends of knowledge, and that they seek it not for pleasure of the mind, or for contention or for superiority to others, or for profit, or fame, or power, or any of these inferior things: but for the benefit and use of life.
>
> BACON, Preface to *The Great Instauration*

As would be expected, the fruits of different species of *Rubus* differ much from one another in edible quality: the best of them, allowed to ripen completely under the protection of a net or muslin, are well worth cultivation in the garden for home consumption. They do not travel well.

Mrs Lankester in Syme's *English Botany* praises the fruit of the dewberry as 'very superior, and larger than any other blackberry', an estimate which one cannot endorse, either as to flavour or size. She is possibly thinking of R. *Balfourianus*, of which one does sometimes meet with a strain producing large fruit of a mulberry flavour, and which Bagnall described as much more delicious even than R. *gratus*. It is perhaps the 'Rubus morus, the *Mulberry Bramble*, so called by the Country People at *Sutton*, in *Essex*'. (Dillenius, *Indiculus*, appended to Ray's *Synopsis* (3rd ed.), extracted from Merret's *Pinax* (1666), p. 106.)

Babington, who cultivated blackberries in the Cambridge Botanic Garden, placed *R. hastiformis* amongst the best flavoured known to him. I have not grown it, but it was chosen at Merton for crossing with *R. inermis* Willd. and a useful fruit has been selected which is being extensively grown.

It is, however, within the power of anyone, without needing to know the name of any bramble, to choose a large fruit that pleases him from a bush on which the fruits are consistently large, sow the seeds at once, and in three years' time begin to enjoy crops of the fruit, probably somewhat improved by cultivation, from his own garden. Not the smallest merit of the blackberry is that it improves the apple pie, pudding or stew, and is itself improved by the association.

One notices that the fruits of those blackberries that bear no stalked glands and have the leaves green beneath are generally, not always, sourish or pleasantly acidulous when raw, improving on cooking; the fruits of those that have leaves white felted beneath but no stalked glands are sweet; those that are abundantly provided with stalked glands usually possess aroma, which varies in quality from species to species and invites exploration and trial.

The flavour of the fruit develops best during dry weather, is spoilt by continued rains, ruined by over-ripeness. Some species, especially those with white felt on the underside of the leaves, crop so heavily that they cannot ripen all their fruit. If they are thought to be of sufficient merit it will pay to remove the smaller buds. Other species ripen all their fruits early and almost simultaneously.

The following species are the best that I have grown for the sake of their fruit:

nitidoides	*gratus*	*oxyanchus*
pyramidalis	*separinus*	*Winteri*
badius	*cyclophorus*	*nemoralis*

Almost as good are:

carpinifolius	*londinensis*	*Balfourianus*
ludensis	*macrophyllus*	*broensis*
Riddelsdellii	*rhombifolius*	*ulmifolius*
pseudo-bifrons	*vestitus*	*fuscoviridis*
foliosus	*largificus*	*scaber*
infestus	*obscurus*	*thyrsiflorus*
insericatus spp. *Newbouldianus*	*Koehleri*	*vallisparsus*

A good selection of manageable sorts for succession and various uses might be:

> *nitidoides* or *carpinifolius*, for early dessert,
> *londinensis*, best for bottling; dessert,
> *gratus*, very large, cooking,
> *separinus*, continuous cropping; for all purposes,
> *Winteri*, the best late for all purposes.

All kinds can be readily reproduced by laying in the tips of the main stem and branches in September. A 2-in. mulch of half-decayed leaves every April is recommended.

XII. Note on the nomenclatural type species for the genus *Rubus* and subgenus *Rubus*

In choosing a type species to represent the genus *Rubus* and subgenus *Rubus* it will be the same species for each group. Obviously the species should be typical of the genus and of the subgenus, that is, it should not in any respect be contrary to the definition of the genus by Linnaeus.

Now Linnaeus himself indicates that two of his species are not typical: 'Rubus Chamaemorus dioicus est; Rubus saxatilis acinis distinctis Baccas profert.' (*Gen. Plant*, 6th ed., p. 254.)

R. saxatilis is atypical also in having a flat, not a conical receptacle, and very narrow, not subrotund petals. *R. arcticus* again must be disqualified, as it has a crumbling fruit like *R. saxatilis*, and a low convex, not a conical receptacle. *R. caesius* also fails to comply with the formula 'Acini Baccarum coaliti sunt nec sine laceratione distinguibiles', for the ripe fruit disintegrates when it is plucked. *R. idaeus* too is thoroughly atypical in its very narrow petals.

J. E. Smith (*Eng. Fl.* (1824), II, 398) adopts for his type of *Rubus fruticosus* L. the specimen no. 5, 'Madera', representing this species in Herb. Linnaeus. Smith's description is *R. ulmifolius* and the Madeiran specimen is *R. ulmifolius*. There need be no hesitation, I consider, in adopting *R. ulmifolius* as the type species of genus *Rubus* and of subgenus *Rubus*; it has none of the disqualifying characters that have been mentioned.

RUBUS L. (type species *R. ulmifolius* Schott f.)

No epicalyx. Fruit a drupe, composed of pulpy drupels containing each a single, woody seed, inserted upon a spongy, flat or convex, conical or cylindrical, pitted receptacle. This runs down into a short central upgrowth from the flattish disc, which extends to the rim of the calyx tube. Calyx tube obsolescent, segments (4) 5 (6–8, 10), imbricate in bud, entire; petals of the same number as the calyx segments, very seldom more than 8 and then superposed, free, inserted on the rim of the calyx tube; stamens numerous, situated on the rim of the calyx tube, free but contiguous at the base, the filaments flattened or subulate, gradually or abruptly narrowed at their apex, the anthers versatile; styles subterminal, deciduous; ovules two, of which only one develops, the seed in most species pitted-reticulate. Pollen grains smooth, subglobose.

Plant perennial, the shoots herbaceous biennial or perennial, usually bearing prickles on all axes and often stalked glands and acicles as well. Cotyledons two, elliptic, ciliate. Leaves alternate, always toothed, the teeth mucronate, stipulate, simple, trifoliate, pinnate or quinate-digitate, or quinate-pedate. Flowers bisexual, rarely dioecious, bracteolate, solitary or panicled, on shoots that grow out from the leaf-stems in their second year.

The floral parts appear to be disposed on the plan of a spiral depressed into a plane. The spiral is for the most part of the 2/5 system, but 3/8 prevails in the terminal flowers of the main axis and in those of the branches of the inflorescence of some at least of the glandular and highly glandular species, giving 5 petals and sepals in the former, and 6–8 in the latter case. The carpels are spirally arranged.

Two stamens stand one on each side of the claw of each petal; they are longer than the rest and are the first to mature. The other, inner stamens (fillers) stand in unequal groups opposite to the base of the calyx segments, and mature in succession. After the anthers dehisce the stamens first stand away from the styles and then generally, even where the stamens are short, come up to the styles again. Where the stamens are very long they invest the young carpels after the petals drop. The receptacle soon lengthens and carries the styles and carpels up above the stamens, which then wither. The fruit ripens in from six to eight weeks.

With respect to the winter buds at the ground level that are to produce the new stems of the following year, the plant is a hemi-cryptophyte; but with respect to the winter buds situated along the first-year stems that are to produce the flowering shoots of the next year, the plant is a phanerophyte. In some species these winter shoot-buds are seriate in the same leaf-axil, the upper one being usually the first to develop. In favourable situations the buds move into growth about mid March, at the same time almost as the seeds begin to germinate. Flowering commences in the earliest species about 8 May, in the latest species not until July is in. Leaf-fall in the deciduous species begins in early October. Other species are more or less wintergreen. The implantation and rooting of stem and branch tips proceeds from mid August until growth is stopped by frost.

The flowers are slightly protogynous, the upper parts of the petals being inflexed upon themselves so that they arch over the anthers whilst exposing the stigmas in a receptive condition. If the styles are red they are at their most brilliant colour at this stage. Nectar is secreted from the surface of the disc in the part lying towards the stamens; it attracts thrips and weevils, bees, butterflies and night-flying moths, whose movements against the erect filaments shake the pollen from the swinging anthers.

XIII. Key to vice-county numbers

England and Wales

1	W. Cornwall	21	Middlesex	41	Glamorgan
2	E. Cornwall	22	Berks.	42	Brecon
3	S. Devon	23	Oxford	43	Radnor
4	N. Devon	24	Bucks.	44	Carmarthen
5	S. Somerset	25	E. Suffolk	45	Pembroke
6	N. Somerset	26	W. Suffolk	46	Cardigan
7	N. Wilts.	27	E. Norfolk	47	Montgomery
8	S. Wilts.	28	W. Norfolk	48	Merioneth
9	Dorset	29	Cambridge	49	Carnarvon
10	I. of Wight	30	Bedford	50	Denbigh
11	S. Hants.	31	Hunts.	51	Flint
12	N. Hants.	32	Northampton	52	Anglesey
13	W. Sussex	33	E. Gloucester	53	S. Lincoln
14	E. Sussex	34	W. Gloucester	54	N. Lincoln
15	E. Kent	35	Monmouth	55	Leicester
16	W. Kent	36	Hereford	56	Nottingham
17	Surrey	37	Worcester	57	Derby
18	S. Essex	38	Warwick	58	Cheshire
19	N. Essex	39	Stafford	59	S. Lancs.
20	Herts.	40	Shropshire	60	W. Lancs.

61	S.E. Yorks.	65	N.W. Yorks.	68	Cheviotland
62	N.E. Yorks.	66	Durham	69	Westmorland
63	S.W. Yorks.	67	S. Northum-	70	Cumberland
64	Mid-west Yorks.		berland	71	I. of Man

Scotland

72	Dumfries	84	Linlithgow or West Lothian	98	Argyll
73	Kirkcudbright			99	Dumbarton
74	Wigtown	85	Fife	100	Clyde Isles
75	Ayr	86	Stirling	101	Kintyre
76	Renfrew	87	W. Perth	102	S. Ebudes
77	Lanark	88	Mid Perth	103	Mid Ebudes
78	Peebles	89	E. Perth	104	N. Ebudes
79	Selkirk	90	Forfar or Angus	105	W. Ross
80	Roxburgh	91	Kincardine	106	E. Ross
81	Berwick	92	S. Aberdeen	107	E. Sutherland
82	Haddington or East Lothian	93	N. Aberdeen	108	W. Sutherland
		94	Banff	109	Caithness
83	Edinburgh or Midlothian	95	Elgin or Moray	110	Hebrides
		96	Easterness	111	Orkney
		97	Westerness	112	Shetland

Ireland

H. 1	S. Kerry	H. 15	S.E. Galway	H. 28	Sligo
H. 2	N. Kerry	H. 16	W. Galway	H. 29	Leitrim
H. 3	W. Cork	H. 17	N.E. Galway	H. 30	Cavan
H. 4	Mid Cork	H. 18	Offaly	H. 31	Louth
H. 5	E. Cork	H. 19	Kildare	H. 32	Monaghan
H. 6	Waterford	H. 20	Wicklow	H. 33	Fermanagh
H. 7	S. Tipperary	H. 21	Dublin	H. 34	E. Donegal
H. 8	Limerick	H. 22	Meath	H. 35	W. Donegal
H. 9	Clare	H. 23	Westmeath	H. 36	Tyrone
H. 10	N. Tipperary	H. 24	Longford	H. 37	Armagh
H. 11	Kilkenny	H. 25	Roscommon	H. 38	Down
H. 12	Wexford	H. 26	E. Mayo	H. 39	Antrim
H. 13	Carlow	H. 27	W. Mayo	H. 40	Londonderry
H. 14	Leix				

Vice-county 1 includes the Scilly Isles; 55 includes Rutland; 69 includes N. Lancs.; 85 includes Kinross; 87 includes Clackmannan; 96 includes Nairn; 102 is the Islay group; 103 is the Mull group; 104 is the Skye group; 110 is the outer group of isles from Lewis to Barra.

XIV. Signs and abbreviations

± more or less

→ common

x set of 7 chromosomes

c. circa, circiter, about

cm. centimetres

m. metres

mm. millimetres

E. England

f. *filius*, son

Herb. Herbarium

H. or Hib. Hibernia

I. Ireland

microg. microgene

N.v.v. Non vidi vivam (plantam), I have not seen the living plant

S. Scotland

ssp. subspecies

t. *tabula*, plate

var. variety

W. Wales in E.W.S.I. (in text), western species (in key).

ARRANGEMENT OF THE GENUS
RUBUS LINN. IN AN
ANALYTICAL KEY

The common species is indicated by →; **N.** = Northern species; **W.** = Western species. **Hib., Penins., Scot., Wales** indicate species that are known from Ireland, the Devon-Cornwall Peninsula, Scotland and Wales respectively.

i. *Stem herbaceous, rootstock creeping and suckering.*
Stipules broad, cauline.

Subgen. **I.** *CYLACTIS*

Shoots annual. Leaf 3-nate. Flowers bisexual. Fruit red to deep purple, drupels hardly coherent, parting from the receptacle. Receptacle flat or convex.

1 saxatilis →   *2* arcticus

Subgen. **II.** *CHAMAEMORUS*

Shoots annual. Leaves simple. Flowers unisexual, dioecious. Fruit red then orange-amber. Receptacle conical.

3 Chamaemorus →

ii. *Stem shrubby. Stipules petiolar.*
a. *Shoots biennial, bark disrumpent. Leaf-fall in autumn.*

Subgen. **III.** *IDAEOBATUS*

Shoots not rooting at tip. Roots suckering. Fruit pubescent, red, coherent, separating whole from the elongate receptacle. Leaves pinnate; stipules small, filiform. Flowers bisexual.

4 idaeus →

Subgen. **IV.** *GLAUCOBATUS*

Shoots decorticating, rooting at tip. Roots not suckering (unless injured). Fruit glabrous, black but pruinose; drupels loosely coherent and loosely adherent to the receptacle. Leaves 3-nate; stipules ovate-lanceolate. Flowers bisexual.

5 caesius →

Subgen. *IDAEOBATUS × RUBUS*

Sect. I. SUBERECTI

Combine characters of *R. idaeus* with characters of Sylvatici and perhaps Discolores. *Stem not rooting*, suberect or high arching, glabrous or slightly pilose. Panicle simple, or hardly compound. Sepals green outside.

Subser.† SUBERECTI VERI

Stem at first erect, later bending but not rooting, hardly branched, glabrous. Leaves 3–7-nate. Panicle usually short racemose. Petals becoming patent, sometimes vaulted before dropping. Fruit deep red or purple, hardly black. Leaf-fall in autumn. Stamens reflexed.

1. *Prickles short*
6 nessensis →
7 scissus →

2. *Prickles and filaments long*
† Filaments exceeding styles
8 sulcatus
9 Bertramii
†† Filaments equalling styles
10 plicatus →
11 ammobius
12 fissus
13 arrheniiformis

Subser.†† SEMISUBERECTI

Habit more bushy, suckering less, tip-rooting more. Stem glabrous or thinly hairy. Leaves less often 6, 7-nate. Panicle more compound. Petals incurved. Fruit black. Leaf-fall in winter. Stamens connivent.

1. *Upper leaves felted*
14 affinis
15 holerythrus **W.**
16 nobilissimus

2. *All leaves green beneath*
17 nitidus
18 integribasis

Subgen. *GLAUCOBATUS × RUBUS*

Sect. II. TRIVIALES

Combine characters of *R. caesius* with those of subgen. Rubus. Most species show the corymbose panicle and the very short-stalked lateral leaflets, and broad stipules of *R. caesius*.

1. *Prickles equal; glands few or none*
† Sepals reflexed; petals pink
19 conjungens →
20 ooliticus **W.**
21 eboracensis **N.**

2. *Prickles equal; glands rather numerous*
† Sepals patent or erect
(a) Petals pink
26 Balfourianus →
27 Warrenii →

3. *Prickles unequal; glands ± numerous, sepals erect*
† Prickles not numerous
32 tuberculatus →
33 Babingtonianus →

†† Sepals reflexed;
   petals white
22 sublustris →
23 latifolius **Scot.**
24 Bucknallii **W.**
   ††† Sepals erect;
   petals white
25 umbelliformis

28 adenoleucus
29 Wahlbergii
   (*b*) Petals white
30 halsteadensis →
   †† Sepals reflexed
31 purpureicaulis

34 rubriflorus
   †† Prickles
   numerous
35 scabrosus →
36 myriacanthus →
37 britannicus →
38 tenuiarmatus **W.**

b. *Shoots biennial or perennial, bark not disrumpent. Leaf-fall in autumn, winter or spring.*

Subgen. **V.** *RUBUS*

Terminal leaflet with *c.* 6–9 secondary nerves on each side of the midrib. Leaves thin or coriaceous. Fruit glabrous at maturity, not pruinose; drupels coherent, entirely adherent to the oblong or elongated, rather pointed receptacle. Leaf 3-nate, or 4–7-nate-pedate, or digitate. Stipules sublanceolate, linear or filiform.

I. *Stem prickles subequal, seated on the angles of the stem. Stalked glands, acicles and pricklets none or rare, or scattered on the stem and more frequent on the panicle.*

Sect. III. SYLVATICI

Stalked glands and acicles none or rare. Stem hairy or glabrescent. Leaves green, greyish or greyish white beneath. Sepals erect, patent or reflexed. Stamens usually long. Flowers longer than 2 cm.

Subsect. **A.** VIRESCENTES

Leaves green beneath, pilose or pubescent, rarely the upper leaves of stem and panicle greyish felted beneath.

Ser. **i.** EUGRATI

Sepals patent to erect. Stem glabrescent or pilose.

I. *Petals pink*
   † Flowers exceeding 2·5 cm.
39 gratus →
40 confertiflorus
41 monensis **Wales**
   †† Flowers *c.* 2 cm.
42 vulgaris
43 latiarcuatus

Ser. **ii.** CALVESCENTES

Sepals ± reflexed. Stem hairy to glabrous.

I. *Petals pink*
   † Prickles broad, some curved
57 nemoralis →
58 Questieri
   †† Prickles slender, subulate
59 oxyanchus
60 cambrensis
61 durescens

EUGRATI (*cont.*)

44 subintegribasis **W**.
45 Averyanus
46 dasycoccus **W**.
2. *Petals pinkish or white*
  † Leaves and panicle pilose
47 calvatus
48 sciocharis
49 nitidoides
50 ludensis **W**.
51 Crespignyanus
  †† Leaves and panicle pubescent
52 carpinifolius →
53 chloophyllus
54 polyoplus
  ††† Petals white. Panicle and sepals glandular
55 horridisepalus
56 glanduliger

CALVESCENTES (*cont.*)

2. *Petals white*
  † All leaves green beneath
62 pullifolius
  †† Upper leaves felted
63 Lindleianus →
64 egregius
65 leucandrus
66 mercicus
67 Maassii
68 Muenteri
69 Sampaianus

Ser. **iii.** PILETOSI

Sepals reflexed or patent. Stem moderate or closely hairy, hair usually persisting.

Subser.† MACROPHYLLI

Prickles long or short. Leaves at first hairy or pubescent beneath, then glabrescent. Sepals ± reflexed. Flowers usually white.

I. *Plants large, large-leaved*
70 macrophyllus →
71 subinermoides →
72 Schlechtendalii
73 Boulayi
74 sylvicola
75 megaphyllus
76 majusculus **Dorset**

2. *Leaves and panicle small*
77 Bakeranus **Surrey**
3. *Leaves and panicle moderate*
  † Prickles small
78 silvaticus
79 chrysoxylon **W**.
80 patuliformis

Subser.†† PYRAMIDALES

Prickles moderate. Leaves thick, soft and often yellowish, pectinately pilose on the nerves beneath. Sepals reflexed, patent or semi-erect. Flowers usually pink.

I. *Petals pink*
  † All leaves green beneath, rarely the upper ones greyish
84 pyramidalis →
85 albionis
86 mollissimus
87 crudelis
88 amphichlorus
  †† Upper leaves or all greyish or grey felted beneath
89 hirtifolius
90 orbifolius
91 obesifolius
92 macrophylloides

†† Prickles moderate
81 amplificatus →
82 eglandulosus **Penins.**
83 belophorus **Lincs.**

2. *Petals white or pinkish*
93 danicus

Subsect. **B.** DISCOLOROIDES

Upper leaves greyish white felted beneath.

Ser. **i.** PROPERI

Stem ± pubescent at first, then glab-rescent. Panicle ample, grey felted and pubescent. Prickles long and moderately numerous or weak and rare. Leaflets convex.

† Prickles long
94 londinensis
95 Libertianus
†† Prickles moderate
96 ramosus
97 Salteri
98 pedatifolius
††† Prickles short
99 poliodes

Ser. **ii.** SUBVIRESCENTES

Stem with crisp patent hair, some-times glabrescent. Panicle villose-hirsute. Prickles strong and long.

1. *Plant and flower large*
100 rhodanthus
101 insularis
102 broensis

2. *Plant and flower small*
103 septentrionalis **Scot.**

3. *Plant and flower moderate*
† Terminal leaflet deeply com-pound-serrate
104 stanneus **W.**
†† Terminal leaflet serrate or serrate-dentate
105 atrocaulis

Ser. **iii.** SUBDISCOLORES

Stem pilose or glabrous. Panicle pilose or also felted, with strong prickles. Mostly somewhat glan-dular.

1. *Eglandular*
114 consobrinus
115 cariensis **Hib.**
116 incarnatus **Hib.**

2. *Slightly glandular*
† Glands very short
117 polyanthemus →
118 rubritinctus →
119 rhombifolius →
120 acclivitatum →
121 prolongatus →
122 alterniflorus →
123 herefordensis
124 obvallatus **Wales**
†† Glands moderately long-stalked
125 separinus
126 septicola

Ser. **iv.** IMBRICATI

Stem glabrous or nearly so, not pruinose. Panicle rachis not or only slightly pilose, with strong and long prickles, subulate patent from a broad base, usually a little bent at the apex. Eglandular.

1. *Petals white*
127 cardiophyllus →

SUBVIRESCENTES (*cont.*)

††† Terminal leaflet unequally serrate

(a) Leaflets acute

106 macroacanthus **Norfolk**

(b) Leaflets long-pointed

107 incurvatus →

108 villicaulis

109 iricus → **Hib.**

110 Langei

111 Favonii **W.**

112 Riddelsdellii **Penins.**

113 lasiothyrsus

IMBRICATI (*cont.*)

128 rotundatus

129 Lindebergii **N.**

2. *Petals pink*

130 errabundus

131 imbricatus

132 Silurum **W.**

Sect. IV. DISCOLORES

Stalked glands, acicles and pricklets absent. Stem hairy or glabrous. Leaves white-felted beneath. Stipules linear. Sepals reflexed. Stamens long or equalling styles. Flowers longer than 2 cm.

A. *Stem arched-procumbent, usually red or purple.*

Ser. **i.** GYPSOCAULONES

Stem pruinose, becoming scaly-waxed on the side turned to the sun. Stamens long or short.

I. *Filaments equalling styles*

133 ulmifolius →

134 pseudo-bifrons →

135 chloocladus

2. *Filaments much longer than the styles*

136 Winteri

137 propinquus **W.**

Ser. **ii.** HEDYCARPI

Stem epruinose or feebly pruinose. Panicle often subpyramidal. Plant often robust. Stamens long.

I. *Petals very narrow*

138 stenopetalus

2. *Petals white*

139 geniculatus

3. *Petals pink; broad*

140 cuspidifer

141 bifrons **W.**

142 vulnerificus **Sussex**

B. *Stem tall, arched, usually furrowed and green.*

Ser. **iii.** CANDICANTES

Stem usually epruinose and glabrous. Leaves silvery greyish-felted and hairy beneath; basal leaflets very short stalked.

I. *Petals pink. Axes glaucescent. Leaves strigose above*

143 neomalacus **Surrey**

2. *Petals pink. Axes not glaucescent. Leaves glabrous above*

144 falcatus

3. *Petals white*

145 hylophilus

146 thyrsanthus

Sect. V. SPRENGELIANI

Stalked glands, acicles and pricklets none or rare. Stem hairy. Leaves green or greyish white beneath. Sepals patent or erect. Stamens shorter than the styles. Flowers less than 2 cm.

1. *Upper leaves felted*
147 Braeuckeri
148 chlorothyrsus

2. *All leaves green beneath*
149 Sprengelii →
150 lentiginosus →
151 permundus
152 Arrhenii
153 axillaris

Sect. VI. APPENDICULATI
Ser. i. VESTITI

Stalked glands, acicles and pricklets very thinly distributed on the stem, more frequent on the panicle. Leaves green, greyish or white beneath. Stem and leaves often hairy.

A. *Armature weak.*
Subser.† NEMORENSES
Slender plants with short prickles which are unequal and pass into acicles, and with sunken, short-stalked glands in the panicle.

154 nemorensis

155 hesperius **Hib.**

B. *Armature strong.*

Subser.†† VIRESCENTES
Leaves green beneath (in some species the upper stem leaves, all branch leaves, and the upper panicle leaves may be grey felted beneath).

1. *Sepals reflexed*
156 fuscicortex **W.**
157 silesiacus
158 eriostachys
159 helveticus

2. *Sepals patent or erect*
   † Terminal leaflet short-pointed
160 lasiostachys →
161 cinerosiformis **Penins.**
162 serratulifolius

Subser.††† HYPOLEUCI
All leaves greyish or white felted and usually pilose beneath.

1. *Sepals reflexed*
   † Terminal leaflet roundish
168 vestitus →
169 podophyllus
170 conspicuus
171 andegavensis
172 salisburgensis
   †† Terminal leaflet not roundish
173 adscitus →
174 leucostachys →
175 scutulifolius

VIRESCENTES (*cont.*)

†† Terminal leaflet long-pointed
163 hebeticaulis **W.**
164 orbus **Penins.**
165 Schmidelyanus
166 flavescens
167 condensatus

HYPOLEUCI (*cont.*)

2. *Sepals patent or erect*
† Terminal leaflet moderate, acuminate
176 Boraeanus →
177 conspersus
†† Terminal leaflet large, round, cuspidate
178 magnificus
179 acutidens **Hib.**
††† Terminal leaflet not round
180 macrothyrsus →
181 criniger →
182 eifeliensis
183 melanocladus **W.**

2. *Intermediate between* **1** *and* **3.** *Stem armature irregular, usually defective on the lower part of the stem. Panicle armature developed much as in* **3.**

Ser. **ii.** MUCRONATI

Stalked glands, acicles (and sometimes pricklets) if present on the stem varying from few to rather many; irregularly distributed on the panicle, sometimes unequal in length, varying in quantity but usually to be seen on the pedicels and calyces.

1. *Stem furrowed*
184 mucronatiformis **W.**

2. *Stem not furrowed, terminal leaflet rounded*
† Anthers pilose
185 mucronifer → **N.**
186 Drejeri
187 badius
†† Anthers glabrous
(*a*) Filaments pink or red
188 Muelleri
189 Briggsii **Penins.**
(*b*) Filaments white
190 cinerosus →
191 atrichantherus

3. *Stem not furrowed; terminal leaflet acuminate*
† Sepals loosely reflexed
192 Lettii **Hib.**
193 Gelertii
†† Sepals patent or clasping
194 Leyanus →
195 hibernicus **Hib.**
196 dunensis **Hib.**
197 hypomalacus →
198 chaerophyllus
199 bracteosus **W.**

Ser. **iii.** DISPARES

Subsessile glands, minute prickles and short acicles are irregularly distributed on stem and panicle, almost wanting in some parts, rather abundant in others, with or without a few longer glands, acicles, gland-tipped acicles and pricklets.

1. *Leaves green beneath*	**2.** *Leaves felted beneath*
200 dentatifolius	*204* taeniarum →
201 ahenifolius	*205* iodnephes
202 Daltrii **Staffs.**	*206* Mercieri **Kent**
203 Gremlii **Middlesex**	

3. *Stem prickles subequal and confined to the angles, but accompanied by crowded glands, acicles and sometimes pricklets, which are either nearly equal, or else equal, but still discrete from the main prickles.*

Ser. **iv.** RADULAE

Stalked glands, acicles and pricklets equal and short on the stem, slightly less so on the petioles, petiolules and panicle.

1. *Sepals reflexed*	**2.** *Sepals patent or erect*
† Glands and acicles all very short	† Leaves felted, glands very short
207 radula →	*218* rudis →
208 sectiramus →	*219* radulicaulis →
209 macrostachys	*220* prionodontus
210 adenanthus **Kent**	†† Leaves mostly green beneath
211 Genevieri	(*a*) Petals white or pale pink
212 crispus **Wales**	*221* granulatus →
213 uncinatiformis **Norfolk**	*222* regillus
†† Glands and acicles rather long, unequal	(*b*) Petals bright pink
214 discerptus →	*223* micans →
215 echinatoides →	*224* pulcher
216 aspericaulis →	
217 malacotrichus **Kent**	

Ser. **v.** APICULATI

Stalked glands, acicles and pricklets numerous, rather unequal. Panicle usually broad, equal and truncate. White flowers with red styles are frequent.

Subser.† FOLIOSI	Subser.†† PALLIDI	Subser.††† SCABRI
Petals white or pink, rarely bright pink. Filaments usually white. Styles often red. Panicle and stem mostly hairy.	As subseries FOLIOSI, but the sepals at length more definitely and uniformly erect.	Flowers as subseries FOLIOSI, usually white. Panicle and stem pubescent or felted. Glands mostly short. Prickles

FOLIOSI (*cont.*)

Sepals reflexed; or partly reflexed and partly patent, sometimes partly erect. Peduncles often bunched or deeply divided.

1. *Leaves distinctly felted beneath*
225 foliosus →
226 rubristylus
227 pseudadenanthus
228 subtercanens
229 sagittarius **Penins.**
230 teretiusculus

2. *Leaves green or slightly greyish beneath*
   † *Filaments slightly shorter than styles*
231 Bloxamii →
232 largificus
   †† *Stem glabrous or glabrescent*
233 cavatifolius
234 corymbosus
235 exsolutus
   ††† *Stem hairy, terminal leaflet roundish (ovate)*
236 fuscus →
237 acutipetalus →
238 trichodes
239 hirsutus
240 hirsutissimus **W.**
   †††† *Stem hairy, terminal leaflet narrow-obovate or rhomboid-ovate*
241 Adamsii
242 Watsonii
243 apiculatiformis
244 racemiger

PALLIDI (*cont.*)

1. *Robust; large leaved*
   † Stem hairy
245 pallidus →
246 drymophilus
247 Loehrii
   †† Stem glabrescent
248 spadix

2. *Leaves small or moderate*
   † Upper leaves felted
249 chlorocaulon
250 Menkei
251 morganwgensis
   Wales
   †† All leaves green beneath; petals pink
252 argutifolius
253 acutifrons **W.**
   ††† All leaves green beneath; petals white
   (*a*) Styles greenish
254 euryanthemus →
   (*b*) Styles red
255 insectifolius →
256 brachyadenes
257 foliolatus
258 acidophyllus
259 erubescens

SCABRI (*cont.*)

usually shortish. Leaflets finely toothed. Sepals patent or erect.

1. *Stem pruinose, hairy*
   † Leaves large or long
260 microdontus →
261 thyrsiflorus
262 derasifolius
263 putneiensis
264 curtiglandulosus **W.**
   †† Petals often minute
265 scaber →
266 tereticaulis
   ††† Petals moderate
267 praetextus
268 derasus **Essex**
269 vallisparsus **Kent**
270 frondicomus **Kent**

2. *Stem hairy, epruinose*
271 longithyrsiger →
272 homalodontus
273 truncifolius
274 concolor **Kent.**

3. *Stem glabrous, pruinose*
275 cyclophorus
276 pseudo-Bellardii
277 luteistylus **Beds.**
278 scaberrimus
279 laxatifrons **W.**

Subser.†††† OBSCURI

Flowers showy. Petals, and usually the filaments and styles also, deep bright pink. Panicle usually pilose-hirsute. The unripe fruit often deep crimson.

1. *Stem glabrous or glabrescent*
280 obcuneatus
281 obscurissimus
2. *Stem ± hairy*
   † Sepals patent or erect; filaments long
   (*a*) Panicle prickles patent
282 entomodontus
283 erraticus
284 adornatiformis
285 obscurus
   (*b*) Panicle prickles slanting, falcate or hooked
286 rufescens →
287 Purchasianus
288 cruentatus
289 grypoacanthus
290 aggregatus
291 fuscoviridis **Penins.**
   †† Sepals patent or erect; filaments short
292 sprengeliiflorus
293 obscuriformis
   ††† Sepals reflexed
294 insericatus
295 rhombophyllus
296 Gravetii

Subser.††††† INCOMPOSITI

Flowers hardly showy, usually white, pinkish or pink. Stem often glabrous or glabrescent, panicle rachis too, often glabrescent below. Leaflets not serrulate. Peduncles not bunched. Prickles sometimes straight and strikingly long and accompanied by stout pricklets. Stalked glands and acicles sometimes few on stem.

1. *Stem glabrous or glabrescent*
   † Leaves felted
297 apiculatus →
298 heterobelus →
299 raduloides
300 melanoxylon
301 pascuorum **W.**
302 longifrons
   †† Leaves green beneath
303 melanodermis
304 inopacatus
305 Reichenbachii

2. *Stem ± hairy*
306 retrodentatus →
307 phaeocarpus →
308 Griffithianus **W.**
309 indusiatus
310 euanthinus **W.**
311 tardus

Ser. **vi.** GRANDIFOLII

Stalked glands and acicles (and pricklets if present) unequal, a few long
ones on the panicle-rachis much
exceeding the rest. Panicle well
developed and elongate, either
pyramidal with racemose lower
branches and ending in a raceme,
or else lax and broad with an abrupt
apex. Mostly luxuriant, stately and
floriferous; later flowering.

1. *Leaves felted*
   † Stem hairy
312 Leightonii →
313 disjunctus →
314 diversus →
315 hostilis
316 Wedgwoodiae
   †† Stem glabrous or glabrescent
317 furvicolor **N.**
318 angusticuspis
319 squalidus

320 pseudoplinthostylus **W.**
321 mutabilis

2. *Stem leaves green beneath*
   † Petals white
322 Turneri **W.**
   †† Petals pink; sepals reflexed
323 Powellii
   ††† Petals pink; sepals patent erect
   (*a*) Terminal leaflet roundish
   (ovate)
324 rotundifolius →
325 rosaceus
326 formidabilis
   (*b*) Terminal leaflet obovate
327 hastiformis
328 Lejeunei
   (*c*) Terminal leaflet ovate or
   elliptic-rhomboid
329 festivus
330 breconensis **Wales**
331 Rilstonei **Penins.**

Sect. VII. GLANDULOSI

4. *Stem prickles very unequally distributed over the sides of the stem, passing into
numerous stalked glands, acicles and pricklets of different lengths.*

Ser. **i.** HYSTRICES

Stem angled, usually strong. The longest prickles strong and broad-based,
usually straight. Panicle with cymose middle branches. Flowers usually
showy, often pink. Petals relatively broad and stamens long.

1. *Flowers 1–1.5 (2 cm.)*
   † Petals white; styles red
332 Murrayi →
333 pygmaeopsis
334 newbridgensis
   †† Petals pink, pinkish or white;
   styles greenish, rarely reddish
   (*a*) Upper leaves felted
335 ochrodermis **W.**

2. *Flowers 2–2.5 (3) cm.*
   † Stem slightly hairy or glabrescent
   (*a*) All leaves green beneath
350 hylocharis →
351 scabripes → **W.**
   apricus var. sparsipilus
352 Koehleri
353 aculeatissimus

40

336 coronatus var. cinarescens

337 adornatus

338 saxicola

339 Chenonii

340 pilocarpus

341 abietinus

342 bavaricus

343 Mikanii

    (*b*) Leaves green beneath; sepals reflexed

344 spinulifer →

    (*c*) Leaves green beneath; sepals erect

345 horridicaulis

346 apricus

347 humifusus

348 plinthostylus

349 rotundellus

    (*b*) Some leaves felted

354 Hystrix →

355 semiglaber **W.**

356 infestus

    †† Stem decidedly hairy

    (*a*) Terminal leaflet coarsely serrate

357 adenolobus →

358 dasyphyllus →

359 Marshallii →

360 Billotii

361 Hartmanii

362 Lapeyrousianus **Surrey**

    (*b*) Terminal leaflet subcompound serrate

363 asperidens

364 emarginatus

    (*c*) Terminal leaflet finely serrate

365 fusco-ater →

366 absconditus

367 oegocladus

368 aristisepalus **W.**

369 hypochlorus

370 hebeticarpus **Penins.**

371 spinulatus **Kent**

Ser. ii. EUGLANDULOSI

Stem rounded or obtuse angled, often pruinose, usually weak and prostrate. Prickles either broad based and curved or narrow-based, slender and weak. Panicle with middle branches racemose, deeply divided, rarely subcymose. Petals moderately broad or narrow, rather small or very small, nearly always white. Styles often red. Stamens often short.

1. *Prickle-base broad, compressed*

372 Schleicheri

373 viridis

374 dissectifolius

375 graciliflorens

2. *Prickle-base narrow, hardly compressed*

    † Glands purple-stalked

    (*a*) Petals white

376 hirtus →

2.†† *Glands yellow or reddish-stalked*

    (*a*) Stem glabrous or glabrescent

385 leptadenes

386 Lintonii

    (*b*) Stem decidedly hairy

    (**i**) Flowers moderate, *c.* 2 cm.

387 elegans →

388 incultus

377 Bellardii

378 pallidisetus **W.**

379 Guentheri

380 perplexus

381 rubiginosus

382 nigricatus **Hib.**

 (*b*) Petals pink

383 purpuratus

384 Durotrigum **Dorset**

 (**ii**) Flowers small, 1–1.5 cm.

389 hylonomus →

390 analogus

391 lusaticus

DESCRIPTIONS

Genus **RUBUS** L.

(Type *R. ulmifolius* Schott f.)

i. Stem herbaceous, rootstock creeping and suckering. Stipules broad, cauline.

Subgen. **I. CYLACTIS** Rafin.

(Type *R. saxatilis* Linn.)

Shoots annual. Leaves 3-nate. Flowers bisexual. Fruit red to deep purple, drupels hardly coherent, parting from the receptacle. Receptacle flat or convex.

1. **R. saxatilis** Linn. 1753, *Sp. Plant.* 494; Syme, 1864, *Eng. Bot.* ed. 3, t. 441.

4x. A small, low plant. Barren stem forming offsets at the tip then dying back nearly to the base, from which rise the flowering shoots in the following year. Flower shoot erect, with a few, weak, straight prickles and simple hairs; terminal leaflet ovate-rhomboid-cuneate, incise biserrate; lateral leaflets subsessile; petiole furrowed. Flowers very few; sepals lanceolate, acuminate, patent; petals 5, glabrous, narrow, erect, white; stamens long, the interior filaments incurved at the apex. Fruit red; drupels 6 in the terminal, 2 in the lateral flower; seed slightly wrinkled or pitted, or nearly smooth. Flowers May–June. Calcicole. Woods and wet rocks.

E., W., S., I. Absent from the midlands and south England, and from central and south Ireland. All Europe except the south-west. Asia to the Altai Mts, Armenia, N. Persia and Kashmir.

2. **R. arcticus** Linn. 1753, *Sp. Plant.* 494; Smith, 1806, *Eng. Bot.* ed. 1, t. 1585.

2x. A very small plant without a barren stem and without prickles. Leaves simple, 3-lobed, or 3-nate and rarely 5-nate; terminal leaflet ovate, cuneate, unequally serrate; lateral leaflets sessile. Flowers 1–3, showy, 1·25–2·5 cm., one terminal and others axillary, long-stalked; sepals and petals 5–7(–10), sepals greenish, glabrous; petals purple or rose, broad ovate, clawed, often toothed; filaments purple, erect, incurved at apex, equalling styles. Carpels pilose. Fruit red to deep purple, drupels numerous, seed faintly wrinkled. Grows usually among *Polytrichum* on gravel or grit covered with bog-earth. Comes up as soon as the snow melts in May and rushes into flower when still only 3–5 cm. high.

E., S. Reported as follows: Ben More, I. of Mull, Rev. Prof. Walker of Edinburgh to J. E. Smith in 1782; high regions of Ben y Glo, Blair, Perth-

43

shire, a specimen sent by Richard Cotton to J. Sowerby after publication of the illustration of *R. arcticus* in *Eng. Bot.* in April 1806; a Yorkshire moor, Watson, *New Botanist's Guide*, 1835; near Kingcausie near Aberdeen, brought unnoticed in peat into J. Irvine Boswell's garden, where it flowered (Syme, *Eng. Bot.*). Possibly brought to Britain in each case as a seed in mud on the leg of a Capercaillie from the Arctic region.

Circumpolar; N. Scandinavia, Finland, Russia, and mountains of central Asia to the Himalayas.

Subgen. **II. CHAMAEMORUS** Focke

Shoots annual. Leaves simple. Flowers unisexual, dioecious. Fruit red, then orange-amber. Receptacle conical.

3. **R. Chamaemorus** Linn. 1753, *Sp. Plant.* 494; Syme, 1864, *Eng. Bot.* ed. 3, t. 440.

8x. Glandular. Prickles absent. Shoots short, erect, all flowering. Leaves reniform with five, obtuse, crenate-serrate lobes; the lower stipules are without leaves, the upper leaves are without stipules. Flowers solitary, terminal, 3 cm., white. Calyx tube obsolescent; sepals ascending. Petals (4) 5 (numerous), hairy, reflexed before they fall. Pedicel and calyx shortly glandular. Receptacle glabrous. Drupels *c.* 20, large. Seed even, punctate. Flowers May–June. Mountain bogs, peaty moors.

N. England, Wales, Scotland, N. Ireland. Circumpolar. N. Europe to S.W. Greenland and Spitzbergen to *c.* 78° 30′ N. lat. N. Asia, N. America.

ii. Stem shrubby. Stipules petiolar.
a. Shoots biennial, bark disrumpent. Leaf-fall autumnal.

Subgen. **III. IDAEOBATUS** Focke

Shoots not rooting at tip. Roots suckering. Fruit pubescent, red, coherent, separating whole from the elongate receptacle. Leaves pinnate; stipules small, filiform. Flowers bisexual.

4. **R. idaeus** Linn. 1753, *Sp. Plant.* 492; Syme, 1864, *Eng. Bot.* ed. 3, t. 442.

2x. Stool shoots with few root suckers but numerous prickles. Leaflets 3, 5 (7), incise-serrate, closely white felted beneath; terminal leaflet often sublobate below the middle; lateral leaflets nearly sessile. Panicle interrupted, with nodding, few-flowered racemes, terminal and axillary. Flowers small calyx-base discoid; petals white, narrow, erect, glabrous; stamens erect, equal, uniseriate or biseriate, equalling the styles. Fruit red or amber; drupels numerous; seeds small, pitted. Flowers mid May to June.

One form has all axes coated with fine felt; another has the same parts glabrous.

forma **obtusifolius** (Willd.) W. Wats.; *R. obtusifolius* Willd. 1811, *Berl.
Baum 2.* ed. 2, 409; *R. idaeus* var. *anomalus* Arrhen. 1840, *Mon. Rub.
Suec.* 14; *R. Leesii* (Bab.) Ed. Lees in Steele, 1847, *Handbk. Field Bot.* 60;
Syme, 1864, *Eng. Bot.* ed. 3, t. 443.

A unisexual male form; it has blunt sepals and blunt leaflets.

Woods and commons, usually near houses. E., I., S., W. All Europe
except the extreme north and south, becoming a mountain plant in the
Mediterranean area.

Subgen. **IV. GLAUCOBATUS** Dumort. pro parte

Shoots decorticating, rooting at tip. Roots not suckering (unless injured). Fruit
glabrous, black but pruinose; drupels loosely coherent, and loosely adherent
to the receptacle. Leaves 3-nate; stipules ovate-lanceolate. Flowers bisexual.

> Up to the present *R. caesius* has been included by all systematists
> except Dumortier in Moriferi. Yet *R. caesius* can be distinguished
> from all Moriferi both by characters of its own and by others that
> are of the same order as, or actually the same as, those that dis-
> tinguish the other subgenera. It has, for example, pruinose fruit
> and a corymbose panicle; it has a decorticating stem and acicular
> prickles like Idaeobatus and *R. saxatilis* (Cylactis); differentiated
> terminal and lateral fruits like *R. saxatilis*; disintegrating fruit like
> both species of Cylactis; finally it has fruit which is partially
> separable from the receptacle, as the fruit is fully separable in
> Idaeobatus. In all these respects *R. caesius* is unlike any morifer.

5. **R. caesius** Linn. 1753, *Sp. Plant.* 493; Syme, 1864, *Eng. Bot.* ed. 3, t. 456.

4x, 5x. Stem flagelliform, greatly branched, terete, pruinose, usually
glabrous, rarely with a few, short glands; prickles few or numerous, weak,
short, straight or curved. Terminal leaflet subcordate, (rhomboid-)ovate,
acuminate, compound-serrate, often ± lobate; lateral leaflets bilobed, very
shortly stalked. Petiole furrowed throughout. Inflorescence consisting of
a terminal and a few axillary 2–3 (5)-flowered corymbs. Flowers 2–2·5 cm.,
pedicels and calyx felted and shortly glandular; sepals clasping; petals ovate
or elliptic, glabrous. Filaments equalling the styles. Carpels and receptacle
glabrous. There are fewer styles in the lateral than there are in the terminal
flowers, which consequently produce many more drupels than the lateral
flowers.

E., W., S., I. Common in wet places, in the open or in woods, and on
disturbed clay or limestone. Europe, middle latitudes south of 62–60°; absent
from the southern mountains, peninsulas and islands. Asia, eastwards to the
Altai Mts and Persia.

var. **grandis** W. Wats. 1956, *Watsonia*, **3**, 285.

5x. A stouter form with large, rugose leaves, closely pubescent beneath, with broad, overlapping petals (often 12–15 in a flower) and broad ovate sepals. Styles and carpels about twice as many as in the typical species. Very common in N.W. Kent; seen in 16, 24, 35, 67.

var. **denticulatus** W. Wats. 1956, *Watsonia*, **3**, 285.

4x. Slender. All leaflets regularly denticulate.

Gravel Road, Addington, Surrey.

var. **pinnensis** W. Wats. 1956, *Watsonia*, **3**, 285.

5x. Robust. Terminal leaflet subcordate-ovate or rhomboid-ovate, gradually acuminate, sharply serrate.

By the Pinn Brook, Swakeley's, Uxbridge, Middlesex.

b. Shoots biennial or perennial, bark not disrumpent. Leaf-fall in autumn, winter or spring.

Subgen. **RUBUS**
(Type species *R. ulmifolius* Schott f.) (MORIFERI Focke)

Terminal leaflet with *c.* 6–9 secondary nerves on each side of the midrib. Leaf thin or subcoriaceous. Fruit glabrous at maturity, not pruinose; drupels coherent, entirely adherent to the oblong or elongated, rather pointed receptacle. Leaves 3-nate, or 4–7-nate-pedate, or digitate. Stipules sublanceolate, linear or filiform.

Subgen. **IDAEOBATUS × RUBUS**

Sect. **I. SUBERECTI** P. J. Muell.

(Type *R. nessensis* W. Hall)

This is a group of species which combine the characters of *R. idaeus* with characters of SYLVATICI and perhaps DISCOLORES. Hitherto they have been classed with MORIFERI, from which strictly their 'idaean' characters should have excluded them. They have now been given an independent place proper to their bi-subgeneric descent.

Subser.† **Suberecti veri** Boul. Eglandular. Stem at first erect, later bending but not rooting, hardly branched, glabrous. Leaves 3–7-nate; terminal leaflet usually cordate, ovate, acuminate; petiole usually channelled throughout. Panicle usually short racemose, flowering in late May or June, followed often in July by a larger, more compound panicle proceeding from a seriate bud in the same leaf axil. Sepals green with a contrasting, white, felted border. Petals becoming patent, sometimes vaulted before dropping. Fruit deep red or purple, hardly black. Leaf-fall in autumn. Suckers arise from a creeping root-stock, or from fibrous roots. Stamens reflexed; or sometimes as far as their length permits connivent after flowering.

Subser.†† **Semisuberecti** W. Wats. Habit more bushy, suckering less, tip-rooting more. Stem glabrous or thinly hairy. Leaves less often 6, 7-nate. Panicle more compound. Petals incurved. Sepals sometimes greyish felted. Fruit black. Leaf-fall in winter. Stamens connivent.

Subser.† **Suberecti veri** Boul.

1. Prickles very short, conical or subulate, 1–1·5 mm. Ripe fruit dark red.

6. R. nessensis W. Hall, 1794, *Edinb. Trans.* **3**, 20; *R. suberectus* Anders. 1815, *Trans. Linn. Soc.* **11**, 218. **Fig. 1.**

4x. Stem erect, up to 3 m., glabrous, glaucescent; *lower prickles reduced to pricklets, purplish-black. Leaves* large, 3, 5 (7)-nate, *shining*, thin, *glabrescent*; petiole faintly channelled throughout, stipules filiform; terminal leaflet ovate, long-acuminate. *Panicle* occupying the whole of the short, distichous, horizontal flowering shoot, *nearly unarmed*, branches sharply ascending. Flowers 2·75 cm.; *petals* white, often flushed red outside, *glabrous*, ovate or elliptic; filaments slightly longer than the styles. Carpels glabrous. Receptacle glabrous. *Flowers c. 20–26 May* to end of June. *The withered stamens hang down* over the reflexed sepals. The fruit tastes of raspberry.

E., W., S., I. Heaths, upland woods and mountains. France, Belgium, Netherlands, Denmark, Scandinavia, central and eastern Europe.

7. R. scissus W. Wats. 1937, *Journ. Bot.* **75**, 162; *R. fissus* auct. mult.; non Lindley, 1835; Sudre, 1908–13, t. 2 (as *R. fissus*).

4x. *Stem inclined*, up to 1·5 m., often only 0·5 m., thinly pilose; prickles numerous, scattered, subulate. *Leaves opaque*, often 7-nate, *petiole distinctly furrowed throughout; leaflets* thick, *densely pubescent beneath*, *imbricate*, plicate, serrate; terminal leaflet cordate-ovate, shortly pointed. Panicle short; flowers 2–2·5 cm., petals often 10–15, pink in bud, pubescent, narrow obovate; calyx usually with one pricklet, sepals patent to erect; *stamens short*, some applied to the styles. Carpels pilose. Receptacle pilose. Begins flowering *c.* 30 May. *Fruit usually partly abortive*, perfect at Rushlye Down, Tunbridge Wells.

E., W., S., I. Heaths, moors, woods and mountains. Belgium, Netherlands, Denmark, S.W. Scandinavia, N. Germany, Russia.

2. Prickles stronger, larger. Ripe fruit black. Petals pubescent or pilose. Affinity with *R. idaeus* less marked.

† Stamens much longer than the styles. Stem furrowed. Basal leaflets stalked.

8. R. sulcatus Vest in Tratt. 1823, *Ross.* **3**, 42; Sudre, 1908–13, 17, t. 3.

4x. *Stem tall, erect*, slightly branched below, *prickles strong*, not numerous. Leaves large, 5-nate, petiole short; *leaflets* contiguous, *pubescent* beneath; terminal leaflet cordate-ovate, long-acuminate. *Panicle* subracemose, lax, long, *nearly unarmed; terminal leaflets narrow, cuneate, deeply incise-serrate.*

Flowers 3 cm.; sepals hairy and felted, loosely reflexed, the leafy tips ascending; petals obovate, pinkish in bud; carpels glabrous; receptacle subglabrous. Fruit oblong, good, seeds large. Flowers from *c.* 12 June. Luxuriant in fresh woods, spiring up to 3 m.

E., W., S. Scarce; 3, 6, 9, 22, 23 (Cogges Wood abundant), 36, 98, 103. Central and east France, Belgium, Holland, Denmark, S. Scandinavia, mid Europe, Switzerland and N. Italy.

9. **R. Bertramii** G. Braun ex Focke, 1877, *Syn. Rub. Germ.* 117; *R. biformis* Boul. in Rouy & Camus, 1900, *Fl. France,* **6**, 40; Coste, 1903, *Fl. France,* **2**, 12, t. 1156.

Stem oblique, up to 1 m.; *prickles slender,* falcate. Leaves large, 5–7–nate, *petiole prickles small, hooked; leaflets* deep green, *glabrescent,* sharply, coarsely or subcompound-serrate; terminal leaflet broad or roundish ovate, acuminate. *Panicle ample,* lax, subcorymbose, with slender, few-flowered branches, *prickles small, sparse, hooked.* Flowers *c.* 2·5 cm.; sepals long-pointed, clasping the base of the fruit; petals pinkish, obovate-cuneate; stamens usually much longer than the styles, anthers often pilose. Carpels with long hairs. Receptacle glabrous. Fruit oblong, good, sweet.

E., W. Scarce; 3, 5, 9, 13, 14, 17, 22, 35, 39, 42, 44, 58. France, Belgium, Denmark, W. Germany, Bavaria.

var. **minor** (Boul.) W. Wats. 1956, *Watsonia,* **3**, 285; *R. biformis* var. *minor* Boul. in Rouy, 1900, *Fl. France,* **6**, 40.

Stem *prickles numerous, stout, hooked.* Leaf smaller. Panicle small, *prickles strong, numerous, hooked.*

England. Scarce; 2, 3, 22, 58. France.

†† Stamens about equalling styles.

10. **R. plicatus** Weihe & Nees, 1822, *Rub. Germ.* 15; Syme, 1864, *Eng. Bot.* 445.

4x. Common. Known by the slender, falcate, yellow and crimson prickles, by the *broad, shortly cuspidate, plicate terminal leaflet,* the abruptly clawed, elliptic petals, the cuspidate, concave-patent sepals, and the strongly hairy receptacle. *Leaflets* 3–5 (6, 7), *imbricate.* Anthers sometimes pilose. Suckers arise close to the stool from underground woody roots. Flowering starts *c.* 13 June.

E., W., S., I. Common on low-lying heaths and in upland woods. All central and northern Europe to Russia.

var. **macrander** Focke, in Aschers. & Graeb. 1902, *Syn. Mitt.-Eur. Fl.* **6 i**, 461.

Stamens much longer than the styles.

England. Very rare; 13 (Blackdown). Germany, Denmark.

var. **micranthus** Lange, 1886, *Dan. Fl.* 770; *R. plicatus* var. *hemistemon* sensu Rogers pro parte, 1900, 22.

A dwarf form with small leaves and small flowers.
England: 14, 16, 36. Denmark.

11. **R. ammobius** Focke, 1877, *Syn. Rub. Germ.* 118.

Stem *c.* 1·75 m., slightly furrowed, glabrescent, glaucescent; prickles yellowish, flattened, long, slender, slightly bent. Leaves 5 (7)-nate, petiole long, deeply furrowed in the lower half, prickles hooked; *leaflets glabrous above, silky greyish felted and pubescent beneath*; terminal leaflet cordate-ovate, prolonged-acuminate, finely, sharply and unequally toothed. *Panicle well developed*, leafy, prickles fine, falcate and hooked. Flowers *c.* 2 cm.; sepals leafy-pointed, loosely reflexed; petals pinkish, obovate, long-clawed; styles yellowish. Carpels and receptacle pilose. Fruit purplish black, thimble-shaped. *The upper leaves of late panicles are grey felted.*

England, only known in 17, a large patch at the roadside on Fairmile Common. Netherlands, N.W. Germany, Denmark.

12. **R. fissus** Lindl. 1835, *Syn. Brit. Fl.* ed. 2, 92; non auct. mult.; *R. Rogersii* E. F. Linton, 1894, *Journ. Bot.* **32**, 213; *R. affinis* ssp. *Rogersii* (E. F. Linton) Sudre, 1908–13, 21, t. 8.

4x. *Stem* 0·5–2 m., becoming *dull, deep purple*, angled to furrowed, erect then arching; *prickles numerous, long*, slanting or bent. Leaves 5–7-nate, small, petiole long, prickles numerous, hooked; leaflets dull and glabrous above, *closely pubescent, appearing felted beneath; terminal leaflet* ovate, gradually acuminate, *serrulate-dentate.* Panicle leafy, subracemose, pedate, patent; rachis flexuose, prickles numerous. Flowers 2–2·5 cm.; sepals triangular-ovate, tips linear, leafy, with pricklets, loosely reflexed; petals pinkish, elliptic-rhomboid; styles yellowish. Carpels and receptacle pilose. Fruit purplish black, small. Flowers *c.* 1 June.

Endemic. E., W., S., I. Widespread, local in the south, frequent in the north, generally in the open among heather.

13. **R. arrheniiformis** W. Wats. 1956, *Watsonia*, **3**, 285.

A low, *small-leaved* bush with bright pink, often semi-double flowers. Stems tufted, arching, shining, brownish, angled, nearly glabrous, shortly branched in late summer; prickles few, shortish, yellowish, patent to falcate from a narrow base. Leaves yellowish green, 5-nate-digitate, petiole prickles small, falcate; *leaflets* remote, glabrescent above, *pubescent and velvety below*, serrate; terminal leaflet ovate-acuminate, base broad and cordate, or narrow truncate, emarginate; basal leaflets very shortly stalked. *Panicle long* with usually one simple leaf, *subracemose with 10–20, long, slender, ascending branches each bearing 1–12, small, falcate prickles.* Flowers 1·75 cm.; sepals acuminate, pubescent, patent to clasping; petals obovate with broad, obtuse apex. Carpels and receptacle usually glabrous. Flowers from *c.* 10 June, the longer panicle in late summer.

E., W., S. Rare; 14, 15, 17, 48 (Wm. Matthews, 1852), 88, 106. France, border of the departments *Vienne* and *Haute-Vienne* (Lamy, 1864)

Subser.†† **Semisuberecti** W. Wats.

1. Upper leaves of stem and panicle pubescent and greenish grey or greyish white felted beneath.

† Stem glabrous or glabrescent. Prickles long, slender, patent or bent, middle panicle branches subcymose.

14. **R. affinis** Weihe & Nees, 1825, *Rub. Germ.* 18; Sudre, 1908–13, 21, t. 7. **Fig. 2.**

4x. Occasionally suckering. Axes becoming red-brown. Stem arching, branching and rooting at the tip, obtuse below to furrowed above. Leaves 5 (6, 7)-nate, petiole faintly furrowed; *leaflets imbricate*, irregularly and coarsely serrate-undulate; terminal leaflet broad cordate-ovate, acuminate. Panicle at first racemose or moderately compound, later enlarged by the growth of long, ascending lower branches, *prickles strong, long, patent and falcate. Flowers 2–3 cm.*; sepals greyish green felted, aciculate, loosely reflexed with the leafy tips rising; *petals large, broad ovate*, pilose, pink (rarely pure white); *filaments a deeper pink, longer than the deep carmine styles*; anthers subpilose. Carpels pilose. *Receptacle densely pilose, especially below the carpels.* Fruit large, of large drupels, but imperfect. Flowers *c.* 20 June, often continuing through July, August and much of September. Leaf-fall in late autumn. Recalls *R. gratus.*

E., W. Rare; in the open on moist heaths and grassland; 3–6, 9–11, 13–17, 19, 22, 25, 27, 34–36, 41, 45, 49, 52, 54. Very rare in Sweden, France, Netherlands, Belgium, Switzerland, more frequent in N.W. Germany from the Rhine to the Elbe.

15. **R. holerythrus** (*holerythros*) Focke, 1895, *Journ. Bot.* **33**, 47; *R. nitidus* ssp. *holerythrus* (Focke) Sudre, 1908–13, 21, t. 6; non Rogers, 1900, 25; *R. affinis* var. *Briggsianus* Rogers, 1900, 24.

A low bush. Stem arching, branching, tip-rooting and suckering, *shining brown.* Leaves moderate, 5-nate; leaflets not imbricate nor undulate, glabrous above, deeply serrate; *terminal leaflet* ovate-oblong, *subobtuse or rather shortly acuminate*, base cordate to subentire. Panicle mostly short and broad and relatively simple, but *sometimes* more compound with subcymose, patent, numerous flowered branches and *powerful, patent and falcate prickles, bracts broad linear, crimson.* Flowers 2·5 cm., incurved; *sepals* often aculeolate, olive green, white-edged, *short-pointed*, incompletely *reflexed*; petals pinkish or pink, roundish; filaments long, pink or white. Carpels glabrous. Receptacle pilose. Anthers sometimes pilose.

S.W. England, Wales and Ireland. Locally frequent; 1–5, 9, 12, 35, 45, 48, 49, 52. Guernsey, Jersey. H. 1, 2, 35, 38. West of France from the Channel to Bordeaux.

†† Stem thinly hairy throughout.

16. **R. nobilissimus** (W. Wats.) Pearsall 1934 (for 1933), *Bot. Exch. Cl. Rep.* **10**, 485; *R. opacus* sensu Rogers, 1900, 23; non Focke.

*4x. *Stem tall*, up to 2 m., not rooting, obtuse-angled, *grooved to the base*; *prickles few, moderate*, broad-based, falcate. Leaves 5-nate, digitate, dull above, grey felted beneath, petiole furrowed; leaflets becoming convex, finely serrate-dentate; *terminal leaflet ovate-oblong, gradually and long acuminate*, base subcordate to entire; intermediate leaflets long. Panicle narrowed, sub-racemose, often with long, numerous flowered lower branches; rachis glabrescent below, felted above, *prickles very few, small*. Flowers 2 cm., incurved; *sepals patent to erect*; petals elliptic or obovate, crumpled, deep pink; filaments white or pinkish, long, anthers subpilose. Carpels and receptacle pilose.

Endemic. E., W., I. Rare; streamsides and watery woods; 3, 6, 9, 11, 17, 41, 69, H. 2.

2. All leaves green beneath.

17. **R. nitidus** Weihe & Nees, 1822, *Rub. Germ.* 49; Sudre, 1908–13, 19, t. 5 *a*.

3x. A *small-leaved*, high bush up to 1 m., suckering close to the stool, but not often, and sometimes tip-rooting. *Stem shining*, reddish brown to *deep violet*; prickles numerous, long, slender, patent or bent up or down towards the tip, *at first bright yellow*. Leaves 5-nate, petiole furrowed, prickles large, hooked, clustered; *leaflets small*, shining above, pubescent, glabrescent beneath *with hard ribs*; terminal leaflet base notched or entire, ovate to oblong-obovate, acute, unequally-serrate. Panicle long, lax, broad, equal with deeply divided, 2–5-flowered branches; rachis, peduncles, pedicels and *calyx base armed with numerous hooked prickles*, at other times almost unarmed. Flowers *c.* 2·75 cm., incurved, deep pink, rarely white; sepals green, patent; *petals roundish, long-clawed*; filaments white or pink, longer than the styles. Carpels and receptacle glabrous. Fruit small. *Flowers hardly before July.*

E., W., S. Widespread but not frequent; wet heaths and streamsides; 3, 4, 6, 9–11, 13, 16–18, 22, 34, 36, 39, 42, 43, 45, 48, 49, 52, 58, 62, 97, 99, 103. Portugal, France, Belgium, Netherlands, Denmark, S. Sweden, Germany, Switzerland, Austria.

18. **R. integribasis** P. J. Muell. in Boul. 1866, *Ronc. vosg.* 23; *R. nitidus* ssp. *integribasis* (P. J. Muell.) Sudre, 1908–13, 20, t. 6.

6x. Stem branched, high arched, up to 2 m., furrowed, glabrescent, *prickles numerous, short, subulate, rather* clustered. *Leaves large*, 5-nate, *petiole long*, prickles hooked and falcate; *leaflets broad, closely pubescent and rather velvety beneath*, irregularly incise-serrate; *terminal leaflet roundish* ovate, *shortly acuminate*. Panicle short and broad, somewhat leafy and compound, lower branches semi-erect, pedicels long, *prickles weak, short, nearly straight.* Flowers 2–2·5 cm.; sepals unarmed, loosely reflexed; petals pink, elliptic-obovate, entire; filaments white, or pink at base, slightly longer than the greenish styles. Carpels subglabrous. Receptacle glabrous.

England, frequent on Surrey commons. France, Belgium, Germany, Switzerland.

Subgen. GLAUCOBATUS × RUBUS:

Sect. II. TRIVIALES P. J. Muell.

These species combine the characters of *R. caesius* with those of MORIFERI. Although no British species has pruinose fruit, one or two Continental species placed here have; and most species show the corymbose panicle and the very short-stalked lateral leaflets and broad stipules of *R. caesius.*

In the following descriptions except where otherwise stated the stem is glaucous, the petioles narrowly grooved throughout, the leaves strigose above, and the stamens about equalling the styles.

1. Prickles equal or nearly equal, stalked glands, if any, few and short.

† Petals pink, sepals reflexed.

19. **R. conjungens** (Bab.) W. Wats. 1935 (for 1934), *Lond. Nat.* 66; *R. corylifolius* var. *conjungens* Bab. 1851, *Man. Brit. Bot.* ed. 3, 129.
4x, 5x. Stem and rachis angled, prickles short, rather strong. Leaflets imbricate; *basal leaflets stalked*; terminal leaflet roundish, cuspidate, not lobed, bluntly serrate. Panicle long and broad, dense at the apex, felted. Filaments white or pink, styles yellowish. Carpels pilose. Fruit shining black, small, partly defective. *Flowering end of May.*

E., W., S., I. Common. Jersey and Guernsey.

20. **R. ooliticus** W. Wats. 1956, *Watsonia,* **3**, 286; *R. corylifolius* ssp. *calcareus* Rogers ex Riddels. 1920, *Journ. Bot.* **58**, 104.

N.v.v. Slender. Axes a light glaucous green, reddening. Stem angled, prickles short, subulate, or broad-based falcate. *Leaves glabrous above,* densely greyish pubescent beneath; *terminal leaflet narrow ovate-elliptic, acuminate-cuspidate,* unequally spreading serrate, *not lobed.* Panicle moderately compound, *rachis densely villose, prickles numerous,* slightly slanting, *pale yellow to bright red.* Flowers 2–2·5 cm.; petals elliptic, narrowed to the claw; *filaments* white, *much longer than the yellowish styles.* Carpels pilose. Receptacle glabrous. Fruit good.

Endemic. Cotswolds and Mendips chiefly.

21. **R. eboracensis** W. Wats. 1956, *Watsonia*, **3**, 286.

Eglandular, felted. Stem sharp-angled, prickles short, patent. *Leaflets subcompound, deeply, coarsely and bluntly serrate, white felted and velvety pubescent beneath*; terminal leaflet moderately broad, subcordate, ovate-rhomboid, shortly acuminate; *basal leaflets sessile and large, imbricate.* Panicle ± floriferous, pyramidal, *branches mostly fascicled*; rachis felted, pubescent, prickles very small on the pedicels; *leaves often 4, 5-nate, the terminal leaflet narrow-rhomboid.* Petals broad ovate. Stamens white, styles greenish. Carpels pilose.

Endemic. England: midlands and the north; 62 (Gormire, 27 July 1937, Watson; also at Sowerby), 67, 55 (Twycross, A. Bloxam, *c.* 1848 as *R. corylifolius* var. *bifrons*).

†† Petals white; sepals reflexed.

22. **R. sublustris** Ed. Lees in Steele, 1847, *Handb. Field Bot.* 54; Syme, 1864, *Eng. Bot.* ed. 3, 455 (as *R. corylifolius*).

5x. *Stem and panicle-rachis round*; *prickles* scattered, *the lower ones very short, deep purple*, the upper ones longer and subulate, glossy dark red in the sun. Leaves 5 (6)-nate, densely, silvery-grey pubescent beneath; *terminal leaflet* cordate, roundish ovate, acuminate, *often lobed above or below the middle*, or rarely divided on one side to the base, *sharply, coarsely and deeply toothed.* Panicle spreading, lax, subcorymbose, the branches divided halfway. Flowers large and showy; petals roundish ovate. Carpels with long hairs. Receptacle pilose. *Flowers mid May.*

E., W., S., I. Common. Jersey.

23. **R. latifolius** Bab. 1851, *Man. Brit. Bot.* ed. 3, 94.

N.v.v. Scotland. Stem furrowed, prickles short, slender, slanting. *Leaves large*, velvety, thickly pubescent beneath; *terminal leaflet* broad cordate-ovate, acuminate, *slightly lobed, compound-serrate*; basal leaflets sessile, imbricate. *Panicle* equal, racemose above, lax, all *branches ascending*, pilose and felted above, with a few, pale, *short, falcate prickles* and a few submerged glands; the upper leaves grey-white felted beneath. Flowers 2·5 cm.; sepals ovate-lanceolate, pubescent felted; petals broad ovate. Carpels and receptacle nearly glabrous. Fruit good.

Endemic. Scotland: 75, 82–84, 88, 90.

24. **R. Bucknallii** White, 1899, *Journ. Bot.* **37**, 389.

Sparsely glandular. *Stem obtuse-angled, greyish to yellow-brown, very hairy*; prickles scattered, numerous, *very slender*, patent or slanting. Leaflets 5, imbricate, pilose beneath; *terminal leaflet* broad cordate-ovate, acuminate, *lobate, closely and sharply serrate*; petiole prickles slanting or hooked; stipules linear, glandular. *Panicle* equal, dense at apex, with short axillary branches, *very hairy and felted*; prickles numerous, slender. *Flowers large*; sepals ovate-attenuate, felted, glandular; petals broad elliptic, pinkish white; filaments longer than the styles. Carpels glabrous.

Endemic. E., I. Rare; 34 (on oolitic hills between N. Nibley and Wotton-under-Edge, White), 36 (fide White (but ?)). H. 12 (Marshall, 1896, ref. no. 1707 as *corylifolius* sp. coll., fide Rogers).

††† Petals white; sepals patent to erect.

25. **R. umbelliformis** Muell. & Lef. 1859, *Pollichia,* 265.

*5x. Kent. Slightly glandular. *Stem with short tufted hair and numerous, scattered, short prickles. Leaves glabrous above,* pubescent beneath; terminal leaflet emarginate, ovate-oblong, acuminate-cuspidate. Panicle well developed with long, widely ascending lower branches; *mid branches deeply divided,* prickles numerous, nearly straight. *Flowers c. 2·5 cm.*; *petals broad oblong, white; stamens white;* styles greenish. Carpels glabrous. Flowers mid June. No doubt derived from *R. caesius × macrophyllus.*

England: frequent on Shooters Hill, W. Kent, with *R. macrophyllus.* Normandy, Paris basin and east of France.

2. Prickles equal or nearly equal, stalked glands rather numerous.

† Sepals patent or erect.

   (*a*) Petals pink or pinkish.

26. **R. Balfourianus** Bloxam ex Bab. 1847, *Ann. Nat. Hist.* **19**, 68; Sudre, 1908–13, 237, t. 209. **Fig. 3.**

4x, 5x. A large, slender, pubescent bramble with large, soft leaves, which are green to greyish beneath, and a large, lax, spreading panicle with *flat, pink flowers 3–5 cm. across,* with *short, pink stamens, hairy anthers* and *red styles.* Calyx and pedicels shortly glandular, the sepal tips long and leafy. Carpels and *receptacle a good deal hairy,* the latter especially so *below the lowest carpels.* Fruit usually imperfect but sometimes large and good, with a mulberry flavour. This is *the first bramble to flower,* sometimes *as early as 8 May, continuing until October.* It probably originated from a cross of *R. caesius* with *R. gratus.*

E., W., S., I. Frequent in open, *oozy spots* in lanes, hedgerows and woods, not shunning clay. N. France and Germany.

27. **R. Warrenii** Sudre, 1904, *Observ. Set. Brit. Rub.* no. 132; *R. dumetorum* var. *concinnus* Warren, 1870, *Journ. Bot.* **8**, 153.

4x. *Leaves small, neat;* terminal leaflet cordate, roundish, cuspidate, evenly and sharply serrate, grey felted beneath. Panicle large, dense-topped, felted, with unequal but not long-stalked glands and acicles, pedicels and calyx-base armed with numerous acicular prickles. *Flowers rather small; petals roundish, erose,* hardly clawed, pinkish. *Styles reddish-based.* Fruit small.

Endemic. England: frequent from Kent and Surrey to Lancs. and Yorks.

28. **R. adenoleucus** Chab. 1860, *Bull. Soc. Bot. Fr.* **7**, 267.

5x. *Usually small in all its parts, and slender. Stem* round, nearly glabrous, *glossy red at the tip*, prickles falcate and recurved. *Leaves* 3 (5)-nate, *glabrescent beneath*, petiole prickles falcate or hooked; terminal leaflet ovate or obovate. Panicle leafy, glandular, *rachis round, nearly glabrous*, prickles acicular. *Flowers rather small*; sepals greenish, soon clasping; petals pinkish, ovate or oblong, shortly clawed; *filaments long*. Carpels glabrous. Fruit small. The glands, deep crimson at first, at length disintegrate into a white matter; hence the name. The species is derived presumably from *R. caesius × rudis*.

England: frequent in the south-east. France, Belgium, Switzerland.

29. **R. Wahlbergii** Arrhen. 1840, *Rub. Suec.* 43; Sudre, 1908–13, 238, t. 209.

Sussex. Robust. Stem sharp-angled to *furrowed*, with scattered, fine, short glands and acicles, thinly hairy; prickles yellowish, subulate, patent and falcate. Leaflets 4–5, almost glabrous above, thinly hairy to greyish felted beneath; *terminal leaflet* rather short-stalked, *cordate, very broad ovate, gradually acuminate*, irregularly and coarsely incise-serrate; *intermediate and basal leaflets broad imbricate. Panicle* compound, apex subcorymbose, thinly hairy *with numerous, straight, longish prickles* and frequent glands. *Flowers c.* 2·5 *cm.*; sepals greyish felted, with prickles and glands; *petals broad ovate*, pinkish. *Stamens long.* Fruit good.

England: 14 (Grisling Common, near Piltdown, and neighbourhood). N. France (one station), N. Germany, Denmark, Scandinavia.

(*b*) Petals white.

30. **R. halsteadensis** W. Wats. 1956, *Watsonia*, **3**, 286; *R. dumetorum* var. *raduliformis* A. Ley, 1904, *Journ. Bot.* **42**, 120; non *R. raduliformis* Sudre, 1910.

★5x. *Axes becoming reddish brown.* Glands and acicles very short. *Stem pilose.* Leaves softly pubescent and grey-white felted beneath; *terminal leaflet* emarginate, *broad ovate, triangular-acuminate* or cuspidate, slightly lobate. *Panicle flowers and fruits congested*, prickles mostly falcate. Calyx broad-based, sepals broad ovate, patent to erect. *Petals moderate, broad ovate, apex notched* or erose. Stamens investing the styles after the petals fall. Carpels glabrous. Fruit suboblong, good.

Endemic. England and Wales: frequent; 2, 6, 13–19, 21, 23, 24, 26, 28, 33–38, 45, 55.

†† Sepals reflexed.

31. **R. purpureicaulis** W. Wats. 1950, *Watsonia*, **1**, 290; *R. corylifolius* var. *purpureus* Bab. 1869, *Brit. Rub.* 267.

★5x. *Axes becoming deep purple. Stem* angled, *furrowed in its upper part*, glaucescent, with a few, fine, tufted hairs at first and a good many short glands, acicles and intermediate pricklets; prickles irregularly distributed, patent or slightly falcate. Leaves 5-nate-subpedate, softly pubescent beneath;

terminal leaflet cordate to entire, *roundish or broad ovate* to elliptic, cuspidate-acuminate; *petiole long, widely grooved towards the base*, prickles falcate, *stipules linear or linear-lanceolate*, glandular-fimbriate. Panicle long and narrow, interrupted, with erect lower branches, felted and shortly glandular; *rachis flexuose*, prickles numerous, unequal, patent and slightly falcate. *Flowers large*; sepals glandular-punctate, ovate, cuspidate; *petals broad ovate, cuneate, erose at the apex, downy, pink; filaments white or pink; styles greenish or pink-based. Carpels closely pubescent.* Fruit moderately developed. The upper leaves of stem and panicle are greyish felted.

Endemic. E., W., I. Probably a good deal more frequent than the following records show, and not recognized; 7–10, 14–17, 21, 22, 26, 34, 36, 40, 62, 67, 70. 'Co. Galway.'

3. Prickles unequal; glands $\pm$ numerous, sepals erect.

† Prickles not numerous.

32. **R. tuberculatus** Bab. 1860, *Fl. Cambs.* App. 5, 306; 1870, *Journ. Bot.* **8**, 170, t. 106. **Fig. 4.**

5x. *Common.* Stem angled, striate, glabrescent, *short-stalked glands, acicles and prickles abundant*; prickles patent. *Leaves* 3, 5-nate, pedate, *softly hairy, pubescent or felted beneath; terminal leaflet roundish with straightish sides* below the middle, *shortly cuspidate*, or cordate, broad or ovate *acuminate, finely dentate-serrate, becoming convex; basal leaflets blunt.* Panicle top dense, corymbose, with long, many-flowered lower branches, and with two or more simple felted leaves; *rachis* angled, striate, felted, $\pm$ hairy and shortly glandular, *prickles subulate, patent. Flower-buds large, depressed*; sepals appendiculate, greenish felted and densely glandular; petals broad ovate or obovate, shortly clawed, pink. Carpels pilose or occasionally glabrous. *Fruit large, rather oblong.*

England: common; 1–3, 6–8, 14–18, 20–23, 26, 27, 29, 30, 33–38, 57, 58, 67. Jersey. Apparently not found in Scotland, Ireland or Wales.

33. **R. Babingtonianus** W. Wats. 1946, *Journ. Ecol.* **33**, 342; *R. althaeifolius* Bab. 1869, *Brit. Rub.* 274; non Host 1823.

*4x. *Habit of a strong R. caesius. Stem obtuse angled*, turning a deep red, *often markedly glaucous*, at first thinly hairy and with a little tomentum which disappears, *stalked glands and acicles short and usually few. Leaves* (3) 4, 5-nate, *petiole prickles straight; terminal leaflet* broad ovate (-rhomboid), acuminate, base shallowly cordate to narrowed entire, *margin lobate compound-serrate, ciliate*, densely and softly hairy as if felted, to glabrescent beneath. Panicle leafy, rather dense, racemose, corymbose at apex, with several, sometimes long, lower corymbose branches; the *pedicels armed with numerous fine prickles.* Flowers rather large; sepals appendiculate; *petals large, broad ovate or elliptic, rounded*, abruptly clawed, some notched, *pink; filaments white, styles yellowish,* *or pink-based*. There is a specimen in Herb. Borrer from the Horticultural Society's garden, dated 1821.

Endemic. Scattered from Kent to S. Uist; 8, 13, 15–17, 21, 29, 35, 38, 39, 59, 62, 102, 110. H. 39.

34. **R. rubriflorus** Purchas, 1894, *Journ. Bot.* **32**, 139 and 187.

Western. Petals, filaments and styles deep pink. Axes and armature crimson. Stem slender, obtuse angled, glabrescent, *with numerous short glands and fewer acicles and minute pricklets; prickles slender*, narrow-based, patent or slanting. Leaves yellowish, 5-nate, densely pubescent as if felted and velvety beneath; terminal leaflet subcordate-ovate or elliptic-acuminate, unequally serrate. *Panicle broad, branches spreading*, felted and pubescent, pilose in the shade; upper simple leaves felted; bracts leafy. Sepals green, ovate-acuminate-appendiculate. Petals ovate or obovate. Stamens longer than the styles. Carpels pilose. I have not seen the ripe fruit, which Purchas says is caesious; and should doubt the correctness of this. Probably derived from a cross between *R. caesius* and *R. dasyphyllus*, as Sudre supposes.

Endemic. Infrequent and confined apparently to the Welsh border and the midlands; 34–36, 39, 40, 55, 57.

†† Prickles numerous, very unequal.

35. **R. scabrosus** P. J. Muell. 1858, *Flora*, 185; Sudre, 1908–13, t. 210. *R. dumetorum* var. *ferox* sensu Rogers, 1900, 93; non Weihe, 1824.

*5x. *Common. Robust.* Stem thinly pilose at first; *prickles very crowded, long, mostly patent.* Leaves (3) 5-nate, *a good deal strigose above*, thickly greyish pubescent to felted beneath; *terminal leaflet widely cordate, roundish ovate, shortly pointed, sometimes slightly lobed*, sharply and unequally incise-serrate-dentate. Panicle broad, round-topped, dense, with 5-flowered, subcymose branches and often a long lower branch; pedicels short, all villose and felted, with numerous submerged glands; prickles numerous, subulate. *Flowers 3 cm.*; sepals broad ovate, cuspidate-appendiculate; *petals roundish, pinkish*. Anthers usually subpilose. Carpels pilose. Fruit good.

E., W. Locally frequent or common from the southern counties northward to Berwick; not reported from Scotland or Ireland. N.W. France, Denmark.

36. **R. myriacanthus** Focke, 1871, *Bremen Abh.*, **2**, 467; *R. dumetorum* var. *diversifolius* Warren, 1870, *Journ. Bot.* **8**, 153, t. 107.

5x. Resembles *R. scabrosus* in the strong, long and patent prickles, but differs in its deep green leaves, rather narrow, oblong or elliptic terminal leaflet, especially on the panicle, the very short-stalked intermediate leaflets, the narrow panicle with remote, upright, short, axillary branches and the very large (3 cm.) white flowers. Petals 5–7, pilose on the margin. Styles long, yellowish. Fruit good.

E., W., S., I. Locally abundant in hedgerows, especially in the west. Belgium, Netherlands, N.W. Germany.

37. R. britannicus Rogers, 1894, *Journ. Bot.* **32**, 49.

6x. *Common. Armature slender, a good many prickles falcate.* Generally keeping to woods or the shelter of bushes, making unusually large roundish leaflets, always green not grey beneath. Known also by its round, hairy stem, flat not channelled petiole, its stalked basal leaflets and filiform stipules. Flowers white, moderate; petals obovate-oblong. Carpels glabrous. Fruit imperfect. Recalls *R. hirtus.* A dwarf form at Snaresbrook, Essex, in 1891.

Endemic. 3, 8, 13–21, 24, 30, 31, 33, 34, 36, 38, 39, 42–44, 49, 55, 86–88.

38. **R. tenuiarmatus** Ed. Lees, 1852, *Bot. Malvern*, 51; *R. dumetorum* var. *triangularis* A. Ley, 1902, *Journ. Bot.* **40**, 70.

Western. Stem roundish, slender, *finely pubescent,* the *largest prickles fragile-pointed in the shade,* yellow to red. Leaves 3 (5)-nate, densely pubescent beneath; *terminal leaflet broad cordate, ovate, gradually acuminate,* lobate, shallowly and obtusely serrate. Panicle large the lowest branch distant, widely spreading, numerous flowered; rachis, peduncles and pedicels densely pubescent and with crowded, yellowish prickles, pricklets and glands. Flowers 2–3 cm., pinkish in bud. *Calyx olive-green, flat-based, sepals broad triangular-ovate, prolonged.* Petals broad ovate, tip erose. Stamens and styles pinkish or white. Carpels pilose. Fruit good. Flowers about midsummer. *The panicle, the terminal leaflet and the sepals are triangular: A. Ley.*

Endemic. E., S. Chiefly in the west of England, in woods; 34, 36, 37, 110.

(**R. pilosus** Weihe ex Lejeune, 1824, *Rev. Spa,* 101 (nomen); *R. dumetorum* var. *pilosus* Weihe & Nees, 1827, *Rub. Germ.* 99.

This has been distributed and recorded as a British bramble. Weihe and Nees define it simply as having stem and leaves hirsute. Sudre says that Weihe's specimens are rather mixed, and have the stem *slightly pubescent;* he believes them to be *R. scaber × caesius.* He also says that Weihe's specimens gathered at Minden, in Herb. Lejeune, look like *R. caesius × scaber,* and have the stem *glabrous,* prickles nearly patent, fruit badly developed, sepals patent or reflexed. Specimens collected by Rogers in Herefordshire and determined by him as *R. dumetorum* var. *pilosus* Weihe are *R. myriacanthus.* His description in the *Handbook* also points strongly to *R. myriacanthus.* There seems no reason for regarding *R. pilosus* as British.)

This ends the list of the known British species of TRIVIALES. Excluded are those purely local crosses of *R. caesius* with MORIFERI or with TRIVIALES, or crosses of TRIVIALES with other TRIVIALES, both whether they are infertile or are partly fertile. It is unlikely that bushes answering to any of these descriptions will be found to possess a distribution area of an extent that cannot be accounted for by tip-rooting, or that they will be found to come up from seed true to type. Such plants are designated *Rubus* — × —, not followed by an author's name, e.g. *Rubus caesius* × *Hystrix*, the names of the two species concerned being placed in alphabetical order. The marks of *R. caesius* will usually be fairly evident (see the description of *R. caesius*); and if it is ascertained what other species are growing in the neighbourhood it will not be difficult to infer the other parent concerned.

There are two items of information in connexion with these *R. caesius* hybrids that one would like to obtain: they are (1) which morifers, if any, will not cross with *R. caesius*, *R. caesius* being the pollen parent; and (2) with which morifers, if any, will *R. caesius* cross, the morifer being the pollen parent.

Subgen. **V. RUBUS** (MORIFERI)

Stem-tip rooting: root not suckering

Sect. **III. Sylvatici** P. J. Muell. Stalked glands and acicles none or rare. Stem hairy or glabrescent. Leaves green, greyish or greyish white beneath. Sepals erect patent or reflexed. Stamens usually long. Flowers longer than 2 cm. (See p. 59.)

Sect. **IV. Discolores** P. J. Muell. Stalked glands, acicles and pricklets *absent*. Stem hairy or glabrous. *Leaves white felted beneath.* Stipules linear. *Sepals reflexed.* Stamens long or equalling styles. Flowers longer than 2 cm. (See p. 97.)

Sect. **V. Sprengeliani** Focke. Stalked glands, acicles and pricklets none or rare. Stem hairy. Leaves green or greyish white beneath. Sepals patent or erect. *Stamens shorter than the styles. Flowers less than 2 cm.* (See p. 105.)

Sect. **VI. Appendiculati** Genev. Stalked glands, acicles and pricklets fairly numerous on the panicle, and usually on the stem; stem prickles confined to the angles and not distributed over the sides. (See p. 107.)

Sect. **VII. Glandulosi** P. J. Muell. Stem prickles very unequally distributed over the sides of the stem, passing into ± numerous stalked glands, acicles and pricklets of different lengths. (See p. 181.)

Sect. **III. SYLVATICI** P. J. Muell.

(Type *R. silvaticus* Weihe & Nees)

Subsect. **A. Virescentes** Genev. Leaves green beneath, pilose or pubescent, rarely the upper leaves of stem and panicle greyish felted beneath. (See p. 60.)

Ser. **i. Eugrati** Sudre. Sepals patent to erect. Stem glabrescent or pilose. (See p. 60.)

Ser. **ii. Calvescentes** Genev. pro parte. Stem thinly hairy to glabrous. Sepals ± reflexed. (See p. 66.)

Ser. **iii. Piletosi** Genev. pro parte. Stem moderate or closely hairy, hair usually persisting. Leaves and panicle closely pilose, the upper leaves sometimes grey-white felted beneath. Sepals reflexed or patent. (See p. 72.)

Subser.† **Macrophylli** Bouv. Prickles long or short. Leaves at first hairy or pubescent beneath, then glabrescent. Panicle branches deeply divided. Sepals ± reflexed. Flowers usually white. (See p. 72.)

Subser.†† **Pyramidales** Bouv. Prickles moderate. Leaves thick, soft and often yellowish, pectinately pilose on the nerves beneath. Sepals reflexed, patent or semi-erect. Flowers usually pink. (See p. 77.)

Subsect. **B. Discoloroides** Genev. Upper leaves greyish white felted beneath. (See p. 81.)

Ser. **i. Properi** W. Wats. Stem ± pubescent at first, then glabrescent. Panicle ample, grey felted and pubescent. Prickles long and moderately numerous, or weak and rare. Leaflets convex. (See p. 81.)

Ser. **ii. Subvirescentes** Sudre. Stem with crisp patent hair, sometimes glabrescent. Panicle villose-hirsute. Prickles strong and long. (See p. 83.)

Ser. **iii. Subdiscolores** Sudre. Stem pilose or glabrous. Panicle pilose or also felted, with strong prickles. Mostly somewhat glandular. (See p. 88.)

Ser. **iv. Imbricati** Sudre. Stem glabrous or nearly so, not pruinose. Panicle rachis not or only slightly pilose, with strong and long prickles, subulate patent from a broad base, usually a little bent at the apex. Eglandular. (See p. 94.)

Subsect. **A. VIRESCENTES** Genev.

Ser. **i. EUGRATI** Sudre

(Type *R. gratus* Focke)

1. Petals pink or deep pink.

† Flowers large, 2·5–4(5) cm. Petals, filaments and styles pink or pink-based.

39. **R. gratus** Focke, 1875, *Alpers. Verz. Gefpfl. Stad.* 26; Sudre, 1908–13, 26, t. 16. **Fig. 5.**

4x. *Robust.* Stem red, arching, furrowed, prickles few, patent. Leaves glabrescent beneath, stipules red; terminal leaflet obovate, base entire, coarsely serrate. *Panicle short, broad, lax, pyramidal, truncate,* pilose, prickles subulate. Sepals greenish grey, white-edged, clasping. *Filaments very long,* at length investing the styles, *anthers pilose. Pollen nearly all perfect.* Carpels glabrous. Receptacle hirsute especially below the carpels. *Fruit large, pleasantly flavoured.* Focke describes the anthers as glabrous, but they are pilose in his specimen in *Rubi Selecti*, no. 42.

E., W., S., I. Frequent; 2, 3, 5, 6, 13–18, 21–24, 27, 30, 34, 36–42, 48, 53, 54, 56–60, 87, 99, 104, 110. H. 23, 37. N. France, Belgium, Netherlands, Denmark, N.W. Germany.

40. **R. confertiflorus** W. Wats. 1935, *Journ. Bot.* **73**, 194. **Fig. 6.**

Robust. Stem rather tall, furrowed, *shining red at the growing point* then opaque; prickles long, often in twos. Leaves large, petiole prickles small, hooked, stipules red; leaflets 5, pubescent and ultimately yellowish beneath; *terminal leaflet roundish to ovate-oblong or elliptic-oblong,* unequal or subcompound serrate-dentate. *Panicle short, subracemose, flowers congested*; rachis pubescent-pilose. Carpels glabrous or nearly so. Receptacle pilose. *The unripe fruit coral-red.*

Endemic. England: rare; 12, 14, 16, 17, 22.

41. **R. monensis** Barton & Riddels. 1932, *Journ. Bot.* **70**, 108; *R. gratus* ssp. *latifolius* sensu Sudre, 1908–13, 26, t. 17.

N. Wales. Robust. Stem nearly glabrous, slightly furrowed. *Leaves* large, *the upper ones greyish felted* beneath; *leaflets imbricate,* irregular to subcompound incise, partly patent-serrate; terminal leaflet subcordate, broad or roundish ovate, acuminate; *basal leaflets very short stalked. Panicle large, broad, compound*; *rachis crooked below, furrowed,* prickles numerous, strong. Flowers large; sepals greyish-felted with yellow hair on the base, ovate, patent; petals broad elliptic; *filaments slightly longer than the styles.* Carpels pilose. Receptacle densely pilose. *Fruit large.*

W., S. Rare; 46, 48, 49, 52, 96. France, *Aisne.*

†† Flowers moderate, *c.* 2 cm. Petals pink, filaments pink or white.

42. **R. vulgaris** Weihe & Nees (exclud. vars. *b, g, d*), 1825, *Rub. Germ.* **38**, 39, t. 14 A.; Focke, 1914, *Rub. Eur.* fig. 46, 47; Sudre, 1908–13, 23, t. 12.

Stem arched, *furrowed,* glabrescent, *blackish purple* and subpruinose; *prickles strong and long, especially on the branches. Leaflets* 5, dull above, *pubescent to greyish felted beneath,* unequally serrate, undulate; terminal leaflet roundish ovate to elliptic-rhomboid, acuminate, base from subcordate to entire. Panicle leafy, interrupted, truncate, rarely very compound; *rachis* pilose, *prickles numerous, strong and falcate*; bracts and calyces finely glandular. *Sepals* grey felted, *patent to loosely reflexed.* Petals elliptic or obovate, notched. Filaments slightly longer than the *reddish or yellowish styles.* Carpels pilose. Fruit subglobose. Flowers *c.* 13 June.

England: local; 11, 16, 17, 20–24, 31 (Holme Fen, Fryer, 1879). N.W. Germany and Belgian Ardennes.

43. **R. latiarcuatus** W. Wats. 1946, *Watsonia*, **1**, 71; *R. vulgaris* var. *mollis* Weihe & Nees, 1825, *Rub. Germ.* 38, 40.

3x. *South-east England. Often robust. Stipules, bracts, pedicels and calyces slightly glandular.* Stem red, angled, slightly hairy; prickles long, straight, numerous, often clustered, occasionally a pricklet. *Leaves large,* yellowish green, glabrescent above, *softly pilose beneath; terminal leaflet roundish or obovate,* like the intermediate and basal leaflets often *subcuneate* below, toothing unequal, coarse, angular and partly patent. Panicle broad and short, upper branches usually one-flowered; *prickles numerous, weak, slanting.* Sepals greenish, white-bordered, pilose, patent to loosely reflexed. *Anthers pilose.* Carpels glabrous. Receptacle strongly pilose. Flowers *c.* 6 June.

England: local; 15, 17, 24. Belgian Ardennes, N.W. Germany.

44. **R. subintegribasis** Druce, 1928, *Brit. Pl. List,* 30; *R. integribasis* sensu Rogers, 1900, 24; not P. J. Muell. 1866.

South-western. Stem high-arched or climbing, much branched, blunt-angled, *spotted red-purple,* thinly hairy; prickles strong-based, slanting, often in twos. *Leaflets 5, not contiguous, becoming convex,* hairy beneath, teeth simple, unequal; *terminal leaflet obovate or elliptic,* acuminate-cuspidate, base ± entire. *Panicle narrow* racemose above, with 1–2, erect, axillary branches; *rachis villous, prickles weak.* Sepals hairy, often 6 or 7. Petals broad obovate. Filaments about equalling the styles. Carpels glabrous. Receptacle slightly pilose. Fruit subglobose, *pale green to coral red* then glossy black.

Endemic. England: 3 (Chagford, Briggs, 1881), 6, 8, 9, 11, 12.

45. **R. Averyanus** W. Wats. 1951, *Lond. Nat.,* Suppl. 97.

Surrey and W. Kent. Plant yellowish green, *densely armed with yellow falcate prickles, rufescent,* eglandular. Stem angled, hairy at first, prickles large, slender, often in twos. Leaflets 5, long-pointed, pubescent and hairy beneath, sharply serrate-dentate, undulate; terminal leaflet of the panicle leaves sub-cordate-ovate to obovate, long-acuminate. Panicle pyramidal, truncate, interrupted below, the *lowest branch usually strongly developed; rachis* hirsute, *densely armed throughout.* Sepals ovate, prolonged into linear leafy tips. Petals lilac, broad ovate or obovate. Filaments slightly longer than the styles. Carpels subglabrous. Receptacle pilose. Fruit ovoid, abundant.

Endemic. England: rare; 16 (Pembury Walk), 17 (Walton Heath; Broadmoor Common; Holmbury Hill, frequent).

46. **R. dasycoccus** W. Wats. 1933 (for 1932), *Rep. Bot. Exch. Cl.* **10**, 21; *R. macrophyllus* var. *amplificatus* sensu Rogers, 1900, 44; *R. pyramidalis* ssp. *amplificatus* sensu Sudre, 1908–13, 47, t. 51; non *R. amplificatus* Ed. Lees, 1847.

N.v.v. Monmouthshire. A very prickly bramble with rather small leaves. Stem sharply angled, furrowed, glabrescent, glaucescent; prickles numerous, rather short. *Leaflets 5, long-stalked,* nearly glabrous above, densely and shortly pilose to felted beneath, *compound incise-serrate; terminal leaflet obovate,* acuminate, *narrowed to the entire base. Flowering branch sharply angled,* glabrescent below; *panicle* nearly equal, with long, erect branches below, the mid branches deeply divided, *leafy to the top;* rachis sharply angled, hirsute above, *prickles numerous, short, yellow. Sepals very prickly,* patent. Petals broad elliptic, clawed. Filaments long. Carpels densely pilose.

35 (around Tintern and Trelleck), 46 (one station). (It has been reported as *R. dumnoniensis* var. *amplificatus* (Lees) Sudre, from Belgium, *Vierset-Barse*.)

2. Petals lilac or pinkish in bud, or pure white. Filaments usually white.

† Leaves, and usually the panicle, thinly or moderately pilose, glabrescent. Sepals green or greyish green.

47. **R. calvatus** Ed. Lees ex Bloxam in Kirby, 1850, *Fl. Leicester*, 42; *R. villicaulis* ssp. *calvatus* (Ed. Lees ex Bloxam) Sudre, 1908–13, 55, t. 61.

Robust. Stem, panicle and sepals slightly glandular and aculeolate. Stem higharched, angled, furrowed, *shining red*, glabrescent, but *remaining hairy below*; prickles numerous, often in twos. Leaves large, prickles hooked; *leaflets* 5, *convex; terminal leaflet* broad ovate-oblong, short-pointed, coarsely serrate-dentate, *the principal teeth jagged*. Panicle broad, lax, nearly equal, few-flowered, leafy and prickly; *rachis crooked, angled. Flowers large, c.* 3 cm.; sepals linear-leafy-pointed; petals ovate or elliptic; filaments pinkish, longer than the yellowish styles. Carpels glabrous. Receptacle moderately pilose. Fruit large, oblong.

Endemic. E., W., I. Sporadic; 3, 7–9, 11, 13, 14, 16, 17, 19, 22–24, 38–40, 50, 53, 55, 57–59, 61. H. 33, 40.

48. **R. sciocharis** (Sudre) W. Wats. 1946, *Journ. Ecol.* **33**, 339; *R. sciaphilus* Lange, 1883, *Fl. Dan.* t. 3026; *R. gratus* ssp. *sciocharis* (Sudre), Sudre 1908–13, 26, t. 17; non *R. sciaphilus* Lef. & Muell. 1859. **Fig. 7.**

Stem much branched and sprawling, *blunt, angled*, hairy; *prickles short* with a very few pricklets. *Leaflets* 3, 4, 5, imbricate, *very short-stalked, petiole also short*, prickles hooked, stipules linear; *terminal leaflet* cordate-ovate, acuminate, *incise-serrate*, softly hairy at first beneath. *Panicle leafy, racemose or short, broad, compound, subcorymbose*; rachis villous, *armed like the pedicels and calyces with weak, yellowish prickles* and a few glands. Flowers 2–3·5 cm.; sepals long-pointed, patent or clasping; *petals pale pink*, elliptic or narrow obovate; filaments long; styles yellowish; *anthers pilose.* Fruit rather large, well developed in the open. In Denmark and Germany the flowers are pure white. On heaths the growth is much restricted and in fruit it resembles *R. plicatus*, with which it sometimes grows. A dwarf form at Seal Chart (v.c. 16), and in Denmark is var. **microphyllus** (Lange) W. Wats.

E., I. Local, generally in woods; 3, 8, 13, 14, 16, 24, 25, 30, 34–37, 39. H. 38. Denmark and Schleswig-Holstein, abundant.

49. **R. nitidoides** W. Wats. 1929 (for 1928), *Rep. Bot. Exch. Cl.* **8**, 786; 1932 (for 1931), loc. cit., 765.

4x, 6x. *Eglandular. Stem nearly erect, seldom rooting,* slightly hairy, becoming *red and ochraceous; prickles numerous, yellow,* hard and very sharp. Leaf-prickles hooked; *leaflets* 5 (6), glabrescent above, pubescent and shortly hairy beneath, *short-pointed, coarsely and slightly compound-serrate; terminal leaflet ovate-elliptic, base entire. Panicle* lax, pyramidal-corymbose, fiercely armed like the stem, the *uppermost branches 1–7-flowered, cymose,* the lower ones semi-erect, long-peduncled, subcorymbose. Flowers, *c.* 2·5 cm.; *sepals shining green, nearly glabrous,* aculeolate, obtuse, often leafy-tipped, patent to erect; *petals long elliptic, incurved, faintly lilac;* filaments sometimes pinkish, longer than the greenish styles. Carpels glabrous. Receptacle pilose. *Fruit abundant, large, oblong-ovoid,* sweet, not very pippy. Flowers about 20 June. Leaf-fall in autumn. First collected by John Stuart Mill on Hayes Common, W. Kent in 1860.

Endemic. Locally abundant in S.E. England; 11, 14, 16–18, 21.

50. **R. ludensis** (*Luda,* Ludlow) W.Wats. 1956, *Watsonia,* **3**, 286; *R. hirtifolius* Rogers pro parte, 1900, 48; *R. sciaphilus* sensu W. Wats. 1929, *Rep. Bot. Exch. Cl.* **8**, 782; sensu Rogers pro parte, 1900, 37; non Lange, 1886–88.

Wales and the Welsh border. Robust. Stem angled, green to fuscous, glaucescent, hairy, scarcely glandular but with a few, small, gland-tipped, as well as the normal broad based, slanting and falcate prickles. Leaves large, subdigitate; *leaflets 5, greyish green and pilose beneath, irregularly not deeply crenate, to subcompound serrate;* terminal leaflet cordate-ovate, gradually acuminate. *Panicle large and compound, pyramidal,* with several ovate and then long linear bracts, the upper branches 1–3-flowered, the mid branches rather oblique, long-peduncled, 3–7-flowered; rachis hairy and glandular, armed like the stem. Flowers 2·5–3 cm.; *sepals* greenish, glandular, prickly, ovate-lanceolate, long-tipped, soon erect, *like the petals 5–7 in number;* petals rather narrow obovate, glabrous on the margin, faintly pink; filaments white, *anthers glabrous,* much longer than the yellowish styles. Carpels glabrous. Fruit ovoid-oblong, rather large.

Endemic. E., W. Locally abundant; 36, 40 (Whitcliff, Ludlow), 42, 43, 49 (Felin Hen, det. *R. sciaphilus* by Gelert and Rogers).

51. **R. Crespignyanus** W. Wats. 1956, *Watsonia,* **3**, 286; *Journ.Ecol.* **33**, 337 (nomen); *R. similatus* sensu W. Watson, 1937 (for 1936), *Rep. Bot. Exch. Cl.* **11**, 444; non P. J. Muell. 1859.

Subglandular. Stem tall, slender, slightly hairy, prickles short, patent to falcate. *Leaflets large,* 5, glabrescent on both sides, finely serrate or serrate-dentate, *rather jagged;* terminal leaflet roundish, cordate-ovate, acuminate. *Panicle* equal with long lower branches, the *mid branches sometimes divided to the base;* rachis slender, flexuose, pilose, prickles short, slanting or recurved.

Flowers c. 2 cm.; sepals long-pointed, *reddish, glandular and aciculate towards the tip*; filaments about equalling the *red styles.* Carpels pubescent. Receptacle pilose. Fruit good but small.

Discovered by de Crespigny on Croham Hurst, Surrey, in 1872: later found by Edward Langley at Bow Brickhill, Bucks., and brought into garden cultivation as the Edward Langley Blackberry.

Endemic. 17 (frequent in woods and on heath over a few square miles in N.E. Surrey), 24.

†† Leaves and panicle pubescent. Sepals grey felted, or green and pilose.

52. **R. carpinifolius** Weihe, in Boenn. 1824, *Prodr. Fl. Monast.* 152; Sudre, 1908–13, 23, t. 10. **Fig. 8.**

4x. *Eglandular. Stem tall, seldom rooting, not much branched,* sparsely hairy; *prickles* numerous, strong-based, *yellow* to brick red. *Leaflets* 5 (6, 7), *pubescent* to felted beneath, *the short hairs semi-applied along the midrib,* unequally to subcompound-serrate, base entire; terminal leaflet broad ovate to elliptic, acuminate. *Panicle mostly racemose or subracemose above,* with spreading, ascending *pedicels armed with numerous straight prickles,* less often very compound; upper leaves grey felted. Flowers rather large and showy; *sepals patent; petals* ovate or elliptic, *long-clawed, incurved, faintly pink in bud,* very rarely fully pink or pure white; stamens long; styles greenish. Carpels glabrous. Receptacle pilose. The glossy, sweet fruits ripen early and ± together. Leaf-fall in autumn. Heaths.

E., W., S., I.(?). Frequent south of Yorks. and Cheshire, not much recorded for Scotland and Wales, doubtful as yet for Ireland.

53. **R. chloophyllus** Sudre, 1907, *Diagn.* 7; *R. silesiacus* ssp. *chloophyllus* (Sudre) Sudre, 1908–13, 45, t. 48.

4x. *High Weald of W. Kent and E. Sussex.* Slightly glandular. *Stem low-arching,* red-speckled, *furrowed to the base,* glabrescent; *prickles small, inclined falcate. Petiole* flat, *prickles small, hooked; leaflets* 5 (6, 7), glabrous and shining above, sharply serrate, jagged; terminal leaflet subcordate, broad ovate to elliptic, acuminate; *basal leaflets subsessile. Panicle short, broad, racemose,* pedicels long, spreading, pubescent, finely glandular; *terminal leaflet obovate-cuneate; rachis* furrowed, *prickles few, small, recurved.* Flowers c. 2·8 cm., incurved, pinkish; sepals green, pilose, leafy-tipped, aciculate, semi-erect; petals roundish to elliptic, clawed, downy; filaments longer than the styles, *anthers pilose.* Carpels and receptacle pilose. *Fruit ovoid-oblong, very good, ripening in late July.*

South-east England, local; 13–16. (Fairly frequent except in 15 (near Tenterden).) France, Seine-Inférieure, Sarthe.

54. **R. polyoplus** W. Wats. 1937 (for 1936), *Rep. Bot. Exch. Cl.* **11**, 220; *R. Salteri* sensu Rogers pro parte, 1900, 45; *R. vulgaris* ssp. *Salteri* sensu Sudre, 1908–13, 25, t. 13; non *R. Salteri* Bab., 1846.

Stem and panicle slightly glandular and aciculate, *prickles yellow, very numerous, short, straight and strong.* Leaves *markedly pedate,* grey beneath; terminal leaflet roundish or elliptic-obovate, base entire, sharply and coarsely serrate; *all petiolules usually long. Panicle long, leafy, compound, intricately branched. Sepals clasping.* Petals pinkish. *Stamens equalling the styles. Carpels densely pilose.*
Endemic. England and Wales: rare; 36, 40, 41.

††† Petals and filaments pure white. Panicle and sepals glandular. Leaves glabrescent.

55. **R. horridisepalus** (Sudre) W. Wats. 1946, *Journ. Ecol.* **33**, 338; *R. chaerophyllus* ssp. *axillaris* microg. *horridisepalus* (Sudre) Sudre, 1908–13, 28, t. 21.

4x. *Robust.* Stem branching, climbing, furrowed, hairy; *prickles* rather strong, *numerous, slanting,* subequal. *Leaves 5-nate, pedate; leaflets imbricate, all short-stalked, the basal ones subsessile,* coarsely, unequally and sharply serrate; *terminal leaflet broad or roundish cordate-ovate, acute* to shortly acuminate. *Panicle short and broad, mostly few-flowered,* lax, leafy, with long lower branches, closely hairy and *with numerous falcate prickles. Calyx* greenish, pilose, *with numerous small prickles and glands,* clasping. Flowers *c.* 2·5 cm. Petals ovate. Stamens long. Carpels a little pilose. Receptacle pilose. Fruit large, ovoid, abundant.
England: rare; 15 (Bigbury), 24 (Lee, near Gt Missenden), 30 (Clophill). Germany (Weissenburg).

56. **R. glanduliger** W. Wats. 1935 (for 1934), *Rep. Bot. Exch. Cl.* **10**, 794; Sudre, 1908–13, t. 11 (as *R. carpinifolius* e).

★4x. W. Sussex. *Axes slender.* Stem arching, furrowed, nearly glabrous, *sparsely glandular;* prickles moderate, subequal, patent to slanting. Leaves 5-nate, digitate, *petiole channelled throughout, prickles small, hooked,* stipules red; leaflets becoming convex and glabrous, unequal or subcompound, angularly serrate or serrate-dentate; terminal leaflet cordate-ovate, acuminate. *Panicle* thinly pilose, *weakly armed, large, lax and broad,* rachis flexuose, *branches 1–6-flowered, pedicels long.* Flowers 3 cm.; sepals greyish green, long-pointed, patent, ascending; *petals glabrous. Receptacle nearly glabrous.* Fruit oblong, drupels numerous.
Endemic. England: 13 (frequent, commons and woods), 16 (Chislehurst).

Ser. ii. **CALVESCENTES** Genev. pro parte

(Type *R. oxyanchus* Sudre)

1. Petals pink or deep pink.

† Panicle prickles strong, broad-based, some falcate or recurved.

57. **R. nemoralis** P. J. Muell. 1858, *Flora,* **41**, 139; non Rogers, nec Genev., nec Focke; *R. Selmeri* Lindeb. 1884, *Herb. Rub. Scand.* no. 33; Trower, 1927–8, *Rep. Bot. Exch. Cl.* **8**, 858 (three plates); *R. vulgaris* ssp. *Selmeri* (Lindeb.) Sudre, 1908–13, 24, t. 13.

4x. *Common.* Stem arched, angled, *furrowed*, becoming *deep purple*; *prickles strong, curved. Leaflets* 5 (6, 7), rather large, *deep green, glabrous above, pubescent to grey felted beneath*, finely, sharply and a little unequally serrate; terminal leaflet long-stalked, base entire or slightly indented, *nearly round*, shortly acuminate-cuspidate. *Panicle* broad, long, leafy, *interrupted, equal*, with simple or compound cymose branches; *rachis crooked, furrowed*, pilose and *dotted with glands. Sepals aculeolate, ending in long, leafy tips*, loosely reflexed then rising. Petals obovate, notched. *Filaments pink, hardly equalling the styles.* Carpels and receptacle pilose. *Fruit large, thimble-shaped*, sweet, abundant. Flowers *c.* 6 June.

var. **microphyllus** (Lindeb.) W. Wats. 1952 (for 1951) *Lond. Nat.* Suppl., 78; *R. pistoris* Barton & Riddels. 1935, *Journ. Bot.* **73**, 127; cf. W. Wats. 1937, *Journ. Bot.* **75**, 160.

A dwarf form found with the typical plant here and on the Continent.

E., W., S., I. Generally distributed and locally plentiful, especially in the north. W. Germany, Denmark, S.W. Norway.

58. **R. Questieri** Lef. & Muell. 1859, *Pollichia*, 120; Sudre, 1908–13, 39, t. 40; Coste, 1903, *Fl. France*, **2**, 36, no. 1164. **Fig. 9.**

Robust. Young shoots bronze. Stem angled, glabrescent; *prickles* brownish, *powerful*, patent to much declining and falcate, a few smaller ones gland-tipped. *Petiole flat above*, prickles strong, falcate, *stipules filiform; leaflets* (3, 4) 5, long, entire-based, soon convex, glabrescent on both sides; *terminal leaflet elliptic or obovate*, acuminate, short-stalked. *Panicle hairy, long, narrow, interrupted*, with sharply rising 1–5-flowered branches, the upper leaves felted beneath. Flowers 3 cm.; sepals long-pointed, loosely reflexed, the tips becoming patent as the fruit swells; *petals* broad obovate, narrowed to the base, *notched*; filaments white or pink-based, slightly longer than the yellowish or pinkish styles. Carpels usually glabrous.

E., I. Rare and very local; woods, and hedges near them; 2, 5, 9 (not 16, 22). H. 1. Jersey. France, Portugal.

R. laciniatus Willd. 1809, *Hort. Berol.* t. 82.

An escape from gardens, bird-sown. Stem furrowed, glabrescent, prickles falcate. *Leaves* 5-nate, digitate; leaflets long-stalked, divided into pairs of *laciniated* segments, glabrous above, glabrescent beneath. Panicle broad, leafy, flat-topped, the upper leaves felted beneath; prickles numerous, short, falcate. *Sepals* felted, *leafy-tipped*, reflexed. *Petals pink, irregularly incised at the apex. Stamens equalling the styles. Fruit long, thimble-shaped*, but variable in different plants. Flowers with *R. vulgaris c.* 13 June.

Not known wild in any country, but is frequently met with near gardens. Described by Philip Miller in 1754 in his *Gardeners' Dictionary*, ed. 4, and figured by Plukenet, 1691, *Phytographia*, t. 108, f. 4.

†† Panicle prickles slender, subulate or a few subfalcate.

59. **R. oxyanchus** Sudre, 1904, *Observ. Set. Brit. Rub.* 18; 1908–13, 38, t. 39, figs. 7, 8 and 11, and 12 (as *R. viridicatus*). **Fig. 10.**

★**4x.** *Robust, forming a high bush.* Eglandular. *Stem crimson,* furrowed; *prickles* hairy, *finely pointed, slanting.* Leaves large, *stipules reddish, narrow-linear,* glandular; *leaflets 3–5, imbricate,* finely and softly hairy to green or greyish felted beneath; *terminal leaflet long-stalked,* cordate, *roundish reniform or broad obovate, truncate, cuspidate,* finely and simply to slightly compound-serrate-dentate. *Panicle* narrow and thyrsoid to broad corymbose, the *lowest branch sharply erect,* the upper branches patent and *terminal flower subsessile;* rachis hairy, *prickles slender, straight,* upper leaves grey-felted, bracteoles glandular. Flowers showy; sepals aciculate; petals narrow obovate, tapered below, margin nearly glabrous; filaments longer than the styles. Carpels and receptacle pilose. *Fruit large, oblong, abundant.*

This plant is called *R. nemoralis* P. J. Muell. by Rogers and *R. nemoralis* Genev. by Focke (1914). These are two different species: the former has been described above, see *R. nemoralis* P. J. Muell. (p. 66); the latter is described below, see *R. propinquus* P. J. Muell. (p. 99).

E., S., I. Scattered, not common; 1–3, 8, 9, 11, 16, 17, 21, 36, 39, 41, 65, 70, 87, 110. France (one locality), Germany (one locality), Belgium (rare).

60. **R. cambrensis** W. Wats. 1956, *Watsonia,* **3**, 286; 1937, *Journ. Bot.* **75**, 196 (nomen nudum); *R. nemoralis* var. *glabratus* sensu Rogers, 1900, 31; non Bab. 1846; nec *R. glabratus* Kunth. 1823.

Western. Axes rufescent. *Stem slender, prickles red-based, short,* subulate. Leaves 5-nate, digitate, *prickles retrorse-falcate; leaflets nearly glabrous above,* pubescent to greyish felted beneath, crenate to slightly compound-serrate; terminal leaflet emarginate, round, cuspidate-acuminate. *Panicle equal, long,* subracemose *with 1–6 roundish-ovate leaves, pedicels long,* prickles short, straight; rachis pubescent and felted, *leaflets serrulate.* Flowers *c.* 2 cm.; *sepals* felted, *gland-dotted and aciculate;* petals broad obovate, notched and clawed. Stamens equalling the styles. Carpels glabrous. Receptacle pilose.

Endemic. E., W., S. 36, 42, 48, 49, 86, 88.

61. **R. durescens** W. R. Linton, 1892, *Journ. Bot.* **30**, 70; *Fl. Derbyshire* (fig. on cover); *R. vulgaris* ssp. *mionacanthus* (Kinscher) H. Hofmann.

Axes deep bright red. Eglandular. Stem suberect, arching, subpruinose. *Leaves* 5-nate, subpedate, *firm; leaflets* plicate, *glabrous above, hard and glabrescent beneath, shallowly and unequally sinuate-dentate-serrate, the principal teeth larger, angular and prominent;* terminal leaflet broad elliptic (-obovate), shortly acuminate, or *broader and cuneate* with the apex only *acute.* Panicle ± equal, the *terminal flower short-stalked,* upper and mid branches 2–3-flowered, deeply divided, erect-patent to patent, *prickles numerous, subulate; rachis crooked, furrowed,* pilose. Flowers 2–2·5 (3) cm.; *sepals* greenish, · felted, aciculate, *loosely reflexed becoming nearly patent; petals bright rose-pink, pilose expanded*, obovate, narrowed below; *filaments pink* or white, longer than the yellowish, reddish-based styles. Carpels densely pilose or glabrous. Receptacle thinly or densely pilose. *Fruit small*, imperfect. Flowers about midsummer.

England: rare; 39, 57. Germany, *Saxony*.

2. Petals white or faintly pink.

† All leaves green beneath.

62. **R. pullifolius** W. Wats. 1948 (for 1946–7), *Rep. Bot. Exch. Cl.* **13**, 328.

Eglandular. Stem round to blunt-angled, glaucescent; prickles moderately slender. *Leaves deep green*; leaflets imbricate, pubescent then hard and glabrescent beneath; *terminal leaflet roundish, short-pointed*, base ± entire, sharply serrate. *Panicle* closely pubescent or felted, broad subpyramidal-corymbose with *usually one large ovate leaf*, lax, branches spreading; *terminal leaflets round, cuspidate or ovate-cuneate, acute.* Flowers about 2·5 cm.; sepals ovate, subcuspidate, reflexed; petals ovate, short-clawed; filaments long; styles pallid. Carpels and receptacle pilose. Fruit subglobose.

Endemic. England: locally abundant; 3 (near Roborough, Briggs, 1865), 9, 11 (Shirley, H. Groves, 1879).

†† Upper leaves of stem and panicle greyish felted beneath.

63. **R. Lindleianus** Ed. Lees, 1848, *Phytol.* **3**, 361; Sudre, 1908–13, 56, t. 63; *R. platyacanthus* Lef. & Muell. 1859, *Pollichia*, 86.

4x. *A common, tall, very prickly, late-flowering bramble with an intricate panicle.* The prickles of the shining stem are strong and large, those of the panicle-branches weak, short and very numerous. *The leaflets are* glabrous above, pubescent and greyish felted beneath, *mostly obovate-cuneate*, a little *jagged*, shallowly and irregularly toothed; *base entire.* Panicle very large and broad, the branches and pedicels long and perpendicularly patent, the pedicels often again branched, densely hairy with a few buried glands. Fruit small. Conspicuously overtopping other bushes, it is proclaimed in July by its show of white flowers after other brambles have gone to fruit. The unclosing white petals are sometimes flushed with red outside.

E., W., S., I. England and Wales in most counties; less general in Scotland and Ireland. France, *Oise*, *Tarn*, Belgium, Netherlands, N.W. Germany.

64. **R. egregius** Focke, 1871, *Bremen. Abh.* **2**, 463; *R. mercius* var. *bracteatus* Bagn. 1894, *Journ. Bot.* **32**, 187; *R. egregius* var. *plymensis* Focke. **Fig. 11.**

4x. Stem arcuate-prostrate or climbing, angled, not pruinose, with few or no stalked glands and acicles; *prickles short, subulate*, yellow to brownish-red, *almost orange.* Leaflets 3 (5), small or large, velvety pubescent beneath, finely to rather coarsely serrate; terminal leaflet roundish ovate-cordate, acuminate. *Panicle long, very narrow, thyrsoid, with sometimes very long, floriferous, ascending lower branches,* and deeply divided, racemose middle branches, armed like the stem; *terminal leaflet* of 3-nate leaves, *obovate-mucronate.* Flowers rather large, incurved; sepals patent with the tips ascending then reflexed; petals glabrous on the tip, roundish ovate or obovate; filaments long. Carpels glabrous. *Fruit small, of 15–20 rather large drupels.* Varies greatly in leaf-form and toothing according to soil and exposure.

var. **pliocenicus** W. Wats. 1952 (for 1951), *Lond. Nat.,* Suppl. 98. **Fig. 11.**

Terminal leaflet narrow-elliptic-oblong, or rhomboid, short-stalked, angularly serrate. Flowers small, petals narrow, distinctly notched at apex; stamens equalling the pink styles.

E., W., I. Rather frequent, woods, heaths and hillsides; 1–3, 9, 13–17, 21–24, 27, 30, 34, 38–40, 48, 49, 53, 55, 58, 59, 67, 69, 70. H. 12. Germany, Denmark, France (rare), Switzerland, Austria.

65. **R. leucandrus** Focke, 1875, *Alpers. Verz. Gefpfl. Stad.* 27; Sudre, 1908–13, t. 53 (as *R. macrophyllus* ssp. *montanus*). **Fig. 12.**

4x. *Eglandular.* A robust, high bush. Stem arching, deep red, shining, sides convex to furrowed; prickles stout-based and strong or subulate. Leaves large, pale green, stipules very narrow; *leaflets imbricate, nearly glabrous above, softly hairy or also grey-felted beneath,* evenly and shallowly crenate-serrate or dentate-serrate; *terminal leaflet* very broad ovate, ± deeply cordate, *gradually acuminate, the other leaflets acute.* Panicle subpyramidal, lax below *with long, sharply rising lower branches, subcorymbose above, often with several long-pointed, narrow, simple leaves;* some of the lower leaves 4, 5-nate; rachis closely pilose above, prickles numerous. Flowers showy; sepals linear-tipped; petals ovate, obovate or elliptic, clawed. Carpels glabrous. Fruit ovoid, of many drupels. *Folds usually appear in pressing the leaves.*

E., S., I. Distributed but infrequent; 2–5, 7–9, 11–13, 16–18, 35, 62, 89. H. 38. Belgium, Germany.

66. **R. mercicus** Bagn. 1892, *Journ. Bot.* **30,** 372.

4x. Slightly glandular and acicular. *Axes purplish red.* Stem angled, glaucescent; *prickles finely pointed* from a wide base, slanting and *pronouncedly falcate. Leaves* (3) 5-nate, pedate, *prickles small, numerous, hooked;* leaflets pubescent beneath, serrate-dentate, obscurely lobed; terminal leaflet roundish, subcordate-ovate-acuminate to obovate-cuspidate. Panicle broad and large, *the terminal flower subsessile or shortly stalked with a long, patent branch beneath,* mid-branches patent with long, patent pedicels; rachis felted, pubescent or villose, prickles as on the stem. Flowers moderate; sepals felted, attenuate, loosely reflexed to subpatent; *petals narrow,* faintly pink with a yellowish base, *becoming patent with the sides reflexed; styles honey-coloured.* Carpels glabrous or pilose. *Receptacle hirsute, especially below the carpels.* Fruit rather large, oblong, of many rather small drupels. Builds up an extensive bush in the open; climbs to 5 m. in woods.

Endemic. E.,W. Local; 7, 16, 26, 29, 30, 38, 39, 49, 53, 55, 67.

67. **R. Maassii** Focke ap. Bertram, 1876, *Fl. Braunschw.* 1, 75; Sudre, 1908–13, 37, t. 38.

Stem tall, arching, much branched, angled; *prickles yellow*, rather few, *slender, slanting and falcate.* Leaves greyish green, long-stalked, prickles falcate; leaflets remote, greyish pubescent and felted beneath, crenate-serrate; *terminal leaflet long-stalked*, subcordate, *oblong-obovate*, acuminate-cuspidate. Panicle pubescent, with 2–3 simple leaves, interrupted, all branches ascending; *rachis crooked, prickles slanting and falcate*, numerous on the long pedicels. Flowers 2–3 cm.; sepals greenish grey, pubescent, aculeate on the base; *petals rather narrow elliptic or attenuate below*, margin glabrous to pilose; filaments longer than the yellowish styles. Carpels glabrous. Flowering throughout July. Winter-green.

England: rare; 11 (Bitterne Common, H. Groves, 1879), 13, 54, 55. N.W. and Central Germany.

68. **R. Muenteri** Marss. 1869, *Fl. Neuvorpom.* 144; *R. Maassii* ssp. *Muenteri* (Marss.) Sudre, 1908–13, 38, t. 39.

Robust, eglandular. *Stem thick*, wide-arching, soon branching, furrowed, green to *deep red; prickles broad-based, ± patent. Leaves rather small*, subdigitate, prickles large, falcate; *leaflets 3–7*, dull above, *soft hairy to felted beneath, serrate-dentate; terminal leaflet often very long-stalked*, cordate, *roundish, somewhat pentagonal.* Panicle nearly equal, mid branches patent, 2–5-flowered, or large, lax, widely spreading, pyramidal-abrupt, the upper branches divided to the base, pedicels long; *rachis felted and villose*, prickles subulate. Flowers 2–2·5 cm., sometimes pink; petals obovate-cuneate, spaced; filaments white or pinkish, long; styles greenish or red. Carpels glabrous. Receptacle pilose.

England. Rare or not generally recognized, apparently mainly a northern bramble; 20 (Mardley Heath), 39 (frequent, Edees), 58 (near Mere). Belgium, north and central Germany.

var. **Robii** W. Wats. 1956, *Watsonia*, 3, 287.

Differs in the smallness of all its parts from the typical plant. Discovered by Miss C. M. Rob in 1937 at Yearsley Moor (v.-c. 62), and thought at first to be *R. furnarius* Barton & Riddels. It has occurred also between Sedbergh and Dent (v.-c. 65) (Rogers 1906, as *R. Godronii* var. *foliolatus*).

69. **R. Sampaianus** Sudre in Samp. 1904, *Rub. Port.* 32; Focke, 1914, *Rub. Eur.* figs. 44, 45; *R. rhombifolius* ssp. *Sampaianus* (Sudre) Sudre, 1908–13, 43, t. 46.

Stem arched, brownish, angled, *rather hairy at first*; prickles subulate. *Leaves rather small; leaflets 5*, entire-based, subglabrous above, slightly *felted* and pubescent beneath, ± contiguous, *sharply incise, finely and unequally to sub-compound-serrate; terminal leaflet* long-stalked, obovate-elliptic, *base rounded; basal leaflets shortly stalked.* Panicle narrow corymbose at the apex, broader below, the terminal flower shortly stalked; rachis crooked, striate, prickles slanting, retrorse-falcate; lateral leaflets gibbous, shortly stalked. Flowers *c.* 2 cm.; sepals closely felted, short-pointed; petals pinkish, elliptic, clawed; filaments white, longer than the greenish styles. Carpels glabrous. Receptacle strongly pilose. *Recalls R. Lindebergii,* under which name it was recorded in the *London Naturalist* for 1927.

England, in two spots at Witley Sandhills (v.-c. 17), where it was seen several times by the author and Mr C. Avery, between 1926 and 1936. Sudre records it for Witley (E. F. Linton), 1908–13, Supplement 1. Belgium, Bavaria, Portugal.

Ser. **iii. PILETOSI** Genev. pro parte

Subser. †**Macrophylli** Bouv.

(Type *R. macrophyllus* Weihe & Nees)

1. Plants large, with large leaves.

70. **R. macrophyllus** (excl. var. *b*) Weihe & Nees, 1825, *Rub. Germ.* 35; Sudre, 1908–13, 48, t. 52.

4x. The upright, bronze shoots soon bend and grow straight forward for perhaps 2–3 m. The stem is slightly furrowed and hairy; *the prickles short, subulate.* The leaves are softly pilose to slightly greyish felted beneath; *the terminal leaflet very large,* cordate-ovate-acuminate, or *with somewhat parallel sides from a rounded base and drawn out into a long angular point.* The panicle is subracemose with deeply divided, spreading mid-branches; *the rachis has* dense spreading hair and *a few stalked glands and weak prickles. Flowers pink,* rarely white; sepals reflexed to patent, sometimes erect on the terminal flower; filaments white, rarely pink, styles greenish. The carpels are slightly hairy, *the receptacle very hairy, especially below the carpels.* Fruit rather large, sweet. Flowering from mid June.

E., W., S., I. Frequent, especially in woods. France, Belgium, Netherlands, Germany, Switzerland, Austria.

71. **R. subinermoides** Druce ex. W. Wats. 1929 (for 1928), *Rep. Bot. Exch. Cl.* **8**, 859; Syme, 1864, *Eng. Bot.* t. 450 (as *R. macrophyllus*). **Fig. 13.**

*4x. Long confused with *R. macrophyllus,* from which it differs in the scrambling, felted rather than hairy, more obtuse angled and furrowed stem, *prickles* shorter, *stouter, recurved; terminal leaflet narrow, long obovate to elliptic, softly greyish white felted* beneath, *becoming very convex; panicle prickles* strong, declining, *weak or wanting in deep shade. Petals white, claw greenish. Carpels* subpersistently and conspicuously *pilose.* Fruit sweet. In the open the leaves are smaller and are white felted though still soft beneath. Flowers from 12 June.

Endemic. England: mostly in woods; 3, 7–9, 11–22, 24, 26, 30, 33, 34, 36, 39.

72. **R. Schlechtendalii** Weihe in Boenn. 1824, *Prod. Fl. Monast.* 152; Sudre, 1908–13, 50, t. 54.

Robust. Stem blunt-angled, furrowed, reddish violet, pruinose, with numerous, sessile glands; *prickles moderate, short-based, slanting or recurved.* Petiole prickles small, curved; stipules glandular. Leaflets 5, base ± entire, greyish white felted and silkily pubescent beneath, *rather shallowly serrate with some mucros patent;* terminal leaflet ovate or obovate, acuminate-cuspidate. *Panicle* long, lax, *leafy, branches spreading, ± compound; rachis hirsute,* prickles of the peduncles and pedicels numerous, short, weak, straight or bent. Flowers *c.* 2·5 cm.; *sepals leafy-tipped, gland-dotted and aculeate, loosely reflexed;* petals usually rose, obovate-rhomboid, hardly clawed; filaments pink or white, slightly longer than the pinkish styles, *anthers pilose.* Carpels and receptacle densely pilose. Fruit large. Distinguish from *R. amplificatus.*

E., W., S., I. Rare; 2 (white flowered), 4, 5 or 6, 14, 23, 30, 50, 87. H. 2. North and central France, N.W. Germany, Austria.

73. **R. Boulayi** (Sudre) W. Wats. 1946, *Journ. Ecol.* **33**, 338.

Hants. and Dorset. *Robust,* slightly glandular. *Stem* very obtuse-angled, striate, *glabrescent,* with some pricklets becoming brownish and dull purple; *prickles numerous, strong-based, subulate,* scattered round the stem; *leaflets 5, entire-based, glabrous above,* pubescent to greyish felted beneath, *serrulate or serrate-dentate,* terminal leaflet elliptic-obovate, cuspidate. *Panicle long,* leafy, *narrow,* interrupted below, *branches short,* deeply divided, *semi-erect; all leaflets obovate-subcuneate, finely serrate-dentate, prickles numerous.* Flowers white or tinged pink. Calyx closely white felted and pubescent, slightly glandular. Carpels pilose. Receptacle pilose.

England: rare; 9, 11. N. France (originally found by Le Jolis at Cherbourg).

74. **R. sylvicola** Lef. & Muell. 1859, *Pollichia,* 113; *R. Schlechtendalii* microg. *silvicolus* (Lef. & Muell.) Sudre, 1908–13, 51, t. 55.

★**4x.** *Robust,* inconspicuously acicular, aculeolate and glandular. *Around Tunbridge Wells.* Stem very obtusely angled, striate, pruinose, *hairy; prickles short, strong-based, recurved.* Stipules linear. *Leaflets 5, imbricate,* glabrescent above, pilose and grey felted beneath, *incise subcompound-serrate, undulate; terminal leaflet roundish-ovate, acuminate-cuspidate, basal leaflets long-stalked. Panicle with very narrow leafy bracts* to the top, pyramidal, with long, numerous flowered, spreading, intricate branches; rachis pilose, prickles numerous, falcate. Flowers 2·5–3 cm. Sepals felted, aculeate, reflexed. *Petals pink, elliptic, narrowed into a long claw.* Filaments white, longer than the greenish styles. *Carpels subpersistently pilose.* Fruit large, ovoid, drupels many.

E., I. Rare; 14 (Benhall Mill Lane), 15 (Tunbridge Wells Common). H. 3 (Skibbereen). France (*Oise, Aisne, Villers-Cotterets* (Questier)).

75. **R. megaphyllus** P. J. Muell. 1859, *Pollichia,* 293, citing Wirtg. *Herb. Rub. Rhen.* **I**, 79 et 80.

Herts. Stem furrowed, glaucescent, dusky and *a good deal hairy; prickles* rather long, *slender, subulate, declining.* Leaves grey felted as well as hairy beneath; *leaflets compound-serrate, petiolules ± glandular; terminal leaflet elliptic-ovate-acuminate. Panicle* rather leafy, broad, long, equal, lax, *glandular; terminal leaflets roundish, incise compound-serrate; prickles various, slender.* Petals pale pink, pilose. Filaments white, longer than the greenish styles. Carpels a good deal pilose.

England. 20 (Sherrardspark Wood, Welwyn Garden City, 1953). France, W. Germany, Bavaria.

76. **R. majusculus** Sudre, 1918, *Bull. Assoc. Pyr.* no. 205; 1908–13, 42, t. 45.

N.v.v. Dorset. Stem sharp-angled, brown, moderately hairy, with numerous sessile glands; *prickles strong,* rather unequal, *slanting,* pilose. Leaves sub-digitate, *leaflets* pubescent to slightly felted beneath, rather finely and unequally serrate, a little dentate; *terminal leaflet* rather broad obovate or ovate-elliptic, *shortly cuspidate-acuminate. Panicle* long, strict, narrow, sub-racemose or with the upper branches half erect, subcymose; rachis felted and pubescent, *prickles* rather few, *much declining to retrorse-falcate.* Flowers up to 2·5 cm.; sepals grey felted, reflexed; *petals narrow elliptic, tapered below,* pink or white, pilose; filaments white, longer than the greenish styles. Carpels pilose or glabrous.

England. 9 (wood north of Verwood Station, E. F. Linton, 1891 (as *R. macrophyllus*)).

2. A high, floriferous bush with small leaves and very numerous, small panicles.

77. **R. Bakeranus** Barton & Riddels. 1935, *Journ. Bot.* **73**, 128; Sudre, 1908–13, t. 39 (as *R. Bakeri*).

*4x. Precocious, eglandular. Stem slightly furrowed, dull; *prickles* patent, subulate to broad-based, falcate, *hairy. Stipules filiform-linear. Leaflets* 5, subcordate, *caudate-cuspidate,* greyish felted and pubescent beneath, finely and sharply serrate; *terminal leaflet* broad elliptic-obovate or ovate, ± *long-stalked.* Panicle simple, or broad pyramidal, compact, mid-branches short, 2–3-flowered, or long-peduncled, half-erect, cymose, with 5–7-flowered lower branches. *Flowers c. 2·2 cm.*; sepals loosely reflexed; *petals bright pink, pilose, roundish, tapered below*; filaments white, longer than the greenish styles, *anther*

sutures pink. Carpels glabrous. Receptacle hirsute, especially below the carpels. Flowering from 30 May, decorating the south metropolitan commons then and in July when the *coral fruits* are showing.

Endemic. E., S. Local; (16, extinct), 17 (frequent on heaths throughout the county), 100 (Arran, Lamlash to Brodick).

var. **milcombensis** Barton & Riddels. 1935, *Journ. Bot.* **73**, 129.

N.v.v. Differs in the leaflets being all subcordate, less deeply toothed, more shortly pointed; and petals, filaments and styles all pink. 23 (Milcombe).

3. Leaves and panicles moderate.
† Prickles small.

78. **R. silvaticus** Weihe & Nees, 1825, *Rub. Germ.* 41; *Fl. Dan.* t. 2904; Sudre, 1908–13, 51, t. 56. **Fig. 14.**

4x. *Hirsute, bright green.* Stem erect, arching, angled. *Leaflets 5, narrow, base entire*, rather coarsely, sharply and simply to ± compound-serrate, green and hairy beneath; *terminal leaflet elliptic, narrowed below, shortly stalked. Panicle dense, often leafy to the top*, long, equal or broader below, *prickles numerous, fine, declining. Sepals pilose, glandular and acicular, leafy-tipped*, loosely reflexed. *Petals* pale pink or white, *elliptic* or obovate, abruptly clawed. Stamens white, much longer than the greenish styles. *Anthers usually pilose.* Carpels pilose. Woods and heaths.

E., W., I. Scattered; 3, 4, 7, 9, 12, 17, 22, 23, 32, 34–36, 39, 49. H. 2, 38. Belgium, Denmark, Germany, Switzerland, Austria.

79. **R. chrysoxylon** (Rogers) W. Wats. 1946, *Journ. Ecol.* **33**, 338; *R. mercicus* var. *chrysoxylon* Rogers in Griffith, 1895, *Fl. Anglesey and Carn.* 42.

Wales and Welsh border. Slender, elegant, prickles weak and short: axes becoming ochreous. Stem furrowed. Leaf-prickles weak, recurved; *leaflets 3, 5,* pubescent to glabrescent beneath, *incise compound-serrate, undulate; terminal leaflet long-stalked*, cordate-oblong or *roundish obovate with a long, abrupt point. Panicle lax with widely spreading branches*, prickly, slightly glandular and acicular, rachis wavy, pedicels shortly woolly-felted. Flowers *c.* 2 cm.; *sepals finely aculeate, patent*, the tips ascending or clasping. *Petals white, narrow elliptic, tapered below.* Stamens long. Carpels glabrous. Receptacle pilose. Fruit subglobose, good.

Endemic. E., W. Infrequent; 35, 36, 42, 49, 52.

80. **R. patuliformis** Sudre, 1906, *Diagn.* 10; *R. silvaticus* ssp. *patuliformis* (Sudre) Sudre, 1908–13, 52, t. 57.

*****4x**. *Robust, very hairy, subglandular.* Stem angled, ± furrowed, *prickles numerous, broad-based, unequal, patent.* Leaves pedate; *leaflets 5, long-stalked*, pilose and greyish felted beneath, *coarsely incise subcompound crenate-serrate;* terminal leaflet broad, subcordate, roundish ovate, acuminate. *Panicle with small leaves to the apex, very broad* pyramidal, truncate, *branches patent*, race-

mose or subpanicled, *densely villose, prickles numerous, very short and weak.*
Flowers *c.* 2 cm.; *sepals loosely reflexed; petals roundish,* pinkish; filaments
white, longer than the yellowish styles, stigmas brownish. Carpels glabrous.
Receptacle pilose. Fruit of numerous drupels.

England: local in several spots around Tunbridge Wells; 14, 16. France
and Belgium (very rare).

†† Prickles moderate.

81. **R. amplificatus** Ed. Lees in Steele, 1847, *Handb. Field Bot.* 58; *R.
incarnatus* ssp. *cryptadenes* var. *stereacanthoides* Sudre, 1908–13, 59.

*4x. *Eglandular.* Stem long, angled to furrowed, striate; *prickles yellow-
brown, subulate, declining from a broad, red base. Leaflets 5, hairy to pubescent and
greyish felted beneath,* unequally and sharply serrate and partly dentate, *base
mostly entire; terminal leaflet* broad to *narrow elliptic-obovate-cuneate, long falcate-
cuspidate; intermediate leaflets often cuneate. Panicle long, narrow, leafy,* lower
branches remote, sharply ascending, upper branches deeply divided; *rachis
hirsute above, glabrescent below, prickles sharply declining from a decurrent base.*
Flowers 2–2·5 cm.; *sepals* greenish, felted, acicles *reflexed;* petals pilose,
pinkish, obovate; filaments long, *anthers pink-sutured.* Carpels glabrous.
Receptacle pilose.

E., W. Rather frequent; woods, hedges and heaths; 2, 3, 5, 7, 10, 11, 13,
15, 16, 18–25, 27, 29, 33, 35–37, 39, 40, 42, 43, 45, 48, 50, 55, 57, 62, 67.
H. 2, 23, 28, 33, 38, 40. France (*Finistère*), Denmark (*Haderslev*).

82. **R. eglandulosus** Lef. & Muell. 1859, *Pollichia,* 164; *R. macrophyllus* ssp.
eglandulosus (Lef. & Muell.) Sudre, 1908–13, 49, t. 53.

Cornwall and Devon. Robust. Stem obtuse-angled to slightly furrowed;
prickles subequal with occasionally a pricklet. Leaves pedate, prickles
falcate or hooked; *leaflets glabrescent on both sides, rather coarsely incise-
subcompound-serrate; terminal leaflet* long-stalked, *cordate,* broad or roundish ovate,
*abruptly long-narrow-pointed. Panicle large, subcorymbose at apex with several
4–10-flowered, oblique, axillary branches;* rachis crooked, hairy, pedicels felted,
prickles numerous, falcate. Flowers 2–2·5 cm., incurved; *sepals* felted and
pubescent, *slightly glandular and aculeate,* loosely reflexed to patent, the tips
erect on the terminal flowers; *petals ovate, faintly lilac;* filaments white,
drying pinkish, slightly longer than the whitish styles. Carpels subglabrous.
Receptacle pilose. *Fruit large.*

> Despite the name, stalked glands occur on the petioles and on the
> panicle, even in Lefevre's specimen in ex Herb. Mueller.

England: scattered in 2 and 3. France (*Aisne, Oise*).

83. **R. belophorus** Muell. & Lef. 1859, *Pollichia,* 111.

N.v.v. *Lincs.* Stem angled to furrowed; *prickles* numerous, often in
twos, *subulate, patent;* leaflets 5, thinly hairy beneath; terminal leaflet base

indented, roundish, cordate-ovate-acuminate, unequally not deeply dentate-serrate, jagged. *Panicle long, lax, pyramidal, with a narrow apex and numerous ovate-lanceolate leaves, prickles numerous, finely subulate, up to 1 cm. long* on the rachis; upper leaves slightly felted. Flowers *c.* 2·5 cm.; sepals felted, leafy-pointed, aculeate, slightly glandular, loosely reflexed; *petals pinkish, ovate.*

England. 54 (Stainton Woods, A. Ley, 1907). France (Oise, rare).

Subser. ††**Pyramidales** Bouv.

(Type *R. pyramidalis* Kalt.)

1. Petals pink.

† All leaves green beneath, rarely the upper ones slightly greyish felted.

84. **R. pyramidalis** Kalt. 1845, *Fl. Aach. Beck.* 275; Sudre, 1908–13, 46, t. 50. **Fig. 15.**

4x. *Panicle glandular. Rufescent.* Stem low-arching, procumbent, hairy to *nearly glabrous,* occasionally with a few glands and acicles; *prickles numerous, gradually narrowed,* straight. Leaves yellow-green, *soft with green felt and shining parallel hairs* on the veins beneath, irregularly and sharply serrate, sometimes jagged; terminal leaflet roundish (-cuneate), ovate or elliptic (-rhomboid), acuminate. *Panicle dense, pyramidal in flower, oblong in fruit,* with the upper branches 1–3-flowered, patent, and usually one simple leaf; rachis sharp-angled, hirsute above, *prickles narrow-based, straight, terminal leaflets cuneate.* Flowers *c.* 2·5 cm.; sepals glandular and acicular, loosely reflexed to patent or ± less erect; petals ovate-elliptic or obovate; filaments white, not much longer than the styles. Carpels ± glabrous. Receptacle pilose. Fruit large. Coming late into flower.

var. **parvifolius** Frid. & Gel. 1887, *Danm. Slesv. Rub.* 87; var. *eifeliensis* sensu Rogers, 1900; 50; non Wirtg. 1862.

A frequent dwarf form. **4x (6x).**

E., W., S., I. Frequent; Scotland, rare. North and central France, Belgium, Netherlands, Denmark, S. Sweden, Germany, Switzerland, Austria.

85. **R. albionis** W. Wats. 1948, *Watsonia,* **1,** 73; *R. Schlechtendalii* var. *anglicus* Sudre, 1908–13, 50.

All axes becoming crimson. Stem sharp-angled, glabrescent, with a pricklet or two; *prickles long-subulate, often bent up at the tip.* Leaflets 5, glabrous above, soft with spreading hairs on the veins beneath, *sharply and unequally or sub-compound-serrate, jagged; terminal leaflet entire-based, elliptic or obovate-cuneate, truncate-cuspidate.* Panicle dense, rather narrow and somewhat diminished to the apex, not leafy, rachis unarmed above, densely felted and hirsute, inconspicuously glandular; prickles acicular on the pedicels. Flowers *c.* 2·75 cm.; sepals ovate-lanceolate, attenuate, reflexed; *petals elliptic-cuneate, entire*; filaments long, pink, anthers usually pilose; styles yellowish. Carpels thinly pilose. Receptacle hirsute.

E., W., S., I. Well distributed in England, rare elsewhere; 3, 5, 7–9, 13, 15–17, 22–24, 36, 38, 40, 41, 45, 47, 102, 103, 110. H. 12, 20. Schleswig.

86. **R. mollissimus** Rogers, 1894, *Journ. Bot.* **32**, 45; *R. danicus* var. *mollissimus* (Rogers) Sudre, 1908–13, 29, t. 23 *b*.

Stem reddish brown, angled, striate, slightly furrowed, hairy, glaucescent; *prickles few*, slender, long patent and ± falcate. Leaves large, digitate, prickles spreading-falcate; *leaflets* 5, becoming convex, *very softly hairy beneath from the parallel spreading hairs* on the veins, *not grey felted*, serrate or finely serrate-dentate; *terminal leaflet* entire-based, *roundish* (-obovate), short-pointed. *Panicle* with trilobed and simple leaves above, *lax, broad, equal and truncate*, lower branches remote; prickles slender, declining and falcate. Flowers *c.* 3 cm.; sepals patent, *petals roundish*; filaments long, pink, rarely white, *anthers pilose*; styles pink-based. Carpels and receptacle pilose.

Endemic. England: rare; 3–5, 9, 14, 16, 17, 22, 24.

87. **R. crudelis** W. Wats. 1933, *Journ. Bot.* **71**, 228.

Stem angled, red-purple, *with rare stalked glands and pricklets; prickles numerous, broad-based, in part curved and confluent, unequal, the intermediate ones gland-tipped*. Petiole short, prickles hooked; *leaflets imbricate*, softly pilose beneath, rather coarse and unequal to subcompound-serrate; *terminal leaflet short-stalked, cordate-ovate, gradually acuminate*. Panicle pyramidal, armed like the stem, its middle branches 3–7-flowered, cymose, the lower branches long and sharply ascending, *the bracts broad, hairy, glandular. Flowers showy*, *c.* 2·5 cm., *the petals, filaments and styles pink or bright pink, the styles sometimes greenish; anthers pilose*. Sepals patent or erect. Carpels glabrous. Receptacle pilose. Fruit rather large.

Endemic. England: rare; 4, 12–14, 17, 22, 24, 39.

88. **R. amphichlorus** (*amphichloros*) P. J. Muell. 1861, *Bonpl.* **2**, 279; *R. pyramidalis* ssp. *amphichlorus* (P. J. Muell.) Sudre, 1908–13, 47, t. 51.

Stem angled, *very hairy, subpruinose, with numerous sessile glands and a few, small pricklets*; prickles slanting from a broad base. *Leaves pedate, leaflets* 5, rather long-stalked, densely and adpressedly pilose, the veins pubescent beneath, unequally and broadly crenate-serrate, *a little jagged; terminal leaflet ovate, oblong or elliptic, acuminate, base emarginate or entire. Panicle equal, rather long and narrow*, the branches mostly 3-flowered, rather deeply divided, weakly armed, *very villose and with sunken glands*. Flowers *c.* 2 cm.; *sepals villose, glandular*, short-pointed, *reflexed; petals* light or deep pink, elliptic-obovate, *glabrous on the margin*; filaments pinkish, pink or red, much longer than the yellowish, reddish-based styles. Carpels glabrous or pilose.

E., I. Rare; 16, 17 (Spring Park, W. Wickham), 39 (roadside near Big Wood, Milwich). H. 12 (near Wexford), 33. France (*Vosges*).

Sylvatici

†† The upper, or all the leaves greyish or grey felted beneath.

89. **R. hirtifolius** Muell. & Wirtg. 1862, *Herb. Rub. Rhen.* ed. 1, no. 173.

Panicle glandular and acicular. Stem low, blunt-angled, becoming deep purple, glaucescent, *considerably hairy*; prickles rather short, subequal, straight, ± patent. *Leaves bright yellowish green, long-stalked*, prickles small, hooked; leaflets velvety, soft beneath with numerous hairs on the veins; *terminal leaflet long-stalked, roundish* (elliptic-) ovate or obovate-*subcuneate*, cuspidate, base entire, *sharply and doubly erect-serrate or jagged*; the *basal leaflet stalks about 5 mm.* Panicle long, equal, leafy with simple leaves to the apex, *lower branches short, erect, remote; rachis prickles numerous, slender*, various; *terminal leaflets obovate, coarsely biserrate.* Flowers 2–2·5 cm.; sepals ovate-acuminate-appendiculate, glandular and acicular, patent or ascending; petals broad obovate; filaments white, *anthers pilose*, slightly shorter than the greenish styles. Carpels glabrous. *Receptacle hirsute, especially below the carpels.* Fruit perfect, moderately large.

E., W., I. Scattered, apparently rare; 9 (Studland Heath), 11 (Shedfield Common), 27 (Mousehold Heath), 34, 41 (Mumbles Road). H. 27, 38. Germany, Belgium and N. France, very rare.

90. **R. orbifolius** Lef. 1877, *Bull. Soc. Bot. France*, **24**, 224; *R. macrophyllus* ssp. *orbifolius* (Lef.) Sudre, 1908–13, 50, t. 53; non Focke, 1914; nec 1925, *Lond. Cat.* ed. 11.

Eglandular. Stem slightly furrowed, glaucescent, a good deal hairy; *prickles numerous, very broad-based, strongly curved.* Leaves digitate, petiole short, *prickles numerous, hooked; leaflets 5, grey felted* and velvety pubescent beneath; *terminal leaflet short-stalked, round*, cuspidate, *base entire*, crenate-serrate. *Panicle* large, pyramidal below, equal and leafy above; rachis felted and pilose, *crowded with yellowish, strongly curved prickles.* Flowers *c.* 2·5 cm.; sepals bearing small, yellow, curved prickles, loosely reflexed to patent; petals broad-ovate or obovate-subcuneate; filaments pink or white, styles slightly greenish. Carpels glabrous. Receptacle hirsute.

Focke confuses this with *R. danicus* and Rogers with *R. mollissimus*. It is the most like *R. obesifolius.*

E., S. Locally frequent; 14 (Chailey, Newick, Piltdown etc.), 17 (Ashtead Forest; Effingham Common), 22 (W. Mortimer Common), 89 (Pitlochry). France (*Saône-et-Loire, Autun* and neighbourhood, frequent).

91. **R. obesifolius** W. Wats. 1956, *Watsonia*, **3**, 287; *R. Grabowskii* sensu Bab. 1847, *Ann. Nat. Hist.* **19**, 83; Bloxam in Kirby, 1850, *Fl. Leicester*, 46; non Wimm. & Grab. 1829; *R. nitidus* var. *rotundifolius* Blox. 1846 (specimen in Fasc.); *R. ? montanus* sensu Rogers pro parte, 1892, *Journ. Bot.* **30**, 142; non Wirtg. 1857.

Stem low, *purple*, angled, hirsute; prickles numerous, long and strong, often biseriate on the angles, patent. *Leaves rather small, pedate*, petiole furrowed throughout, prickles curved; *leaflets 3, 4, 5, grey felted and pilose* beneath; *terminal leaflet cordate, reniform* with a short, broad point, crenate-dentate. *Panicle* long, equal, dense above, with one, felted, ovate leaf, hirsute, sparsely glandular, *crowded, with strong, curved prickles*; mid and upper branches 3-flowered. Flowers *c.* 2·5 cm.; *sepals* felted, prickly and glandular, *soon erect*; petals ovate or obovate; filaments white, hardly longer than the greenish styles, *anthers pilose*. Carpels glabrous. Receptacle hirsute, the hairs protruding between and below the carpels. *Fruit perfect, ovoid.*

In *British Rubi* Babington confused this plant with *R. Grabowskii* of Henfield, Sussex, but his original description relates exclusively to the Cadeby plant. Rogers dropped the name *R. montanus* Wirtg. as being too doubtfully British, in *London Catalogue*, ed. 9.

Endemic. England: local; 39 (Meerbrook, Belmont to Cheddleton), 55 (Cadeby), 58 (Mate's Lane, Malpas).

92. **R. macrophylloides** Genev. 1868, *Mem. Soc. Akad. Maine-et-Loire*, **24**, 172; *R. Schlechtendalii* microg. *macrophylloides* (Genev.) Sudre, 1908–13, 50, t. 55.

N.v.v. Stem brown, obtuse-angled, striate, pilose, minutely and sparsely glandular; *prickles moderate, falcate or slanting. Petiole short*, prickles hooked, small. *Leaflets 5, short-stalked*, large and long, narrow, *sharp and unequal to subcompound-serrate*, glabrescent above, grey with fine felt and *very soft with spreading hairs on the veins beneath*; terminal leaflet base entire, elliptic-obovate, long-acuminate. Flowering branch long with some 4, 5-nate leaves; panicle short, narrow, subracemose above, pilose, *sparingly glandular; rachis prickles few, falcate, weak.* Flowers 2–2·5 cm.; sepals felted, pilose, *slightly aculeate* and glandular, reflexed; *petals elliptic; filaments pink, longer than the pink styles. Carpels glabrous.*

E., S., I. Rare, woods; 7 (Ramsbury), 54 (Apley), 88 (Killin). H. 23 (Crooked Wood, near Deveragh). France (rare).

2. Petals white or pinkish.

93. **R. danicus** Focke ex. Frid. & Gel. 1887, *Danm. Slesv. Rub.* 71; Sudre, 1908–13, 29, t. 23; *R. purbeckensis* Barton & Riddels. 1932, *Journ. Bot.* **70**, 190.

Stem arched, angled, *hairy*, sessile glands numerous; prickles rather long, subequal, slanting or falcate. Leaves subdigitate, *petiole prickles large, falcate*; leaflets softly pilose on the veins beneath, *sharply and unequally serrate*, undulate, *the principal teeth salient and sometimes jagged; terminal leaflet long-stalked*, roundish or obovate, acuminate, base nearly entire. Panicle long, narrow, subpyramidal, upper branches spreading, 1–2-flowered, lower branches 3–5-flowered, semi-erect; rachis felted and hirsute, glandular, *prickles slender, straight and falcate, lateral and basal leaflets of the lower leaves subsessile.* Flowers showy, *sepals and petals 5–8; sepals* grey-felted, *glandular and aculeolate*, patent; petals ovate-rhomboid; *filaments long*, drying *pinkish*, *anthers pilose*; styles greenish. Carpels ± pilose or nearly glabrous. Receptacle pilose. Fruit rather large.

E., S. Principally in Scotland; 9, 11, 13, 23, 65, 68, 74, 77, 86–89, 92, 96–98, 103, 106, 108, 110 (not 55). N. Germany, Denmark.

Subsect. B. DISCOLOROIDES Genev.

Ser. i. PROPERI W.Wats.

(Type *R. pedatifolius* Genev.)

Name from *R. properus* Chab., M.S., from its alleged very early flowering. Focke founded this group to take Chaboisseau's species for no other reason than that he was unable to decide to what existing group it belonged.

† Prickles long and moderately numerous.

94. **R. londinensis** (Rogers) W. Wats. 1938 (for 1937), *Rep. Bot. Exch. Cl.* **11**, 442; *R. imbricatus* var. *londinensis* Rogers, 1903, *Journ. Bot.* **41**, 89; *R. Daveyi* Rilst. 1950, *Journ. Linn. Soc.* **53**, 413. **Fig. 16.**

Robust. Stem thick, tall, *blunt-angled*, pubescent to glabrous, *subperennial; prickles* partly in twos, *slender, patent, bent or slightly sigmoid.* Leaves large; leaflets long-stalked, thinly pubescent to greyish felted beneath, broadly serrate; *terminal leaflet subcordate, roundish, subquadrate to oblong, cuspidate. Panicle* pubescent and felted, *broad and open, truncate, the mid branches long, cymose-partite, 5–7-flowered; rachis* blunt-angled, *prickles rare or none, small subulate. Flowers c. 3·5 cm.*, incurved; *petals* broad ovate or obovate, cuneate, *deep pink*, rarely white; sepals pubescent with linear, leafy tips, loosely reflexed; *filaments long*, white or pink. Carpels slightly pilose. Receptacle pilose. *Fruit large.*

Endemic. England: locally frequent; 1, 2, 16–18, 21, 24, 55.

95. **R. Libertianus** Weihe in Lej. & Court. 1881, *Comp. Fl. Belg.* **2**, 163.

Stem angled, glabrescent, becoming reddish purple, glaucescent; *prickles subulate*, mostly *slanting*, rather many, *yellowish*, subequal. Petiole pubescent, prickles hooked; *leaflets* 5, *subimbricate*, grey felted and pubescent beneath, *often rather cuneate; terminal leaflet ovate, acute* to elliptic cuspidate, compound or shallowly crenate-serrate. *Panicle large, long and dense*, leafy, apex subcorymbose, *branches all ascending*, pubescent and hairy above, *prickles subulate.* Flowers 2–2·5 cm.; sepals leafy-tipped, loosely reflexed or patent; *petals obovate, pinkish to white, fimbriate; filaments white, long.* Carpels glabrous. Fruit good. *All axes and the head of the panicle gland-dotted.*

England: rare; 22 (Boars Hill, Foxhill), 30 (Heath and Reach, and near Woburn). Belgium, France (one station).

†† Prickles of moderate length.

96. **R. ramosus** Briggs, 1871, *Journ. Bot.* **9**, 330.

Mainly western. Stem striate, *furrowed*, glabrescent, glaucescent, *prickles* scattered, *brownish*, patent and slightly declining. *Leaves tough, pedate; leaflets* 3, 5, *smallish, becoming convex, shining above*, closely grey felted and pilose beneath, irregularly serrate; *terminal leaflet* base emarginate, *oblong-obovate, short-pointed*, rather long-stalked. *Panicle slightly glandular and acicular above*, large, lax *with very long, sharply ascending, numerous flowered, axillary branches*, upper branches patent, 3 (5)-flowered; rachis crooked, felted and pilose, *prickles rather few*, weak. Flowers *c.* 2 cm.; sepals felted, patent during and after flowering; *petals pink*, broad ovate, abruptly clawed, or obovate-cuneate; filaments white, hardly longer than the *light red styles*. Carpels pilose. Receptacle pilose. *Fruit subglobose, usually imperfect.*

Endemic. E., W., I. Frequent in the south-west; 1, 2, 3, 7, 9, 11, 21, 49. H. 1, 21.

97. **R. Salteri** Bab. 1846, *Ann. Nat. Hist.* **17**, 172; non sensu Rogers, 1900, 45.

Stem tall, arching, furrowed, glabrescent, prickles brown, slender, few. Leaflets 5, glabrous above, softly greyish pubescent beneath; *terminal leaflet* subcordate, elliptic or oblong-obovate, cuspidate, *compound serrate*. Panicle leafy, large, broad pyramidal, abrupt corymbose, terminal flower shortly stalked; mid branches spaced, patent-cymose, 3–5-flowered; axillary branches spreading, closely pubescent and greyish throughout, slightly glandular, prickles tiny. Flowers 2·5 cm., cupped; sepals felted and pubescent, appendiculate, patent-ascending after flowering; *petals* remote, elliptic, *notched, glabrous, white*; filaments white, slightly longer than the yellowish styles. *Carpels glabrous.* Receptacle pilose. Fruit oblong.

Endemic. England: 10 (frequent in the south-east of the island, Apse Castle Wood and the downs to the south), 36 (Winforton Wood).

98. **R. pedatifolius** Genev. 1860, *Mem. Soc. Acad. Maine-et-Loire*, **8**, 71, 93, **10**, 85; *R. clethraphilus* Genev. 1869, *Mon.* 167; Sudre, 1908–13, 25, t. 14; *R. properus* Chab., M.S.

Subeglandular Slender. Stem brown and shining, angled, slightly furrowed, virtually glabrous; *prickles yellowish*, rather few, subulate from a broad flattened base. Leaves, digitate, petiole flat or nearly so, glabrescent; leaflets 3, 5, rather long-stalked, glabrous above, velvety and shortly pilose and on the upper leaves greenish grey felted beneath, *simply, irregularly, open serrate;* terminal leaflet elliptic-obovate, shortly and abruptly pointed. Panicle moderately long with 1–2 simple leaves below, broad, lax, the upper and middle branches 2–3-flowered, spreading, with numerous, weak, straight prickles throughout, and rare glands and acicles. Flowers *c.* 2·5 cm.; sepals greyish felted and pilose, appendiculate, loosely reflexed to patent; *petals 5, elliptic, fimbriate, rosy pink*; filaments white, drying pinkish, longer than the greenish styles. *Carpels densely and shortly pilose. Receptacle hirsute, especially below the carpels.*

England: very rare; 40 (Wyre Forest (in mid-flower 15 July 1953)). W. France, Pyrenees and N. Italy.

††† Prickles short.

99. **R. poliodes** W. Wats. 1956, *Watsonia*, **3**, 287.

Robust, building up a tall bush in open situations, climbing in woods. *Stems stout,* blunt-angled, furrowed, glaucescent, *with short divergent hair, prickles short or very short, sharply declining.* Leaves large, digitate; *leaflets convex, broad,* emarginate, pubescent to grey felted beneath, serrate-crenate to sharply serrate; shortly cuspidate; *terminal leaflet* short-stalked, *roundish subquadrate or obovate. Panicle* leafy, pubescent below, felted above, *lax, compound,* with spreading 3–7-flowered middle branches, *prickles few, fine, short and straight.* Flowers 2·5–3 cm.; *sepals* felted, *short-pointed,* patent with ascending tips; petals ovate or obovate, pinkish; filaments white, much longer than the greenish styles. Carpels glabrescent. Receptacle hirsute. *Fruit deep blood red to black.*

Endemic. England: scattered; 22 (Boars Hill, (holotype)), 24 (Bow Brickhill), 27 (Mousehold Heath and Ringland Heath), 30 (Kings Wood, Heath and Reach).

Ser. **ii.** SUBVIRESCENTES Sudre

(Type *R. villicaulis* Koehl.)

I. Robust, large-flowered plants, leaflets rather evenly serrate.

100. **R. rhodanthus** W. Wats. 1933, *Journ. Bot.* **71**, 224; *R. carpinifolius* var. *roseus* Weihe & Nees, 1825, *Rub. Germ.* 36; *R. rhombifolius* sensu Focke, Rogers, Sudre (1908–13, t. 45) et auct. mult.; non Weihe, 1824; *R. Banningii* Focke pro parte, *Syn. Rub. Germ.* 261; *R. rhombifolius* var. *megastachys* W.-Dod, 1906, *Journ. Bot.* **44**, 64.

Stem angled to furrowed, becoming deep red; *prickles purple-based* from the first, *long subulate.* Petiole prickles large, hooked; leaflets 5, softly pubescent and felted beneath; *terminal leaflet* subcordate or emarginate, ovate to *ovate-lanceolate (or subrhomboid), acuminate,* short-stalked, *usually almost regularly serrate or serrate-dentate. Panicle slightly glandular,* nearly equal, or moderately narrowed, *sometimes very broad with numerous flowered cymose lower branches, prickles large, slanting or curved. Flowers c. 3·5 cm.;* sepals greenish, thinly felted and pilose, usually aciculate, acuminate with leafy tips, reflexed to subpatent; *petals deep pink, rather narrow elliptic* or obovate, tapered below; *filaments* pink or white, *very long, anthers reddish, pilose;* styles greenish or reddish. Carpels subglabrous. *Fruit large.*

E., S., I. Moderately frequent in the south; 6, 7, 13–19, 22–24, 30, 34–36, 38, 57, 95. H. 27. Guernsey. France (rare), Belgium, Denmark, Germany, Switzerland (rare), Bavaria.

101. **R. insularis** Aresch. 1881, *Skannes Fl. et Bot. Notiser*, 158; *Fl. Dan.* t. 2414.

Stem deep-red, angled to furrowed, thinly hairy; *prickles powerful*, few or many, long, patent, slightly bent towards the tip, or falcate, hairy. Leaves olive-green, 5-nate, subpedate; *leaflets* densely and *softly pilose* to greyish felted *beneath, rather unequally* and sharply serrate, *or evenly serrate-dentate*, or a little jagged; *terminal leaflet* rather long-stalked, *roundish*, broadly ovate or elliptic, broadly and rather abruptly acuminate. Panicle lax, broad, narrowed to a truncate apex, leafy, mid branches sometimes long-peduncled, 3–7-flowered, cymose; stipules, bracts and bracteoles broad and green, rachis armed like the stem. *Flowers showy, up to 3–4 cm.*, incurved; sepals reflexed, the long tips spreading; petals rather narrow, diminished to the base, deep pink; filaments pink or white, long, *anthers sometimes pilose*; styles pink. Carpels usually pilose. Receptacle pilose. *Fruit* large oblong, soon turning *deep blood red*.

E., W., S., I. Rather rare in S. England; 6, 9, 18, 22, 24, 27, 42, 43, 49, 72, 87, 95–97, 106–108. H. 22. Common on the eskers, W. H. Mills. Sweden, Denmark.

102. **R. broensis** W. Wats. 1946, *Journ. Ecol.* **33**, 338; *R. umbraticus* Lindeb. *Herb. Rub. Scand.* 11; non P. J. Muell. 1859; *R. villicaulis* f. *silvestris* Frid. & Gel. *Bot. Tidsskr.* **16**, 69.

Stem angled, *glabrescent*; prickles numerous, strong, flattened. Leaves digitate; leaflets pubescent and closely greyish green felted beneath, finely and unequally patent-serrate; *intermediate leaflets narrow-oblong; terminal leaflet* emarginate, *rather narrow elliptic-obovate*, cuspidate-acuminate. *Panicle large and long, narrow* and dense and leafy above, broader and laxer below, *prickles strong, numerous, retrorse-falcate. Flowers 3–4 cm., opening widely; sepals conspicuously white-edged*, short-pointed, loosely reflexed; *petals* broad obovate, some notched, *pink; filaments white, very long; styles greenish.* Carpels glabrous. Fruit oblong. Late summer shoots are glandular and aculeolate.

England. Open heaths and woods, rare: 12, 17, 22, 55. H. 33. Sweden (Bro (Areschoug)).

2. A small bramble, with small, faintly lilac flowers.

103. **R. septentrionalis** W. Wats. 1946, *Journ. Ecol.* **33**, 338; *R. confinis* Lindeb. *Herb. Rub. Scand.* 12; non P. J. Muell. 1859.

4x. *N.v.v. Scotland.* Stem arched, prostrate, slightly furrowed, a good deal pilose, glaucescent; *prickles crowded, subulate, mostly patent, pilose. Leaves rather small*, 5-nate pedate, *stipules large, linear-lanceolate; leaflets* greyish green beneath, *sublobate, sharp lyserrate; terminal leaflet roundish-reniform* to broad ovate, elliptic or obovate, acuminate, *long-stalked*. Panicle villose, equal, interrupted below, the lower branches remote, racemose, prickles falcate; *bracts and bracteoles sublanceolate, large*. Sepals loosely reflexed. Petals obovate. *Filaments white, long. Styles yellowish.*

Scotland: rare; 94, 96, 97, 104, 105, 110. W. Sweden (Bahusia, rare).

3. Plant and flowers of moderate size.
† Terminal leaflet deeply and sharply compound-serrate to sublaciniate.

104. **R. stanneus** Barton & Riddels. 1934, *Journ. Bot.* **22**, 231.

Western. Eglandular. Stem angled, furrowed, pilose, becoming deep purplish glaucescent; *prickles rather clustered.* Leaflets 5, greyish felted and pubescent beneath, late summer leaves frequently becoming nearly regular serrate. *Panicle floriferous, with patent, long-stalked, cymose mid branches,* villose; prickles subulate, on the pedicels acicular. Flowers 2–2·5 cm.; *sepals* felted and villose, *patent; petals pale pink or pure white,* obovate; filaments white, longer than the whitish, *sometimes reddish-based styles. Carpels strongly pilose.* Receptacle pilose. Fruit oblong or subovoid, freely produced.

Endemic. E., S. Scarce, except in the south-west; 1–3, 5, 17 (one bush), 110 (S. Uist).

†† Terminal leaflet evenly serrate or serrate-dentate.

105. **R. atrocaulis** P. J. Muell. 1859, *Pollichia*, 163; non auct. dan.; *R. stereacanthos* P. J. Muell. ex Genev. 1868, *Mem. Soc. Acad. Maine-et-Loire*, **24**, 189.

Axes, bracts and sepals gland-dotted. Stem becoming *purple-black in the sun,* furrowed, with scattered, short, clustered hairs; *prickles numerous, broad-based. long, subulate, patent. Petiole* long, *channelled throughout; leaflets 5, subglabrous above,* felted and pubescent beneath; *terminal leaflet* cordate-ovate, acuminate to *suborbicular, shortly pointed.* Panicle very broad, subcorymbose, the primary flower short-stalked, the upper branches cymose; *prickles hard, very long, various;* pedicels and calyces felted; sepals finely aculeolate, loosely reflexed with the long tips ascending; *petals rose-pink, roundish-cuneate,* apex notched or erose; filaments pinkish, slightly longer than the yellowish, reddish-based styles. Carpels usually pilose. Receptacle pilose.

England: locally frequent; 5, 9, 16–18, 20, 21, 24, 39, 57. E. France, W. Germany.

††† Terminal leaflet unequally toothed.
(*a*) Flowers smaller than 2 cm. Leaflets acute.

106. **R. macroacanthus** (*macroacanthos*) Weihe & Nees, 1825, *Rub. Germ.* 44, t. 18; non auct. brit.; cf. W. Wats. 1948, *Watsonia*, **1**, 74.

Norfolk. Stem short, arched, angled, furrowed, red and very hairy at the base, subglabrous upwards, glaucescent; *prickles* flattened, numerous, *biseriate, very strong and long, patent.* Leaves pedate, petiole prickles numerous, strong, recurved; *leaflets 5, imbricate,* nearly glabrous above, felted and pilose beneath, unequally crenate-serrate, obsoletely lobate; *terminal leaflet shortly stalked,* subcordate, broad obovate, *acute.* Panicle long, leafy, subcorymbose at the top, interrupted below, the lower axillary branches erect. *Flowers small, c. 1·8 cm.;* sepals shortly pointed, loosely reflexed to patent; *petals round to broad obovate, deep rose-pink;* filaments deep pink, slightly longer than the yellowish styles. Carpels pilose. *Fruit well-formed, ovoid, rather small.*

England. 27 (Sprowston and Mousehold Heath). Germany (*Minden*).

(*b*) Flowers broader than 2 cm. Leaflets long-pointed.

107. **R. incurvatus** Bab. 1848, *Ann. Nat. Hist.* Ser. ii, **2**, 36.

4x. *Stem* angled, furrowed, hairy especially at the base, purplish-glaucescent; *prickles crimson-based.* Leaves pedate; *leaflets 3, 5, imbricate, yellowish green, thick, softly pilose to greyish felted beneath,* ± coarsely and doubly serrate-dentate, undulate. *Panicle very long and narrow, with dense, short cymose branches*; rachis crooked and furrowed, felted and villose, with numerous falcate prickles. Flowers large, showy, incurved; sepals linear-tipped, grey felted and *yellowish hairy,* aculeate, reflexed with the tips ascending; *petals* pink, pinkish or white, *broad-elliptic-obovate, cuneate*; filaments white or pink-based, longer than the pink-based styles. Carpels pilose. Fruit moderate. A dwarf form occurs.

E., W., S., I. Well distributed, but more frequent in the west than in the east; 4, 6, 9, 13, 15–17, 19, 22, 23, 36, 39, 41, 42, 48–52, 57, 58, 66, 67, 69, 74, 88, 104, 105, 110. H. 2, 12, 23, 33, 35, 40. Jersey. ?Portugal.

108. **R. villicaulis** Koehl. ex Weihe & Nees, 1825, *Rub. Germ.* 43; Sudre, 1908–13, 54, t. 55; *R. castrensis* W.-Dod, 1906, *Journ. Bot.* **44**, 63. **Fig. 17.**

4x. *Robust, shaggy.* Stem arching and procumbent, angled to furrowed, patently pilose; prickles hairy, straight, mostly slanting, occasionally a pricklet. Leaves 5-nate, prickles strong, falcate, stipules red, linear; leaflets hairy to pubescent or greyish felted beneath, unequally serrate-dentate or coarsely serrate; terminal leaflet cordate, roundish ovate, acuminate to rather narrow elliptic, base entire. *Panicle large, broad,* terminal flowers subsessile, upper branches cymose, *lower axillary branches rather numerous, long, oblique; prickles numerous, straight, strong, slanting.* Flowers 2–2·5 cm.; sepals long-pointed, usually aciculate, loosely reflexed; petals rather broad, rose or pink, rarely white; filaments pink or white; styles greenish or pinkish. Carpels glabrous or pilose. *Receptacle hirsute, especially below the carpels.* Fruit moderate. Our form is var. **marchicus** (Krause) Sudre, described above. It sometimes has a few glands and pricklets in the panicle. Koehler's typical plant has been found in N. Essex.

E., S., I. Scattered; 2–4, 7–9, 13, 15–17, 19 (Marshall), 22, 23, 38, 39, 58, 67, 98. H. 1 (Dingle Peninsula). Jersey. Netherlands, Germany, Austria. Absent from France, Belgium, Denmark and Scandinavia.

109. **R. iricus** Rogers, 1896, *Journ. Bot.* **34**, 506.

*4x. Ireland. *Robust, hirsute*. Stem angled, striate, hardly furrowed, *purple, hairy; prickles numerous, large, subulate. Leaf-petiole short*, stipules red, long-linear; *leaflets 5, subimbricate*, dull and glabrescent above, hairy and felted beneath, *rugose; terminal leaflet* broad-*elliptic-rhomboid, short-stalked, deeply and sharply subcompound-serrate. Panicle floriferous*, dense, broad, pyramidal with rounded apex, becoming rather corymbose from *the lower branches drawing up*, middle and upper branches cymose; *rachis hirsute, prickles long-subulate. Flowers c. 3 cm., showy*; sepals grey-villose, leafy-tipped, aciculate, loosely reflexed; petals obovate, rose-pink; filaments rose-pink, at any rate at the base, much longer than the reddish-based styles. Carpels subglabrous. Fruit rather large but imperfect.

Endemic. Ireland: frequent; H. 1–3, 5, 9, 16, 26–28, 35, 39.

110. **R. Langei** G. Jens. ex Frid. & Gel. 1877, *Rub. Dan.* 67; *R. atrocaulis* auct. dan.; non P. J. Muell. 1859; *R. rectangulatus* Maass, nomen.

*4x. Stem furrowed, brown, pilose; *prickles hairy, long, subulate from a narrow base*. Leaves yellow-green, softly pilose to felted beneath; *leaflets 5, sharply unequal or doubly serrate*; terminal leaflet cordate, broad ovate or roundish to elliptic, acuminate-cuspidate. *Panicle* narrowed to the leafy apex; rachis crooked, villose and slightly glandular, *prickles* long, *slender*, various; pedicels and calyces pilose, aculeolate and glandular. Sepals loosely reflexed. *Petals narrow-obovate, long-clawed, deep pink*. Filaments white or deep pink, far exceeding the green or deep red styles, *anthers usually pilose*. Carpels pilose. A dwarf form, with terminal leaflet finely toothed and apex more produced, is var. *parvifolius* (Jens.) Frid. & Gel.

England: rare; 9, 11, 17 (with the dwarf at Horsell Common), 24, 34, 62. Denmark, N. Germany.

111. **R. Favonii** (*Favonius*, the west wind) W. Wats. 1946, *Journ. Ecol.* **33**, 338; *R. pyramidalis* microg. *dumnoniensis* var. *eupectus* et var. *transiens* Sudre, 1908–13, 47, t. 51*b*; *R. pyramidalis* sensu Rogers pro parte, 1900, 50; *R. Buttii* Riddels. 1948, *Fl. Gloucester*, 148.

Western. Stem angled, *furrowed*, thinly pilose, *deep purple*, glaucescent; *prickles* long, *slender, patent to recurved*. Leaves yellow-green, pedate, stipules linear-filiform; *leaflets 3–5*, broad, very *softly and densely yellowish pilose beneath*, rather *coarsely to sublobate serrate*; terminal leaflet very broad, cordate-ovate, broadly acuminate. *Panicle large, floriferous*, leafy, *lax, pyramidal the lowest branches widely spreading*, from patent to half-erect; *rachis prickles fine but large, patent to hooked*. Flowers *c.* 2·5 cm.; sepals leafy-pointed, glandular and aciculate, patent to semi-erect; *petals lilac*, obovate, long-clawed; *fila-*

ments white, *barely longer than the greenish styles.* Carpels pilose. Fruit rather large, ovoid. *Stipules and bracts red-veined.*

E., W., S., I. Well distributed in the west; 9 (Bere Wood), 33, 35, 36, 38, 40 (Wyre Forest), 49 (Llanberis, Babington, 1860, as *villicaulis*), 102, 104, 110. H. 23. France (*Valois*), Belgium.

112. **R. Riddelsdellii** Rilst. 1950, *Journ. Linn. Soc.* **53**, 415.

Devon and Cornwall. Usually robust, eglandular. Stem angled, deep brownish red to *violet*, glaucescent, hairy; *prickles numerous, hairy, long and subulate.* Petiole prickles hooked, *leaflets 5, rather shortly stalked,* glabrescent above, felted and velvety pubescent beneath, *teeth* unequal *ending in long mucros*, undulate, convex; *terminal leaflet elliptic or oblong-obovate,* base subentire, acuminate-cuspidate. *Panicle* large, *long*, leafy, *equal and dense above, broader below*, most of the branches axillary; *rachis very hairy and prickly, some leaves 4, 5-nate. Flowers c. 2·5–3 cm.*; sepals ovate-cuspidate, grey felted and aculeate, reflexed; *petals long, oblong or rhomboid-elliptic, pale pink; filaments white, long; styles yellowish. Fruit large, thimble-shaped.* It seems to combine the characters of *R. villicaulis* and *R. propinquus*, two frequent Cornish species.

Endemic. Frequent in east Cornwall; 1–4.

113. **R. lasiothyrsus** Sudre, 1908-13, 61, t. 68; *R. villicaulis* Boreau, 1857, *Fl. Cent. France*, ed. 3, no. 762; non Koehl. 1825.

Stem angled to furrowed, pruinose, *a good deal hairy,* rarely a pricklet or acicle; prickles rather long, straight, hairy. Leaves digitate; *leaflets 5, glabrous* above, *greyish and softly hairy beneath,* unequally incise-serrate, often a little jagged; terminal leaflet roundish-oblong, cuspidate, to *elliptic-obovate acuminate. Panicle short, branches 1–3-flowered, spreading; rachis* angled, *villous,* prickles falcate; middle segment of the trifid bracts, and the sepal tips, green, leafy; pedicels and sepals slightly glandular. Flowers *c.* 2·5 cm.; sepals felted, reflexed; *petals obovate, pinkish or white, sometimes narrow attenuate to the base; filaments white, equalling the greenish styles.* Carpels pilose. Receptacle pilose. Fruit rather large, perfect. In woods.

England: very rare; 16, 17 (Spring Park, West Wickham). Belgium, N. France (*Normandy, Finistère*) and west of France to the Pyrenees.

Ser. **iii. SUBDISCOLORES** Sudre

(Type *R. rhombifolius* Weihe)

1. Eglandular.

114. **R. consobrinus** (Sudre) Bouv. 1903, *Rub. Anj.* 676; *R. villicaulis* ssp. *consobrinus* Sudre, 1899, *Rub. Pyr.* 46; *R. argenteus* ssp. *consobrinus* (Sudre) Sudre, 1908–13, 58, t. 65.

Robust, very prickly. Stem glaucescent, slightly pilose at first; *prickles crowded, long, straight or falcate,* often in twos. *Leaflets 3, 4, 5, imbricate, sub-*

acuminate; terminal leaflet roundish, ovate or obovate, subcordate. Panicle broad, leafy; *rachis* clothed with numerous, spreading, short hairs and shorter felt, *prickles long, slender, nearly equal, crowded.* Sepals ovate-cuspidate, greenish grey with a narrow, white-felted border, patent to erect. *Petals* ovate-oblong, long-clawed, *pink. Filaments white, or occasionally pink*, longer than the styles, which are greenish, or sometimes red-based. *Anthers pilose.* Fruit rather small.

England: infrequent; 14, 16, 17, 34 (near the Buckstone). France, Belgium, Pyrenees.

115. **R. cariensis** Rip. & Genev. 1872, *Mem. Soc. Acad.* **28**, 55; non sensu Rogers 1900, 25.

Ireland. Robust. Stem obtuse, red to blackish purple, subpruinose, *glabrescent. Prickles strong*, rather long, falcate *from a compressed, broad base.* Leaflets 5, subglabrous above, grey-white felted beneath; *terminal leaflet subcordate, ovate-acuminate*, teeth erect, sharp and deep subcompound. *Panicle large, pyramidal*, apex narrow, truncate, lower branches numerous flowered, mid branches *c.* 5-flowered, cymose; *prickles numerous, strong, recurved*, not large. Flowers showy, incurved; sepals felted, aculeolate, loosely reflexed to patent; *petals rose, broad ovate*, tapered below; filaments white, slightly longer than the greenish styles. Carpels pilose. *Fruit perfect, abundant.*

Ireland: H. 2 (roadside near the entrance to Torc Wood, Muckross). France (*Loire-Inférieure*).

116. **R. incarnatus** P. J. Muell. 1859, *Pollichia*, 95; *R. argenteus* ssp. *incarnatus* (P. J. Muell.) Sudre, 1908–13, 59, t. 65, figs. 16–18, not 19, 20.

Ireland. Stem red-brown, arched, angled, slightly furrowed, *densely pubescent; prickles* numerous, long, *narrow-based*, slanting. *Leaves short-stalked, petiole prickles numerous*, strong, falcate; *leaflets 5, glabrescent above*, felted and pubescent beneath, finely serrate, a little dentate and jagged; *terminal leaflet short-stalked, roundish or broad ovate with a long falcate point and often broad truncate base. Panicle long, dense, narrow*; rachis pubescent-felted, *prickles numerous*, strong, slanting and recurved, leaflets undulate, plicate, long-pointed. Flowers *c.* 2·5 cm.; sepals felted, subpatent; *petals rosy-pink*, broad ovate, notched, margin villose; filaments pink below, longer than the styles, *anthers pilose.* Carpels pilose. Fruit subglobose.

Ireland. H. 2 (near the Post Office, Muckross, 1935). France (*Valois*), Belgium, Germany (*Alsace*). Known apparently only in a single station in each of these countries.

2. Stem and panicle slightly glandular.

† Glands very shortly stalked, inconspicuous.

117. **R. polyanthemus** (*polyanthemos*) Lindeb. 1883, *Bot. Notiser*, 105; Trower, 1929 (for 1928), *Rep. Bot. Exch. Cl.* **8**, 857, no. 29 (plates); Sudre,

1908–13, 61, t. 67; *R. pulcherrimus* Neum. 1883, *Oefvers. Vet. Akad. Foes-handle. Stockh.* 65; Syme, 1864, *Eng. Bot.* t. 446 (leaf and stem only) as *R. rhamnifolius.*

Stem angled, moderately pilose, prickles moderate, stronger on the branches, yellow to red. *Leaflets 3–6 (7), dull above, grey felted and pubescent beneath, entire-based, becoming convex,* nearly simply-serrate or serrate-dentate; *terminal leaflet* roundish to *broad or rather narrow obovate, cuspidate. Panicle long, slightly narrowed upwards,* numerous flowered, mid branches cymose; rachis with numerous, strong, slanting prickles, felted-pubescent, *glandular with sometimes a few pricklets below. Sepals* greyish felted and pilose, *glandular-punctate and aculeolate,* reflexed. Petals broad obovate, pink, claw yellowish. Filaments pink or white, much longer than the greenish or rosy-based styles. Carpels pilose. Receptacle hirsute. Fruit of numerous, small drupels. Luxuriant late-produced shoots are often a little heteracanth.

E., W., S., I. Generally common. Denmark, S. Sweden, N. Germany.

118. R. rubritinctus W. Wats. 1956, *Watsonia,* **3**, 287; *R. cryptadenes* Sudre, 1904, *Bull. Soc. Etudes Sci. Angers,* 31; non *R. cryptadenus* Dumort. 1863; *R. erythrinus* sensu Briggs, 1890, *Journ. Bot.* **28**, 204; non Genev. 1868; *R. bipartitus* Boul. & Bouv. ex Genev. pro parte, 1880, *Mon.* 254.

4x. *Rufescent. Stem glabrescent becoming crimson and deep purple on the upper side,* green and *glaucescent beneath.* Leaves pedate, *leaflets* 5, glabrescent above, yellowish, shortly hairy and slightly grey felted beneath, *with rather shallow, spreading, unequal teeth; terminal leaflet* elliptic or broad *obovate, often rather truncate, with a long point,* base subcordate to nearly entire. Panicle lax, with long, and often one particularly long, lower branches, succeeded now and then by a still lower axillary, barren, trailing shoot; *rachis* felted and pilose, slightly glandular and aciculate above, *glabrescent below. Sepals* felted, aculeate, *reflexed. Petals pink, obovate, often notched, glabrous on the margin.* Filaments pink or white, long. Styles whitish or reddish. Carpels glabrous or ± hairy. *Fertile fruit rather large.*

E., W., I. Commonest in the south-west; 1–7, 9, 11–17, 22–26, 30, 34–42, 44–46, 57, 60, 67. H. 33. Guernsey, Jersey. France.

119. R. rhombifolius Weihe in Boenn. 1824, *Prodr. Fl. Monast.* 151; non sensu Rogers et auct. mult.; *R. argenteus* Weihe & Nees, 1825, *Rub. Germ.* 45; Sudre, 1908–13, 56, t. 64; non auct. brit.; cf. W. Wats. 1949, *Watsonia,* **1**, 74; *R. incurvatus* var. *subcarpinifolius* Rogers & Riddels. 1925, *Journ. Bot.* **63**, 13. **Fig. 18.**

4x. *A high bush. Stem* angled, furrowed, glabrescent, *dark red;* prickles large, long, rather numerous. Leaves rather large; *leaflets 5, imbricate, glabrous above,* silkily-whitish felted and pubescent beneath, unequally and finely serrate-dentate; *terminal leaflet short-stalked,* broad-based, or truncate, *ovate, gradually acuminate; basal leaflets subsessile. Panicle* large, leafy, *pyra-*

midal, with long-stalked cymose branches, the terminal leaflets of the uppermost 3-nate leaves rhomboid-acuminate, prickles numerous, nearly straight. *Sepals* grey-green, felted, aciculate or aculeolate and *glandular, patent. Petals downy, pink, roundish-ovate, cuneate.* Filaments pink or pinkish, much longer than the styles. *Anthers pilose.* Carpels pilose. *Fruit very good.*

E., W., S., I. Rather frequent; 3, 13–24, 29, 30, 33, 35, 36, 39, 42 (Glanau), 49, 55, 68, 110. France, W. Germany, Austria.

120. **R. acclivitatum** W. Wats. 1946, *Journ. Ecol.* **33**, 339; *R. argentatus* var. *clivicola* A. Ley, 1896, *Journ. Bot.* **34**, 158; Sudre, 1908–13, 57; *R. gymnostachys* Genev. pro parte, 1880, *Mon.* 223.

Stem angled, hairy; prickles strong-based, declining. Leaflets 5, grey felted and hairy beneath, undulate, *openly or ± compound-serrate; terminal leaflet long-stalked, broad obovate, cuspidate,* narrowed to the entire base. *Panicle broad, lax and long, mostly without any simple leaves,* with 1–3 (5)-flowered branches, *prickles numerous, short, falcate or hooked; terminal leaflets obovate-cuneate.* Flowers incurved; sepals glandular-punctate, aculeate, patent or loosely reflexed; *petals* pale pink, *obovate-cuneate;* filaments pink, long; styles yellowish. The whole plant becomes very hairy when growing in the shade: on sunny slopes the leaves are closely white felted beneath.

E., W., I. Most frequent in Wales and the Welsh border counties; 3, 5, 7, 8, 16, 20, 23, 24, 33–37, 42, 43, 47. H. 1, 2, 38. France (*Maine-et-Loire,* Couboureau).

121. **R. prolongatus** Boul. & Letendre ex Lef. 1877, *Bull. Soc. Bot. France,* **24**, 221; *R. lasiothyrsus* microg. *prolongatus* (Boul. & Let.) Sudre, 1908–13, 62, t. 68; *R. micans* sensu Rogers pro parte, 1900, 48; non Gren. & Godr. 1848; *R. griseoviridis* Barton & Riddels. 1933 (for 1931–2), *Proc. Cottesw. Field Cl.* **24**, 204.

4x. *Hirsute, small-leaved. Stem* sharply angled, *densely hairy,* glaucescent; *prickles yellowish brown to bright red,* robust, broad-based, slanting and falcate, pilose. *Leaflets 3* (4, 5), dull and glabrescent above, grey felted, pubescent and hairy beneath, *sharply compound-serrate; terminal leaflet short-stalked, elliptic-obovate, cuneate,* cuspidate-acuminate, *base subentire. Panicle large, lax, pyramidal, branches compound, spreading at right angles to the rachis,* with several simple leaves above; prickles short, broad-based, declining or recurved. Sepals greyish felted and hairy, leafy-pointed, reflexed. *Petals pink, narrow elliptic to obovate.* Filaments white, long. Styles greenish. A few stalked glands are to be found in the panicle and on the petioles.

E., W., I. Fairly frequent in the south and the west; 1–4, 7, 9, 14, 16, 17, 34, 35, 36, 41, 42, 45, 46, 49, 52. H. 20. Jersey. France (*Normandy*).

A large-fruited fertile hybrid of *R. prolongatus* with *R. ulmifolius* grows in an extensive patch on Tunbridge Wells Common. What appears to be a hybrid of *R. prolongatus* with *R. propinquus,* forming large but comparatively few drupels, is spread over a large area of W. Cornwall and has been described under the name of R. *pydarensis* Rilst. (see under R. *propinquus*).

122. **R. alterniflorus** Muell. & Lef. 1859, *Pollichia*, 160; Sudre, 1908-13, 63, t. 70. **Fig. 19.**

*4x. *Stem becoming glaucescent and blackish purple,* furrowed, *with a few short hairs, minute stalked glands and pricklets; prickles firm, straight, yellow, rufescent.* Leaves long-stalked, leaflets 5, *glabrous above,* ± felted beneath, *finely serrate-dentate, sometimes jagged; terminal leaflet long-stalked,* roundish, ovate, elliptic or obovate, *short-pointed. Panicle large, lax, pyramidal,* the lower branches long, oblique, the mid branches patent or ascending; rachis flexuose, intricately pilose and minutely glandular; *prickles very numerous, various, generally short, firm and fine.* Sepals glandular-punctate, aculeate, reflexed. Petals rather large, oblong or pointed ovate, *deep rose, pink or pinkish.* Filaments white or pinkish, longer than the greenish or reddish-based styles. Carpels glabrous or pilose. Receptacle glabrescent or pilose. Not averse to clay or chalk.

E., W., S., I. Fairly well distributed but not frequent; 2–5, 14–17, 21, 24, 33, 35, 39, 48, 69, 106, 110. H. 12, 20, 33. France, Belgium.

123. **R. herefordensis** Sudre, 1904, *Observ. Set. Brit. Rub.* 135; R. *alterniflorus* microg. *herefordensis* (Sudre) Sudre, 1908–13, 64, t. 71; R. *pubescens* sensu Rogers, *Set. Brit. Rub.* no. 115 (Caplar); non Weihe, 1824.

Western. Hirsute. *Stem pentagonal, red-brown,* hairy *and* felted, glaucescent, pricklets and stalked glands minute and rare; *prickles moderate, declining, slightly unequal, hairy.* Leaves pedate, *leaflets 3–5, glabrescent above, softly hairy* to *pubescent and felted beneath, veins yellowish, unequally angular serrate to sublobate; intermediate leaflets short-stalked;* terminal leaflet cordate, roundish or rhomboid-ovate, cuspidate. Panicle equal, interrupted below, with several simple leaves above; rachis hairy and felted with sunken, short glands. Flowers *c.* 2·5 cm.; sepals long-tipped, loosely reflexed or partly patent; *petals deep pink, roundish ovate; filaments deep pink,* longer than the pink or greenish styles. Carpels thinly pilose. *Bracteoles and stipules conspicuously glandular-fimbriate.*

This has the appearance of being a stable derivative of R. *propinquus* × (perhaps) *vestitus.*

England: principally in the south-west and western midlands; 1–3, 6, 7, 20, 22, 23, 33, 34, 36–38, 40. Guernsey (*St Andrews*).

124. **R. obvallatus** Boul. & Gillot, 1873, *Assoc. Rub.* 35; Genev. 1880, *Mon.* 175; R. *alterniflorus* ssp. *obvallatus* (Boul. & Gillot) Sudre, 1908–13, 66, t. 72.

Merioneth. Stem glaucescent, *angled with the sides concave,* thinly pilose and glandular; prickles subequal, moderate, yellowish, slightly declining. Leaflets

4, 5, softly greyish felted and pubescent beneath, *incise, ± compound serrate-dentate, all the foliar prickles weak, straight or spreading-falcate; terminal leaflet with base entire, obovate, acuminate.* Panicle broad and rather short, racemose or more compound with the branches widely spreading, divided near the apex; prickles weak, patent, unequal. Flowers *c.* 2·5 cm.; sepals greyish felted, gland-dotted, reflexed; petals pinkish to white, *broad elliptic to obovate*; filaments white, long; styles greenish. Recalls *R. egregius. The petioles are a good deal glandular and a little aciculate.*

Wales: 48 (frequent around Dolgelley). France (Brittany, the centre and south), Switzerland.

†† Stalked glands more conspicuous, especially on petiolules, pedicels and sepals.

125. **R. separinus** Genev. 1860, *Mém. Soc. Acad. Maine-et-Loire,* **8**, 71, 90, 111; *R. alterniflorus* ssp. *separinus* (Genev.) Sudre, 1908–13, 64, t. 71; *R. Gelertii* sensu Rogers, 1900, 56; non Frid. 1886; *R. cissburiensis* Barton & Riddels. 1931, *Journ. Bot.* **69**, 238.

4x. *Stem blunt-angled, glabrescent,* with a few stalked glands and acicles, *glaucous, spotted with red in semi-shade;* prickles strong, slanting. *Leaves pedate; leaflets 3–5, tough, glabrous above,* grey or white felted beneath, unequally serrate-dentate; *terminal leaflet broad obovate or elliptic-oblong, cuspidate. Panicle large, leafy, pyramidal, corymbose,* with several, long, ascending, numerous flowered axillary, and long-peduncled, cymose middle branches; *rachis and branches closely felted,* slightly hairy, *with* subulate prickles and *a few conspicuous glands.* Flowers *c.* 2·5 cm.; sepals grey felted, short-pointed, patent with the points erect; *petals pink, elliptic-obovate, clawed;* filaments pinkish or white, longer than the greenish, pink-based, rosy or red styles. *Carpels always conspicuously pilose, turning fuscous* before they become black. *Fruit* excellent, *abundant, large* and sweet. *An aggressive species, usually in quantity where it is found.*

England: in woods, hedges and open scrub, frequent on clay; 7, 11–17, 20, 21, 24, 33. West and south France, Austria.

126. **R. septicola** (Sudre) W. Wats. 1956, *Watsonia,* **3**, 288; *R. alterniflorus* ssp. *septicolus* Sudre, 1909, *Rub. Eur.* 65, t. 72.

Stem climbing, shallowly furrowed, *hairy, with a few small pricklets and short glands;* prickles small, declining. *Leaves large, tough; leaflets 5, glabrous above,* with thin, short hair or felt beneath, rather sharply and unequally serrate or crenate-dentate; *terminal leaflet base subentire, elliptic-obovate, short-pointed and short-stalked. Panicle very lax, interrupted,* long and narrow, leafy, the lower *branches erect, long-peduncled, panicled, the middle and upper ones in short fascicles;* rachis felted and pubescent, prickles hooked, finely glandular and aciculate. *Flowers 2 cm. or less;* sepals furnished like the pedicels with numerous weak prickles, loosely reflexed or patent; petals pink, obovate; *filaments white, equalling the greenish styles.* Carpels pilose. Fruit subglobose, small. *The very large leaves combined with particularly small flowers and fruit and short recurved prickles, make recognition very easy.*

England: only known in the hedges east of the Tunbridge Wells cemetery; 14 and 16. France (in a wood near Angers).

Ser. **iv. IMBRICATI** Sudre

(Type *R. imbricatus* Hort.)

1. Petals white, or at first faintly pink.

127. **R. cardiophyllus** Lef. & Muell. 1859, *Pollichia*, 86; *R. rhamnifolius* ssp. *cardiophyllus* (Lef. & Muell.) Sudre, 1908–13, 68, t. 76; *R. rhamnifolius* sensu Rogers pro parte, 1900, 29; Syme, 1864, *Eng. Bot.* t. 446 (flowering pieces only); non Weihe & Nees; 1822–27. **Fig. 20.**

4x. *Young shoots red. Stem sharp-angled, furrowed, glabrescent, bright red,* glaucescent; *prickles* not many, *long subulate.* Leaflets 5 (6), felted beneath, *teeth sharply incised,* somewhat compound-serrate; *terminal leaflet unusually long-stalked,* widely cordate, roundish ovate-acuminate. Panicle equal, rather short and broad, armed with short red prickles, rachis thinly felted. *Flowers c. 2·5 cm., incurved;* sepals grey felted, reflexed; *petals hairy, roundish, abruptly clawed, inflexed at the apex,* faintly pink to white; filaments longer than the styles. Carpels glabrous. Receptacle hirsute. *Fruit moderate, subglobose.* The leaflets of the flowering branch leaves become concave.

var. **fallax** W. Wats. 1937, *Journ. Bot.* **75**, 161.

Differs in being far smaller in every part, the leaflets nearly equally serrulate, the terminal leaflets of the upper 3-nate leaves of the flowering branch being obovate-cuneate.

> *R. furnarius* Barton & Riddels. 1935, *Journ. Bot.* **73**, 129; is a semi-dwarf variety of *R. cardiophyllus,* differing from it only in its rather smaller size. N. England; 65, ? 64, ? 69.

E., W., S., I. Frequent south of Lancs. and Yorks., rare elsewhere. Jersey, Guernsey. France, Denmark, N. Germany. The var. *fallax*; 3, 9, 10 (Bembridge, Salter, 1844), 12, 13, 17, 33, 39, 42, 62. H. 16.

128. **R. rotundatus** P. J. Muell. ex Genev. 1869, *Mon.* 177; *R. Questieri* ssp. *calvifolius* microg. *rotundatus* (P. J. Muell. ex Genev.) Sudre, 1908–13, 41, t. 41, figs. 6–9, t. 51, fig. 1; *R. dumnoniensis* Bab. 1890, *Journ. Bot.* **28**, 338; Rogers, 1900, 32; *R. cariensis* sensu Rogers, 1900, 25; non Rip. & Genev. 1872; *R. altiarcuatus* Barton & Riddels. 1933 (for 1931–2), *Proc. Cottesw. Field Cl.* **24**, 198.

4x. *Robust, very large-flowered. Stem procumbent when unsupported, furrowed, deep purple,* glabrescent; prickles long, mostly patent, sometimes in twos.

Leaves large, especially in shade, digitate; leaflets grey or white felted beneath, rather coarsely, unequally and towards the point subcompound-serrate; *terminal leaflet roundish-ovate or ovate-deltoid to oblong-obovate, shortly acuminate*; basal leaflets large. Panicle large, long, equal to subpyramidal, dense above, branches semi-erect, short, usually subcompound; *rachis* with clustered hairs and *felt, prickles straight and falcate, mixed. Flowers 3–4 cm.*; sepals loosely reflexed; *petals roundish(-obovate), apex notched or erose, ciliate*; filaments white, long; styles occasionally reddish based. *Carpels bearded or glabrescent.* Receptacle hirsute. *Fruit large, oblong.* A few short stalked glands and pricklets are sometimes to be found in the panicle, and sessile glands on the sepals.

E., W., S., I. Frequent; 1–9, 11–18, 22–24, 34, 36, 38, 39, 41–46, 49, 52, 57, 58, 69, 96–98, 101–104, 106, 110. H. 2, 11, 12, 26. Channel Isles. France, Belgium.

129. **R. Lindebergii** P. J. Muell. 1859, *Pollichia*, 292; Sudre, 1908–13, 84, t. 87; *R. lacustris* Rogers, 1907, *Journ. Bot.* **45**, 9.

4x. *Stem high-arching, then procumbent, angled, somewhat furrowed, thinly hairy, glaucescent, soon branching; prickles red-based, strong,* patent or falcate. *Petiole long, prickles spreading, hooked, large. Leaflets rather small, not contiguous, greyish above,* softly pubescent and felted beneath, *serrulate, mucros longish, curved; terminal leaflet base entire, rather narrow elliptic-obovate, shortly cuspidate-acuminate, very long-stalked.* Panicle narrow, long, leafy below, some leaves 5-nate, the *middle branches deeply divided or clustered, 2–3 together* in the same leaf or bract axil; rachis crooked angled, prickles numerous, strong, hooked. Flowers 2·5–2·8 cm.; sepals ovate, acute, white felted, reflexed; *petals narrow obovate, tapered to base*; filaments long. Carpels glabrous. Fruit moderate, roundish.

E., W., S., I. Frequent in the north, rare or absent in the south; 22, 34–36, 39, 40; 42, 43, 45, 50, 55, 57–60, 62–67, 69, 70, 75, 85, 87, 88, 90, 110. H. 38. Denmark, S. Sweden.

2. Petals pinkish, pink or rose.

130. **R. errabundus** W. Wats. 1946, *Journ. Ecol.* **33**, 339; *R. Maassii* ssp. *Scheutzii* sensu Sudre, 1908–13, 38, t. 39; *R. Scheutzii* sensu Rogers, 1900, 31; non Lindeb. 1885.

*4x. *Mainly northern.* Stem angled, subpilose, glaucescent; *prickles yellow, rufescent, straight,* long and strong with slender sharp points. Petiole prickles hooked. *Leaflets 5, all broad imbricate,* softly pubescent beneath, *serrulate-dentate*; terminal leaflet rather long-stalked, roundish or subquadrate, apiculate. Panicle equal, dense at the apex, leafy, branches deeply divided, slightly glandular, *very prickly*; terminal leaflets of the upper leaves cuneate, serrulate, *the lower leaves very long-stalked. Flowers up to 3·5 cm.*, strongly incurved; sepals greyish green, aculeate, loosely reflexed or subpatent; *petals lilac-pink,* *roundish obovate,* downy; *filaments pink, deeper at the base,* longer than the yellow styles, *anthers reddish, usually pilose.* A dwarf form in v.-c. 87.

Endemic. E., W., S., I. Scattered in the south and midlands, more frequent in the north; 12, 13, 15, 17, 23, 34, 49, 52, 58, 60, 67, 70–74, 86–88, 96, 97, 99, 110. H. 37–39.

131. **R. imbricatus** Hort. 1851, *Ann. Nat. Hist.* **4**, 113; Sudre, 1908–13, 67, t. 74.

4x. *A rather small-leaved bush with small dense panicles.* Stem biennial, soon throwing out *flagelliform branches,* blunt-angled to furrowed, glabrescent, glaucescent, reddening; prickles rather short, few. *Leaflets* yellowish green, *imbricate,* convex, subcordate, dull above, pale and shortly hairy to grey felted beneath, *unequally incise-serrate, the principal teeth larger and prominent; terminal leaflet broad, cordate, obovate, acuminate; lateral leaflets short-stalked; basal leaflets subsessile.* Panicle leafy, narrow, dense, equal, racemose above, interrupted, the lower branches sharply ascending; *rachis* hairy above, *glabrescent below,* prickles few, short, pedicels felted. Flowers 2–3 cm.; *sepals* felted, cuspidate, *sharply reflexed; petals roundish-obovate, notched,* pink or deep pink; filaments white, slightly longer than the yellowish or rosy-based styles. Carpels pilose. Fruit small, subglobose.

E., S. Mainly in S. England; 2–6, 9, 11, 14, 16, 17, 22, 34–36, 38, 110. N.W. France.

132. **R. Silurum** (A. Ley) W. Wats. 1937, *Journ. Bot.* **75**, 197; *R. nemoralis* var. *Silurum* A. Ley, 1894, *Journ. Bot.* **32**, 142.

Eglandular, pubescent and felted, not pilose. Stem angled, ± furrowed, becoming yellowish brown to deep red, slightly and finely hairy then glabrescent; *prickles rather short,* slanting, tending to be distributed in small groups on the angles of the stem. *Leaves digitate, pubescent and* on the upper leaves greenish grey *felted beneath,* nearly evenly, rather finely *dentate-sinuate-serrate; terminal leaflet roundish to elliptic-obovate, base entire,* apex gradually acuminate. Petiole-prickles rather numerous, rather small, recurved-falcate. *Panicle,* whether short or long, always *pyramidal,* with 1–2 simple leaves; *rachis sharply-angled, furrowed, pubescent to felted above,* prickles moderate, subulate, patent or slanting. *Flowers incurved, c.* 2·5 cm.; sepals felted to pubescent, patent to loosely reflexed; petals elliptic to roundish, white or pale lilac; filaments white, longer than the sometimes reddish styles. Carpels glabrous. Receptacle hirsute. Fruit subglobose.

Endemic. E., W., S. Local, mainly in or near Wales; 6, 8, 23, 34–36, 41–44, 46, 47, 49, 60, 71, 104 (Rhum).

Sect. **IV. DISCOLORES** P. J. Muell.

(Type *R. ulmifolius* Schott. f.)

A. Stem arched-procumbent, usually red or purple.

Ser. **i. GYPSOCAULONES** P. J. Muell. Stem pruinose, becoming scaly-waxed on the side turned to the sun. Stamens long or short. (See p. 97.)

Ser. **ii. HEDYCARPI** Focke. Stem epruinose or feebly pruinose. Panicle often subpyramidal. Plant often robust. Stamens long. (See p. 100.)

B. Stem tall, arched, usually furrowed and green.

Ser. **iii. CANDICANTES** Focke. Stem usually epruinose and glabrous. Leaves silvery greyish white felted and hairy beneath; basal leaflets very shortly stalked. (See p. 103.)

Ser. **i. GYPSOCAULONES** P. J. Muell.

(Type *R. ulmifolius* Schott. f.)

I. Filaments equalling styles.

133. R. ulmifolius Schott fil. in Oken, 1818, *Isis*, fasc. 5, 821; Sudre, 1908–13, 69, t. 77; Syme, 1864, *Eng. Bot.* 171, t. 447 (as *R. discolor*).

2x. Stem furrowed, clothed with numerous, minute, tufted, semi-appressed hairs; prickles robust, broad-based, patent to recurved, finely felted. Leaves 3–5-nate, pedate; *leaflets small, becoming convex and subcoriaceous,* glabrous and usually shining above, closely white felted (in some forms also pubescent) beneath, variously toothed; terminal leaflet variously shaped. Panicle long, subequal, for the greater part without leaves, branches patent, pedicels long, all densely felted and armed as the stem. Flowers showy, of moderate size; sepals white, felted, reflexed; *petals roundish, cuneate, crumpled;* anthers glabrous, pollen grains nearly all conform. *Carpels densely pilose.* Fruit roundish, perfect, of many small drupels. Flowering after midsummer.

Usually the petals, filaments and styles are all rose-pink; rarely the petals are a paler pink, filaments white and styles greenish. When the anthers are pilose or the pollen grains difform and the fruit imperfect, a hybrid origin may be suspected. But very fertile hybrids also are frequent, and are to be known by the mixture of characters and especially the *ulmifolius* character of the fruit. The following natural fertile *ulmifolius* crosses have been identified, mostly in south-east England viz:

subinermoides	*pseudo-bifrons*	*villicaulis*	*radula*
egregius	*prolongatus*	*vestitus*	*scaber*
longifrons	*spinulifer*	*hostilis*	

Many subspecies and varieties occur, especially in west and south England.

E., W., S., I. Channel Islands. Generally distributed, but infrequent north and just south of the Scottish border. Prefers open sunny situations, including those on clay and chalk. Western Europe, N.W. Africa and the east Atlantic isles.

134. **R. pseudo-bifrons** Sudre ex Bouv. 1907, *Rub. Anjou*, 32.

4x. Stem glabrescent, prickles numerous, strong, patent, declining or slightly falcate from a broad, crimson base. *Leaves* deep green, *villose as well as white felted beneath*; terminal leaflet long-stalked, base ± entire, oblong-obovate, acuminate-cuspidate, subplicate, margin undulate, *shallowly and unequally divaricate-dentate. Panicle long, dense, felted and pilose*, prickles numerous, robust, falcate; *pedicels very short, flowers and fruit consequently dense. Sepals pale greenish white, slightly aculeolate, felted and pubescent*, loosely reflexed. Petals contiguous, roundish oblong, abruptly and shortly clawed, pink. *Filaments white. Styles yellowish.* Carpels and receptacle pilose. *Fruit* rather large, *ovoid-oblong*, sweet; usually not all ripened.

E., W. Abundant in the south-eastern counties; 7, 8, 13–22, 24, 35, 40 (Wyre Forest). W. France.

135. **R. chloocladus** W. Wats. 1956, *Watsonia*, **3**, 288; *R. pubescens* Weihe in Boenn. 1824, *Prod. Fl. Monast.* 152; non Rafin. 1811.

4x. *Stem arching and sprawling*, somewhat *furrowed, thinly and minutely felted and hairy*, becoming purple and *pruinose; prickles strong*, declining to strongly *recurved*. Petiole prickles hooked. Stipules filiform. Leaflets 5, closely pubescent and greyish white felted beneath, unequally serrate-dentate; *terminal leaflet* subcordate-ovate-acuminate to *elliptic or ovate-lanceolate, prolonged. Panicle* long, narrow, equal, with short, spreading, deeply divided, cymose branches; rachis and branches felted and pubescent, *prickles long-based and strongly curved*; lower leaves 5-nate. Sepals reflexed; *petals white or pink*, broad ovate, clawed; *filaments white*, long or short. Carpels long-pilose. Fruit globose to oblong.

> Focke and Sudre give the stem as epruinose. I have collected *R. chloocladus* at Aachen with the lower part of the stem pruinose, as it is at Clophill. Weihe and Nees describe the stem as covered with a glandular substance; this was perhaps *pruina*.

England: rare; 30 (Clophill). N. and S. Germany, Austria, S. France, Portugal.

2. Filaments much longer than the styles.

(For *R. chloocladus* (*pubescens*), see above.)

136. **R. Winteri** P. J. Muell. ex Focke, 1877, *Syn. Rub. Germ.* 196; *R. Godronii* ssp. *Winteri* (P. J. Muell. ex Focke) Sudre, 1908–13, 78, t. 80. **Fig. 21.**

4x. *Robust. Stem* angled, purple, *pilose at the base*, furrowed and with *semi-appressed hairs above; prickles strong, slender, long*, broad-based, deep purple. *Leaflets 5*, tough yellowish green, white felted and softly hairy beneath, sharply and unequally to *subcompound-serrate, undulate; terminal leaflet long-stalked*, ovate to obovate-*cuneate*, base rounded, entire. Panicle lax, long, much narrowed, pilose, prickles long, falcate and hooked, a few much smaller; peduncles spreading, *pedicels short*. Flowers *c.* 2·5 cm.; *sepals* loosely reflexed with long spreading points, *usually aculeolate*; petals pink, roundish obovate, clawed; *filaments white or deep pink, anthers usually pilose*. Carpels pilose. One of the last of the brambles to come into flower, its late fruit large and excellent.

E., W., I. Scattered, locally frequent, does not shun clay; 2, 3, 11, 13, 14, 16, 17, 21, 22, 24, 25, 35, 39, 49, 53, 55, 56, 57. H. 1, 2. Alderney. France, W. Germany, Austria, Switzerland.

137. **R. propinquus** P. J. Muell. 1859, *Pollichia*, 88; *R. Godronii* ssp. *propinquus* (P. J. Muell.) Sudre, 1908–13, 79, t. 81; *R. nemoralis* sensu Genev. 1880, *Mon.* 205; *R. lamburnensis* Rilst. 1950, *Journ. Linn. Soc.* **53**, 415.

Western. Young shoots reddish bronze. Axes and leaves covered with sessile glands when young. Robust. Bushy. Stem angled, sides flat to slightly furrowed, *with tufted, large and minute hairs, becoming blackish purple* and glaucescent; *prickles hard*, long, hairy, ± patent. *Leaflets 5*, subglabrous above, pubescent and white felted beneath, the *veins yellow*, deeply, nearly simply and sometimes finely serrate, the margin crisped; *terminal leaflet roundish or broad ovate-acuminate*, or slightly obovate-cuspidate. Panicle broad, nearly equal, or pyramidal-oblong, dense, *middle and upper branches spreading*, 1–5-flowered, *peduncles long, pedicels short*; rachis angled, striate, felted and hirsute, prickles various, lower leaves 5-nate. Flowers 2·5–3 cm.; sepals felted and hirsute, reflexed; petals roundish ovate or obovate, clawed, rose, rose and white (parti-coloured), pink or (rarely) white; filaments long, white, anthers subpilose; styles reddish-based. Carpels thinly pilose or bearded. Receptacle pilose. Fruit roundish, of moderate size, rather dry, late. *Flowering July.*

The size and breadth of the leaflets and the character of the toothing, varies as much here as in France, according to the conditions of the habitat. Specimens having rank leaves with broad leaflets, sent by Chaboisseau to Genevier from Pindray, the same place from which the specimens that were named *R. propinquus* were obtained, Genevier described and named as a separate species, *R. crassifolius*, which Sudre rightly reduced to a synonym of *R. propinquus*.

E., W., S., I. Common in some places in the west of Britain; 1–4, 34–36, 38, 41, 42, 48, 49, 58, 100, 110. H. 2, 21, 23, 27. Guernsey. France (very common in the centre and the west, rare in the north), W. Germany (Bavaria).

Sudre states that *R. propinquus* produces numerous hybrids, of which he describes eleven in *Rubi Europae*. In England *R. herefordensis* Sudre,

1904, *Observ. Set. Brit. Rub.* 135, and *R. Riddelsdellii* Rilst. 1950, *Journ. Linn. Soc.* **53**, 415, are fertile and stable, and have attained a fair area of distribution. *R. londinensis × propinquus* (*R. carnkiefensis* Rilst. 1950, *Journ. Linn. Soc.* **53**, 417) is fertile; it occurs with both parents at Carnkief, and thence for seven miles towards Truro. More local are scattered bushes of *R. prolongatus × propinquus* (*R. pydarensis* Rilst. 1940, *Journ. Bot.* **78**, 13), semi-infertile in v.-c. 1; *R. propinquus × rufescens*, good fruiting, Riggs Wood, Sellack, v.-c. 36; *R. propinquus × stanneus* and *R. propinquus × rubritinctus*, semi-fertile, near Port Isaac, v.-c. 2.

Although such plants may properly be taken into the private herbarium and studied, it seems very desirable that great restraint should be observed in describing, naming and publishing them as independent species, to take their place beside such well-distinguished fertile and widely ranging forms as *R. ulmifolius*, *R. propinquus*, *R. prolongatus*, *R. rufescens* and *R. londinensis*, which have produced them.

Ser. **ii. HEDYCARPI** Focke

(Type *R. procerus* P. J. Muell.)

1. Petals very narrow. Panicle prickles small, falcate and hooked.

138. **R. stenopetalus** Lef. & Muell. 1859, *Pollichia*, 194; *R. pubescens* ssp. *aduncispinus* microg. *stenopetalus* (Lef. & Muell.) Sudre, 1908–13, 86, t. 89.

Wiltshire. Stem deep red, furrowed, glabrescent; prickles strong, recurved. Petiole prickles strong, *hooked. Leaflets* 5, glabrous above, white felted and pubescent beneath, *sharply unequal,* in part patent-, *dentate-serrate; terminal leaflet base rounded, broad obovate-cuspidate.* Panicle generally narrow, axillary branches sharply ascending, upper branches 1–3-flowered; *terminal leaflets narrow oblong-obovate or subcuneate*; rachis villous to the base, *prickles falcate or hooked.* Flowers moderate; sepals felted and villous, reflexed; *petals pale lilac,* elliptic or obovate, *often very narrow, tapered* to the base; filaments white, longer than the greenish styles.

England: apparently very rare; 8 (Everleigh Ashes). France (Paris basin).

2. Petals white. (Cf. *R. cuspidifer.*)

139. **R. geniculatus** Kalt. 1845, *Fl. Aach. Beck.* 267; Sudre, 1908–13, 87, t. 90; *R. falciferus* P. J. Muell. 1859, *Pollichia*, 84; *R. cerasifolius* Lef. & Muell. 1859, *Pollichia*, 98.

Bucks. Known by its flat-sided, glabrous, red stem, elliptic-rhomboid terminal leaflet, which is glabrous above and white felted beneath, entire-based; a large, long, tapering panicle armed with numerous, long, strong prickles, all straight on rachis, peduncles and pedicels; and finally by its showy, white, elliptic petals and long, white stamens. Fruit rather large, fertile.

England: 24 (College Wood, Little Horwood, E. F. Linton, 1892, as *R. rhamnifolius*, with doubt).

3. Petals pink, broad.

140. **R. cuspidifer** (*cuspidiferus*) Muell. & Lef. 1859, *Pollichia*, 89; Sudre, 1908–13, 82, t. 84. **Fig. 22.**

*4x. *Robust, showy.* Stem red, furrowed, *with a few, small, tufted hairs; prickles numerous, strong-based,* mostly long and slender. *Leaflets 5,* long-stalked, felted and pilose beneath, *coarsely and deeply serrate, undulate; terminal leaflet roundish-ovate-acuminate,* base rounded, truncate or cordate. *Panicle leafy, broad, pyramidal,* blunt, *densely villous,* the lower branches often very long and leafy; prickles numerous, long and strong. *Flowers c. 3 cm.*; sepals reflexed; petals pink or white, roundish, obovate or elliptic, shortly clawed; filaments white, pink, or deep pink at the base, far longer than the rosy-based styles. Carpels densely pilose. A dwarf form occurs on Midhurst Common, W. Sussex.

England: locally frequent; 2, 3, 13, 14, 16, 17, 22, 30, 36. France, Bavaria.

141. **R. bifrons** Vest in Trattinick, 1823, *Rosac. Mon.* **3**, 28; Sudre, 1908–13, 80, t. 82; Coste, 1903, *Fl. France,* **2**, 38, t. 1173; *R. nemoralis* var. *cornubiensis* Rogers & Riddels. 1925, *Journ. Bot.* **63**, 14; *R. cornubiensis* (Rogers & Riddels.) Rilst. 1950, *Journ. Linn. Soc.* **53**, 413.

A much-branched low bush, subevergreen. *Stem* stout, dull, red-brown, glabrescent, *obtuse-angled* with flat striate sides; *prickles subulate. Leaflets* 3–5, deep green, glabrous above, *all long-stalked,* green and shortly hairy to closely *chalky white felted beneath, finely serrate-dentate, jagged;* terminal leaflet *roundish or broad* with a rather truncate base and sometimes truncate apex, *shortly acuminate or cuspidate,* or oblong-acuminate. Panicle not leafy, ± long, moderately broad, equal, or slightly narrowed to the rounded apex, densely and shortly hairy above, felted below; *the lower rachis prickles long subulate* or a few bent, *the upper ones* and those on the branches *acicular.* Sepals felted and pubescent, gland-dotted, aculeolate; *petals* deep or pale pink, *very broad ovate-cuneate, margin crisped;* filaments white or pinkish, long; styles greenish or reddish. Carpels thinly pilose. Fruit moderate, shining. *On the stem, and especially on the panicle, a few minute shortly stalked glands and sometimes a few acicles may be found.*

Boulay says that in France *R. bifrons* crosses freely with any species with which it grows. Focke crossed it with *R. gratus* and obtained a fertile seedling. *R. Tresidderi* Rilst. 1950, *Journ. Linn. Soc.* **53**, 44, appears to combine the characters of *R. bifrons* and *R. londinensis,* in the neighbourhood of which it frequently grows around Lambourne and Per-ranzabuloe Church, v.-c. 1. The author of *R. Tresidderi* himself remarks on its affinity with *R. cornubiensis* (= *bifrons*).

E., W., S. Local; 1–4, 35, 71. South central and western Europe east-wards to Silesia. Rare in N. France.

142. **R. vulnerificus** Lef. ex Corbiere, 1893, *Fl. Norm.* 201; *R. cuspidifer* microg. *vulnerificus* (Lef. ex Corb.) Sudre, 1908–13, 83, t. 85.

E. Sussex. Stem furrowed, with short tufted hairs; prickles yellowish, *very long,* ± patent. *Leaves* digitate, *prickles subulate to slightly spreading-falcate; leaflets 5,* felted and pubescent beneath, *subcompound serrate-dentate; terminal leaflet obovate-cuspidate,* much *narrowed to the entire base. Panicle long, lax, pyramidal,* mostly without leaves, *branches nearly patent,* cymosely divided near the apex; rachis felted and pubescent, *prickles numerous, subulate, up to 11 mm. long,* pricklets few, slender. Flowers *c.* 2·5 cm.; sepals broad ovate-cuspidate, reflexed; petals rose-pink, broad obovate-cuneate, notched; filaments white, longer than the greenish styles, anthers subpilose. Fruit good.

England: 14 (between Uckfield Rocks and Buckham Hill, near the turn to Shortbridge). France (Normandy and the north-west).

(**R. procerus** P. J. Muell. 1864, *Ronc. Vosg.* 7; Sudre, 1908–13, 87, t. 92; *R. armeniacus* Focke in Bremen, 1874, *Abh.* **4**, 183; *R. macrostemon* Focke, 1877, *Syn. Rub. Germ.* 193; *R. thyrsoideus* var. *viridescens* Rogers & Riddels. pro parte 1925, *Journ. Bot.* **63**, 14; *R. oplothyrsus* Sudre, 1904, *Observ. Set. Brit. Rub.* 134.

4x, 7x. *Very large and robust. Stem* light brown on the underside, purple above, *thinly hairy at first,* glaucescent, *slightly furrowed;* prickles broad-based, gradually narrowed. *Leaves very large, digitate, prickles hooked; leaflets* glabrescent above, pilose as well as felted beneath, *unequal to subcompound incise-serrate; terminal leaflet* roundish or *oblong, shortly acuminate-cuspidate, base truncate, emarginate. Panicle large,* lax, com-pound, *pyramidal-truncate,* mid branches long cymose with rather long pedicels, villous. *Flowers c. 3 cm.;* sepals reflexed; petals rose-pink, rarely pure white, broad ovate, cuneate; filaments white, long, anthers pilose; styles greenish. Carpels subpilose. *Fruit very large.*

This was collected by Griffith near Holyhead in 1895 and was identi-fied by Focke as *R. thyrsoideus;* it was collected by Rogers at the Lizard in 1904 and was identified by him as *R. thyrsoideus* Wimm. sp. coll., changed later to *R. thyrsoideus* var. *viridescens* Rogers & Riddels.; there is a specimen of it in Herb. Rogers, marked '*R. Schlechtendalii,* a very strong form', from South Island, New Zealand.

The plant is a native of Transcaucasia and Armenia and is distributed right across southern Europe to France. I have met with it in many parts of England in the past 20 years, growing as an 'escape' from gardens. It is spoken of as 'American Blackberry' (? for Armenian Blackberry) and 'Himalayan Giant'. It is unsafe to assume that it is native in any place in these islands.)

Ser. iii. CANDICANTES Focke

(Type *R. candicans* Weihe)

1. Petals pink. Axes glaucescent. Leaves strigose above.

143. **R. neomalacus** Sudre ap. Gandog. 1905, *Nov. Conspect.* 144; *R. thyrsoideus* ssp. *neomalacus* (Sudre) Sudre, 1908–13, 90, t. 94; *R. incarnatus* sensu Genev. 1880, *Mon.* 306; non P. J. Muell. 1859; *R. bipartitus* Boul. & Bouv. pro min. parte.

★4x. *Stem* slender, furrowed, striate *with scintillating particles*, glabrescent; prickles long, straight or slightly bent. *Leaves* 5-nate, *pedate, rather small; leaflets subimbricate, rugose, velvety pubescent and felted* beneath, incise, sharply and unequally to subcompound-serrate, *undulate*; terminal leaflet base emarginate, roundish ovate, *obovate or oblong*, long-acuminate or cuspidate; *basal leaflets short-stalked. Panicle narrow, dense, short, terminal flowers subsessile,* rachis and pedicels villous. Flowers up to 2·5 cm.; incurved; sepals reflexed; petals broad obovate, notched, shortly clawed; filaments pink-based, slightly longer than the greenish styles. *Carpels densely pilose.* Receptacle hirsute. Recalls *R. imbricatus* and *R. rubritinctus.*

England: rare; 17 (locally frequent on Holmbury Hill and throughout the River Wey basin), 22 (Aldworth). Central and north-central France.

2. Petals pink. Axes not glaucescent. Leaves glabrous above.

144. **R. falcatus** Kalt. 1845, *Fl. Aach. Beck.* 266.

5x. *Stem* glabrous, *lustrous*, furrowed, prickles strong-based, patent to falcate. Leaflets grey-white felted and pubescent beneath, *subcompound incise serrate*; terminal leaflet emarginate, broad or roundish, cordate-ovate-acuminate to elliptic-acuminate. Panicle short, broad and pyramidal-corymbose, or *elongate, equal, leafy*; rachis and branches pubescent, *prickles hooked and falcate. Flowers* 2–2·5 cm., *in usually short-peduncled cymes*; sepals felted and pubescent, reflexed; *petals ovate or elliptic*, abruptly clawed; filaments white, slightly longer than the greenish styles. Carpels pilose. Fruit of moderate size.

England: rather frequent in the midlands; 21–23, 29, 30, 33, 34, 36, 38, 39, 52, 55, 56, 67.

Focke says that the two bushes known to Kaltenbach near Aachen disappeared many years before 1877. He considered that it was an abnormal form, probably hybrid, because the petals were described as incurved at the apex, like a hood. Sudre, without ever having seen a specimen, improves upon this by placing it among the hybrids of *R. candicans*. In 1937 on going to Kaltenbach's locality Mr N. D. Simpson and I found a mutilated bush or two bearing normally developed fruit. I do not know of its occurrence elsewhere on the Continent, and it looks as if the Aachen locality represents an attempted colonization from Britain.

3. Petals white. Axes not glaucescent.

145. **R. hylophilus** Rip. ex Genev. 1880, *Mon.* 238; *R. Brittonii* Barton & Riddels. 1931, *Journ. Bot.* **69,** 191. **Fig. 23.**

★3x. *Stem suberect, arching, furrowed to the base,* thinly pilose; *prickles* strong-based, *subulate. Leaves rather large, olive-green;* leaflets 5, convex, glabrous above, velvety grey felted and pubescent beneath, *nearly simply and bluntly serrate;* terminal leaflet emarginate, broad ovate-oblong, suddenly shortly pointed. *Panicle broad,* equal or subpyramidal, truncate, *with 1–4 ovate leaves,* middle and upper branches oblique, 3-flowered, the lowest branches long and erect, villous, *prickles falcate, rather slender.* Flowers 2·2–2·5 cm.; sepals short-pointed; *petals ovate or broad elliptic,* with a slender claw and a truncate, erose and fimbriate apex; filaments slightly longer than the styles. Carpels a good deal pilose. Receptacle pilose. Fruit variable: sometimes the terminal fruits are thimble-shaped and perfect whilst the other fruits are poor; or all the fruits are good but small and subglobose; or all fruits are semi-infertile. The pollen is 3 per cent fertile, but with some giant grains.

England: confined to the south-east, rather rare except in Surrey; 13, 16, 17, 21–24. France (Normandy and the north-west), Switzerland (Vaud), Bavaria.

146. **R. thyrsanthus** Focke, 1877, *Syn. Rub. Germ.* 168; *R. thyrsoideus* ssp. *thyrsanthus* (Focke) Sudre, 1908–13, 91, t. 94.

4x, 3x. *Robust and large-leaved.* Axes becoming dark purple. *Stem grooved to the base,* glabrous or with a few stellulate hairs, arching, procumbent and rooting; prickles rather few, *very strong-based, some recurved.* Leaves digitate; leaflets glabrous above, felted and pubescent beneath, unequally or compoundly angular, divaricate serrate, with longish mucros, margin undulate; *terminal leaflet base entire, broad elliptic-ovate, cuneate, shortly acuminate. Panicle large and long, numerous flowered, dense, leafy, narrow above,* pyramidal below, *hardly armed above,* felted and shortly and closely pilose. *Flowers small, often 1–1·5 cm.;* sepals reflexed; *petals roundish, cuneate, faintly pinkish white;* filaments slightly longer than the greenish styles. Carpels bearded. Fruit semi-infertile, year after year.

E., I. 16 (Hayes Common, one bush), 35. H. 33. Denmark, S. Germany to Silesia, Switzerland, Bavaria, Bohemia.

(*R. thyrsoideus* var. *viridescens* Rogers & Riddels. consists of two different brambles; *R. procerus* P. J. Muell. and *R. holerythrus* × *ulmifolius.*)

Sect. **V. SPRENGELIANI** Focke

(Type *R. Sprengelii* Weihe)

1. Upper leaves greyish white felted and hairy beneath.

147. **R. Braeuckeri** G. Braun, *Herb. Rub. Germ.* 85, 197; T. Braeucker, 1882, *Deutsche Rub.* 292.

*3x. *Petals pink. Stem weak, climbing, sharply angled, strongly pruinose; prickles numerous, often confluent,* long patent. Terminal leaflet subcordate, oblong, short-pointed. *Panicle slightly glandular, very large, lax, prickly; flowers 1–2 cm., stamens slightly shorter than the styles. Fruit small, ovoid, imperfect.* Anthers and carpels pilose.

England: local; 14, 16 (frequent from Uckfield to Westerham). West Germany.

148. **R. chlorothyrsus** (*chlorothyrsos*) Focke, 1871, *Abhandl. Nat. Ver. Breme*, **2**, 462; Sudre, 1908–13, 36, t. 35; *R. phyllothyrsus* Frid. 1896, *Rub. Gall.* Exsic. no. 81; non *R. Babingtonii* var. *phyllothyrsus* sensu Rogers, 1900, 70.

4x. *Petals white.* Stem low-arching, prostrate, *densely hairy,* slightly glandular, and acicular; *prickles hairy, numerous, rather small, slender,* subequal, slanting or falcate. Leaves 5-nate; *leaflets pectinate on the veins to greyish felted beneath,* coarsely unequal or subcompound-serrate; terminal leaflet short-stalked, ovate-acuminate or broadly elliptic with a rounded base; *basal leaflets subsessile. Panicle slightly glandular, large, lax, equal, with lanceolate leaves to the top, very prickly* and floriferous, branches spreading, 1–7-flowered, cymose. *Flowers 1–2 cm.;* sepals grey-white felted and hairy, long narrow ovate with linear leafy points, *at length patent-erect;* petals remote, elliptic, clawed; filaments white, slightly shorter than the greenish styles. Carpels pubescent. Fruit small, perfected.

E., I. Rare; 16 (Tunbridge Wells Common, in one spot). H. 16 (Maam Bridge, *Simpson*). Denmark, N.W. Germany, Austria.

2. All leaves green beneath.

149. **R. Sprengelii** Weihe, 1819, *Flora*, **2**, 18; Sudre, 1908–13, 32, t. 29. **Fig. 24.**

4x. *Common.* A low bush. Stem roundish, hairy, with an occasional pricklet and stalked gland; *prickles slender, falcate to subuncinate. Leaflets* 3–5, *hairy beneath,* unequally incise and sharply serrate; *terminal leaflet elliptic-obovate,* acuminate. *Panicle slightly glandular, subcorymbose,* short, broad, lax, *branches and pedicels widely spreading, prickles short, hooked. Flowers c. 2 cm.;* sepals felted and pilose, long-tipped, loosely clasping; *petals* narrow, obovate-oblong, crumpled, *bright rose-pink;* filaments pink or pink-based, about equalling the styles. Carpels pilose. Fruit rather small.

E., W., S., I. Frequent in woods and among heather and gorse; 3–24, 34–40, 42, 46, 48, 49, 51, 55, 57–60, 62–64, 73, 74, 104. H. 11, 12. Guernsey, Jersey. N. France, Belgium, Netherlands, Denmark, Germany.

150. **R. lentiginosus** Ed. Lees in Steele, 1847, *Handb. Field Bot.* 60; *R. cambricus* Focke in Griffiths, *Fl. Anglesey & Caern.* 46.

4x. *Stem furrowed*, hairy at first with numerous sessile glands and occasional pricklets and stalked glands; prickles numerous, yellow, long-based, declining and falcate. *Leaves 5 (6, 7)-nate, petiole long; leaflets concave, glabrous above, glabrescent beneath, incise and sharply subcompound-serrate;* terminal leaflet ovate to rhomboid or obovate, acuminate, base truncate, emarginate. Panicle lax, subracemose or pyramidal, the long lower branches sharply ascending, the upper ones patent, very prickly, hairy and slightly glandular. *Flower-buds ovoid, pointed; flowers 1–1·8 cm.; sepals aculeate, glandular, erect; petals pinkish, elliptic, crumpled, inflexed at the apex,* recurved at the sides; filaments white, barely equalling or rarely longer than the styles. *Anthers sometimes pilose.* Carpels pilose. Fruit oblong, pale green to coral, then deep vinous red to glossy black.

Endemic. E., W. Chiefly Wales and the west midlands; 3, 8, 11, 22 (Owlsmoor), 27, 28 (Mousehold Heath), 34–40, 49, 51, 58.

151. **R. permundus** W. Wats. 1946, *Journ. Ecol.* **33**, 337; *R. mundiflorus* W. Wats. 1938 (for 1937), *Rep. Bot. Exch. Cl.* **11**, 445; non Sudre, 1901.

4x. *Surrey, Bucks. A rather small, upright bush,* axes deep purple. *Stem stout, blunt-angled,* hairy, slightly glandular; *prickles moderate,* subulate, declining. *Leaves 5-nate, pedate,* yellowish green; *leaflets short-stalked,* dull and glabrescent above, roughly hairy beneath, rather *coarsely, bluntly and irregularly serrate; terminal leaflet emarginate, roundish ovate, broadly acuminate. Panicle large, leafy, broad and abrupt above, broader still below,* the lower branches panicled, numerous flowered, the middle and upper *branches spreading, 4–12-flowered, cymose-partite; rachis* and branches *rigid* pubescent, *mostly unarmed.* Flowers 1·5–2 cm.; *sepals* greyish green, *leafy-tipped, patent; petals roundish,* clawed, opening flat, *bright pink;* filaments white, sometimes pink at the base, shorter than the styles. Carpels pilose. Fruit moderate, subglobose.

Endemic. Rare; 17 (plentiful over a large area under the bracken on Netley Heath, where it was discovered by Mr C. E. Britton; Mare Hill, Witley), 24 (Burnham Beeches, *G. C. Druce*).

152. **R. Arrhenii** Lange, 1864, *Haandb. Dansk. Fl.* 386; *Fl. Dan.* t. 2720 et 2833; Sudre, 1908–13, 31, t. 27.

4x. *Stem slender,* not much branched, *round,* pilose; *prickles numerous, small,* abruptly narrowed from a broad, yellow base, declining. Leaves digitate, bright green, stipules filiform-linear; leaflets 3, 5, pubescent beneath, regularly incise-serrate; *terminal leaflet elliptic, acuminate,* base subcordate to rather cuneate. *Panicle hanging, very long, not dense, straight, subracemose,* the numerous branches semi-erect, hairy and *inconspicuously glandular*; rachis hirsute. Flowers 1·5–2 cm.; *sepals ovate-lanceolate to lanceolate, attenuate*, some with linear, leafy tips, patent to erect; *petals glabrous, pink*, pinkish, or white, roundish or *obtuse at the apex*, narrowed below, *slow in falling off*; filaments white or pink, shorter than the styles; *styles subpersistent.* Carpels pilose or glabrous. Pollen almost perfect.

England: very rare; 8 (Longleat Park Wood), 15 (wood N.E. of Tenterden), 22 (Wokingham Road west of Crowthorne), 37 (Wyre Forest, *Gilbert* 1857). Denmark, N.W. Germany, and Silesia.

var. **polyadenes** Gravet ex Focke ap. Aschers & Grab. 1902, *Syn.* **6**, 530.
Stalked glands and acicles much more numerous and more unequal, some longer than the hair.

England: very rare; 39 (Bailey's Hill, Biddulph, in two spots). Belgian Ardennes.

153. **R. axillaris** Lej. in Lej. & Court. 1831, *Comp. Fl. Belg.* **2**, 166; *R. Leyi* Focke, 1877, *Syn. Rub. Germ.* 268; *R. chlorothyrsus* ssp. *Leyi* (Focke) Sudre, 1908–13, 36, t. 36; *R. scanicus* Aresch. 1881, *Skanes Fl.* ed. 2, 270; cf. H. Allander, 1934, *Bot. Not.*

4x. With coarsely toothed, roundish, subcordate terminal leaflet and pyramidal panicle like *permundus*, and stalked glands and flowers like *R. Arrhenii*, but stamens about equalling the styles. The petals are roundish and pink, not elliptical and white, as stated by Lejeune. I have seen it in the Ardennes.

England: 39 (near the gates of Belmont Hall, Constall, 1950). Belgian Ardennes, Denmark, S. Sweden, Germany (in the north-west and in Saxony).

Sect. **VI. APPENDICULATI** Genev.

Ser. **i. Vestiti** Focke. Stalked glands, acicles and pricklets very thinly distributed on the stem, more frequent on the panicle. Leaves green or white beneath. Stem and leaves often hairy. (See p. 108.)

Subser.† **Nemorenses** Sudre. Slender plants with short prickles which are unequal and pass into acicles, and with sunken, short-stalked glands in the panicle. (See p. 108.)

Subser.†† **Virescentes** Sudre. Leaves green beneath (in some species the upper stem leaves, all branch leaves and the upper panicle leaves may be greyish felted beneath). (See p. 109.)

Subser. ††† **Hypoleuci** Sudre. All leaves greyish or white felted and usually pilose beneath. (See p. 114.)

Ser. **ii. Mucronati** W. Wats. Stalked glands, acicles (and sometimes pricklets) if present on the stem varying from few to rather many; irregularly distributed on the panicle, sometimes unequal in length, varying in quantity but usually to be seen on the pedicels and calyces. (See p. 121.)

Ser. **iii. Dispares** W. Wats. *Subsessile* glands, *minute* pricklets and *short* acicles are irregularly distributed on stem and panicle, almost wanting in some parts, rather abundant in others, *with or without a few longer* glands, acicles, gland-tipped acicles and pricklets. (See p. 127.)

Ser. **iv. Radulae** Focke. Stalked glands, acicles and pricklets *equal and short* on the stem, slightly less so on petioles, petiolules and panicle. (See p. 129.)

Ser. **v. Apiculati** Focke. Stalked glands, acicles and pricklets numerous, *rather unequal.* Panicle usually broad, equal and truncate. White flowers with red styles are frequent. (See p. 137.)

Subser.† **Foliosi** W. Wats. Petals white or pink, rarely bright pink. Filaments usually white. Styles often red. *Panicle and stem mostly hairy. Sepals reflexed; or partly reflexed and partly patent,* sometimes partly erect. Peduncles often bunched or deeply divided. (See p. 137.)

Subser.†† **Pallidi** W. Wats. As subseries *Foliosi,* but the sepals at length more definitely and uniformly erect. (See p. 144.)

Subser.††† **Scabri** W. Wats. Flowers as subseries *Foliosi,* usually white. Panicle and stem pubescent or felted. Glands mostly short. Prickles usually rather short. Leaflets finely toothed. Sepals patent or erect. (See p. 150.)

Subser.†††† **Obscuri** W. Wats. Flowers showy. Petals, and usually the filaments and styles also, deep bright pink. Panicle usually pilose-hirsute. The unripe fruit often deep crimson. (See p. 159.)

Subser.††††† **Incompositi** W. Wats. Flowers hardly showy, usually white, pinkish or pink. Stem often glabrous or glabrescent, panicle rachis too, often glabrescent below. Leaflets not serrulate. Peduncles not bunched. Prickles sometimes straight and strikingly long, and accompanied by stout pricklets. Stalked glands and acicles sometimes few on the stem. (See p. 166.)

Ser. **vi. Grandifolii** Focke. Stalked glands and acicles (and pricklets if present) unequal, *a few long ones on the panicle-rachis much exceeding the rest. Panicle well developed and elongate,* either pyramidal with racemose lower branches and ending a raceme, or else lax and broad with an abrupt apex. Mostly luxuriant, stately and floriferous; later flowering. (See p. 172.)

Ser. **i. VESTITI** Focke

(Type *R. vestitus* Weihe)

Subser.† **Nemorenses** Sudre

(Type *R. nemorensis* Lef. & Muell.)

154. **R. nemorensis** Lef. & Muell. 1859, *Pollichia,* 198; Sudre, 1908–13, 52, t. 59.

W. Kent to E. Sussex border. Stem blunt-angled, striate, thinly pilose, with scattered *fine acicles* and short glands; *prickles fine,* numerous, scattered, unequal, *narrow-based.* Leaves pedate, with fine prickles, glands and acicles on the petiole and petiolules; leaflets densely pubescent beneath, incise sub-compound serrate, or sublobate; *terminal leaflet subcordate, broad ovate,* or *subrhomboid-ovate, cuspidate-acuminate.* Panicle pyramidal, obtuse, inter-rupted, *the uppermost leaves lanceolate and linear lanceolate,* greyish felted beneath, branches 3–5-flowered, cymose; rachis pubescent and felted, striate, armed like the stem. Flowers 2–2·5 cm.; sepals glandular, reflexed; *petals pinkish-white, narrow elliptic, pointed; filaments long,* white; styles greenish. Carpels pilose.

England: 14 and 16 (around Hawkenbury). France, Belgium, Switzer-land, Bavaria.

155. **R. hesperius** Rogers, 1896, *Journ. Bot.* **34**, 504.

Ireland. N.v.v. Stem reddish brown, sharply angled, glabrescent, *glauces-cent, with a few pricklets* and sessile and short-stalked glands; *prickles numerous,* rather scattered, short, unequal, often in twos, *strong-based,* declining or fal-cate. Leaves pedate, short-stalked; *leaflets 5, broad, imbricate, rugose,* unequally and coarsely incise-serrate, undulate, felted and pubescent or glabrescent beneath; *terminal leaflet broad, cordate, ovate, acuminate.* Panicle long, leafy, lax below, narrowed upwards, middle branches 1–3-flowered, spreading—ascending; rachis flexuose, pubescent, prickles numerous, short, unequal, with numerous glands and acicles sunken in the hair. Flowers *c.* 2 cm.; sepals greenish, felted and pilose, long linear pointed, patent; *petals white, elliptic, long-clawed; filaments white, hardly longer than the styles.* Carpels pilose. Dis-covered by E. S. Marshall. Gilbert's 'R. hesperius' found at Tunbridge Wells was *R. nemorensis.*

Endemic. Ireland: H. 16 and 26.

Subser.†† **Virescentes** Sudre

(Type *R. Schmidelyanus* Sudre)

1. Sepals reflexed.

156. **R. fuscicortex** Sudre, 1904, *Observ. Set. Brit. Rub.* 26; *R. mucronifer* ssp. *fuscicortex* (Sudre) Sudre, 1908–13, 114, t. 112; *R. podophyllus* sensu Rogers pro parte, 1900, 67; non P. J. Muell. 1861.

Western. Plant small, axes deep crimson. Stem low, glaucescent, pubescent, with few short glands and acicles; *prickles rather short, slender. Petiole prickles straight or* spreading, falcate. *Leaflets* 3 (4, 5), glabrescent above, softly pubescent beneath; *finely serrate-dentate; terminal leaflet rather broad, oblong, cuspidate.* Panicle small, *subracemose,* felted and pubescent, finely glandular and aciculate, *prickles numerous, short, weak, straight. Flowers 1·5–2 cm.* Sepals felted and pubescent, glandular. *Petals pink or pinkish, subglabrous, obovate-lanceolate or slightly rhomboid;* filaments white, longer than the greenish styles. Carpels glabrous. Receptacle hirsute. *Fruit large.*

Endemic. E., W. Rather frequent on slate in N. Wales, rare elsewhere; 3 (not 16), 42, 48, 60, 67.

157. **R. silesiacus** Weihe in Wimm. & Grab. 1829, *Fl. Sil.* I, 53; Sudre, 1908–13, 44, t. 47.

4x. *W. Kent. All axes slender.* Stem arched-prostrate or climbing, *much branched*, angled, thinly pubescent, glandular and aculeolate; prickles yellowish, moderately declining. Leaves rather large, pedate; *leaflets 5,* pubescent beneath, unequally serrate, a little sinuate-dentate, *sometimes slightly lobate; terminal leaflet oblong-obovate, long-acuminate.* Panicle long, narrowed above, *dense, intricately branched,* the branches deeply divided, or shorter and pyramido-corymbose; rachis felted and pubescent, glandular and acicular, prickles numerous, slender. Flowers *c.* 2·5 cm.; sepals narrow, greyish felted, pubescent, aculeate and glandular; *petals narrow elliptic, white; filaments white, equalling the greenish styles.* Carpels glabrous. Fruit roundish ovoid, moderate.

England: 16 (Shrewsbury Park, Shooters Hill). South-central Europe.

158. **R. eriostachys** Muell. & Lef. 1859, *Pollichia,* 225; *R. Colmannii* ssp. *eriostachys* (Muell. & Lef.) Sudre, 1908–13, 118, t. 117.

Tunbridge Wells. Stem obtuse-angled, pilose, slightly glandular and acicular; *prickles short, unequal, not strong.* Leaves rather large; leaflets 3 (4, 5), glabrescent to thinly felted beneath, *coarsely and angularly serrate* with some points patent; *terminal leaflet broad, obovate, rounded or somewhat truncate-cuspidate,* base subcordate or subentire. *Panicle rather short, narrow subracemose with one or numerous simple leaves,* or lax pyramidal, truncate *with a few, spreading 3–7-flowered branches; rachis villose or hirsute* and glandular, *prickles yellowish, weak.* Flowers *c.* 2·5 cm.; sepals greenish felted with short prickles and short glands; *petals ovate-oblong, white; filaments white, longer than the reddish-based styles. Anthers pilose.* Carpels glabrous or pilose.

England: 16 (Tunbridge Wells Common, near the cricket ground), 39. France, Belgium.

159. **R. helveticus** Gremli, 1870, *Beitr. Fl. Schw.* 36.

Stem green, blunt-angled, thinly hairy, with few glands and slender pricklets; *prickles yellow, very unequal, abruptly narrowed close above the base,* some glandtipped, *straight. Leaves large, deep green, prickles all straight; leaflets 5, coarsely and unequally serrate,* finely long-cuspidate, greyish green felted beneath; *terminal leaflet round, base entire. Panicle large but short,* lax, pyramidal, *with simple* grey felted *leaves nearly to the top;* rachis felted rather than hairy, with unequal subulate prickles and a few stalked glands. Sepals felted and pilose. *Petals narrow, white. Filaments white, equalling the greenish styles.* Carpels pilose. *Fruit roundish, drupels large.*

England: 17 (Putney Heath, near the Telegraph Arms). Switzerland, Bavaria.

2. Sepals patent or erect.

† Terminal leaflet shortly pointed, roundish.

160. **R. lasiostachys** Muell. & Lef. 1859, *Pollichia*, 107; *R. leucanthemus* sensu Rogers, 1900, 51; non P. J. Muell. 1859; *R. surrejanus* Barton & Riddels. 1932, *Journ. Bot.* **70**, 189; *R. hirtior* W. Wats. 1932 (for 1931), *Rep. Bot. Exch. Cl.* **9**, 764.

Stem blunt-angled, striate, *densely hairy*, with a few short glands; *prickles* broad-based, *short*, mostly patent. *Leaflets 3, 5, broad, subimbricate, densely and shortly hairy beneath*, unequally and towards the apex rather compound-serrate; *terminal leaflet* subcordate, *roundish, shortly acuminate; basal leaflets long-stalked. Panicle leafy, long, narrow*, equal, lax below, *mid branches short*, semi-erect, about 3-flowered; *rachis hirsute, with buried glands, prickles subulate, short*. Sepals patent or loosely reflexed; petals broad elliptic, suddenly clawed, white; *filaments white, equalling the greenish styles*. Carpels glabrous. Fruit subglobose.

England: locally common in the south-east; 13–17, 21–24. France (*Valois*, one locality).

161. **R. cinerosiformis** Rilst. 1940, *Journ. Bot.* **78**, 165; *R. lentiginosus* ssp. *devoniensis* Focke, 1914; *R. macrophyllus* var. *devoniensis* Focke ex Rogers, 1892, *Journ. Bot.* **30**, 205; *R. fusco-ater* sensu Briggs pro parte, 1880, *Fl. Plym.* 124; *R. mucronatus* sensu Riddels. pro parte, 1939, *Fl. Devon*, 263.

Devon and Cornwall. Stem deep purple, the *sides hollow, hairy*, with rather frequent short glands and acicles; *prickles* mostly *small, subulate, declining. Leaves 3 (5)-nate, large in the shade; leaflets pectinately pilose on the veins beneath*, finely and unequally serrate or dentate-serrate, undulate; *terminal leaflet subcordate, reniform, roundish, obovate or oblong*. Panicle leafy, equal, lax below, dense above, rounded at the top, *branches fascicled*, 1–6-flowered, strongly ascending; rachis densely and shortly hairy with some sunken glands and acicles, prickles slender, declining. Flowers *c.* 2 cm.; sepals felted, glandular, ovate-cuspidate, leafy-tipped; *petals broad, elliptic-rhomboid, pink or pinkish; filaments white, hardly equalling the reddish-based styles*. Carpels and receptacle pilose. Fruit fertile.

Endemic. England: 1–3 (frequent on hill slopes and in woods).

Briggs was familiar with this bramble, which is frequent around Plymouth. He sent seeds of it to Focke, who grew it at Bremen and named it, as he informed R. P. Murray in 1882, 'Rubus devoniensis Focke, Herb.' (unpublished).

162. **R. serratulifolius** Sudre, 1909, *Rub. Tarn.* 31; *R. gratiosus* ssp. *serratuli-folius* (Sudre) Sudre, 1908–13, 122, t. 121.

W. Kent and E. Sussex. Stem sharply angled, moderately pilose with scattered fine acicles and glands; *prickles rather short*, declining and falcate. *Leaves pedate, large, petiole prickles small, hooked; leaflets glabrous above,* pubescent beneath, the upper ones greyish felted, *serrulate; terminal leaflet cordate, roundish or broad obovate.* Panicle rather large, often leafy, lax, broad pyramidal, truncate, with 3–7-flowered cymose branches; rachis pilose, glandular and finely aculeate. Flowers showy, *c.* 2·5 cm.; *sepals* greenish, white *felted, narrow, linear-tipped, clasping; petals* obovate-elliptic, cuneate, *pink; filaments* white, *deep pink based,* slightly longer than the *greenish or red styles.* Carpels densely pilose.

England: very local; 14 and 16 (around Tunbridge Wells), 17 (Watts Hill, The Chart, Limpsfield). France, Belgium (one station), Germany (one station).

†† Terminal leaflet moderate or long-pointed, ovate, ovate-elliptic, or obovate-oblong.

163. **R. hebeticaulis** (*hebecaulis*) Sudre, 1900, *Rub. Pyr.* 63, 1908–13, 123, t. 122; *R. Thurstonii* Rilst. pro parte, 1950, *Journ. Linn. Soc.* **53**, 416. **Fig. 25.**

Western. Vigorous and villous. Stem trailing, deep purple, *blunt angled,* glaucescent, *villous,* with rare glands and pricklets; prickles long and slender, slightly unequal, straight. *Leaves* 3, 4, 5-nate, *pedate; leaflets* closely and very softly *yellowish hairy,* or the upper ones greyish felted and pubescent beneath, unequally and finely incise-serrate; terminal leaflet base entire or indented, broadly *elliptic-oblong to obovate,* acuminate. *Panicle* with several ovate leaves and long linear bracts, *very long, dense thyrsoid,* or shorter, lax pyramidal, the lower branches semi-erect; *rachis villous,* sparingly glandular, *terminal leaflets often cuneate.* Flowers *c.* 2·5 cm.; *sepals* villous, glandular, aculeolate, *ending in long linear curving points; petals rosy-pink,* rarely white, obovate-cuneate; *filaments slightly longer than the styles.* Carpels glabrous or glabrescent, rarely densely pilose. Fruit perfect, ovoid.

E., W., S., I. Frequent in the west; 1–6, 33, 36, 37, 45, 48–50, 70, 110. H. 2, 21, 26, 34, 38. Bloxam and Hort collected it in 1847, Ed. Lees in 1848, Babington, at Torc, in 1858. It has been called *hirtifolius, pyramidalis, silvaticus* and *villicaulis.* France (Belgium as var.), W. Germany, Switzerland, Bavaria.

164. **R. orbus** W. Wats. 1946, *Journ. Ecol.* **33**, 339; *R. iricis* var. *minor* Rogers & Riddels. 1925, *Journ. Bot.* **63**, 14.

Devon. Cornwall. Stem angled, *hairy,* sparsely glandular and acicular; *prickles subulate.* Leaflets 5, thinly pilose and slightly greyish felted beneath; terminal leaflet emarginate or subcordate, *oblong with a twisted point, coarsely jagged, with smaller intermediate teeth lying in a sinus; basal leaflets short-stalked.* Panicle not very large, equal or pyramidal, with spreading, 1–2-flowered branches above, and axillary, 3–6-flowered branches below; *terminal leaflets oblong, obtuse or acute;* rachis felted and villous, glandular, the *prickles subulate,* the pedicels with extremely fine prickles. *Flowers c. 3 cm., showy;* sepals narrow ovate, linear-pointed; *petals narrow obovate*, diminished to the base, *red; filaments long*; styles greenish or pinkish. Carpels glabrous.

Endemic. England: frequent in 1–4.

165. R. Schmidelyanus Sudre, 1904, *Bull. Soc. Bot. Fr.* **51**, 21; 1908–13, 119, t. 118.

Stem distinctly angled, *very hairy, with pricklets but few or no glands*; prickles strong, unequal, some long, patent. *Leaves moderate*, leaflets green or greyish pubescent beneath, unequally and rather angularly serrate; *terminal leaflet long-stalked*, base nearly entire, *roundish ovate*. Panicle equal, rather short, usually with only one simple leaf, mid and upper branches 1–3-flowered; *rachis hairy, prickles numerous, very slanting and falcate*, with smaller ones and some acicles intermixed; *terminal leaflets broad, ovate, short-pointed*. Flowers *c.* 2 cm.; sepals long-tipped; *petals rather narrow, obovate-oblong*, white, or slightly rosy in bud; filaments white or pinkish, longer than the greenish styles. Carpels ± pilose. Receptacle hirsute. Fruit globose-ovoid.

England: very rare; 14 (Hindleap, Ashdown Forest), 19 (Pods Wood, Tiptree), 39 (Swynnerton). France, Switzerland, Belgium, Germany, Bavaria.

var. **longiglandulosus** Sudre, 1904, *Bat. Eur.* 26; 1908–13, 119, t. 118, figs. 10, 11.

5x. *A larger plant.* Stem and panicle more glandular and aciculate, *a few of the stalked glands very long. Prickles longer, more slender* and not broad-based.

England: 17 (Littleheath Wood, Selsdon). N. France, Germany.

166. R. flavescens Muell. & Lef. 1859, *Pollichia*, 195; *R. Colmanni* ssp. *flavescens* (Muell. & Lef.) Sudre, 1908–13, 118, t. 117.

Kent. Surrey. Yellowish green. Stem rather blunt-angled, hairy, with a few small glands, acicles and pricklets; *prickles numerous, yellow, rather short* and unequal, slanting and recurved. Petiole prickles hooked. Leaflets 3, 4, 5, hairy beneath, unequally to subcompound-serrate; *terminal leaflet ovate-oblong to elliptic-acuminate. Panicle lax, broad*, leafy above, *all branches patent, long-peduncled, few-flowered*; rachis and branches villous, with a few glands and numerous, *short, yellow prickles*. Flowers *c.* 2 cm.; *sepals armed with numerous yellow prickles*, patent; *petals ovate-elliptic, pink*, pilose, apex notched or erose; filaments white, longer than the *reddish-based styles*. Carpels pilose. Fruit subglobose.

England: local; 13, 16 (Tunbridge Wells Common, Crockham Hill Common), 20. France (*Valois*, scarce).

167. R. condensatus P. J. Muell. 1858, *Flora*, **41**, 167; *R. densiflorus* Gremli, 1870, *Beit. Fl. Schw.* 36.

Middlesex. Stem low-arched, procumbent, brownish-red, blunt-angled, shortly hairy, with some scattered short glands; *prickles rather short*, broad-based, *subulate, declining. Leaves large*; leaflets 3–5, harsh and shortly hairy beneath, bluntly serrate; terminal leaflet cordate, broad elliptic-oblong or roundish, acuminate; other leaflets very shortly stalked. *Panicle* compound, *with a dense terminal agglomeration of simple or cymose branches, and similar axillary clusters below*, the lowest branches spreading and remote; the terminal leaflets oblong-apiculate; rachis crooked, some prickles retrorse-falcate. *Flowers c. 1·5 cm.*; sepals patent-erect; *petals roundish-obovate, notched, white, veiny; filaments white, equalling the greenish styles*. Carpels pilose. Fruit globose. Strong panicles have more advanced armature, recalling *R. fuscus* and *R. Bloxamii*.

England: 21 (Harrow Weald Common, north end). France (in the Seine basin, Alsace), Switzerland, Austria.

Subser. ††† **Hypoleuci** Sudre

(Type *R. vestitus* Weihe)

1. Sepals reflexed.

† Terminal leaflet roundish, shortly pointed.

168. **R. vestitus** Weihe ap. Bluff & Fingerh. 1825, *Comp. Fl. Germ.* **1**, 684; Sudre, 1908–13, 101, t. 101; *R. leucostachys* sensu Rogers et auct. mult.; non Smith, 1824.

4x. *Stem* angled, *violet purple, hirsute and felted*, with a few glands and pricklets; *prickles deep purple, long subulate*. Leaves moderate; *leaflets* 3–5, dull above, *strongly pectinate-pilose on the veins beneath*, with underlying felt, *evenly serrate-dentate*, undulate; *terminal leaflet* base entire and subcordate, *round, shortly cuspidate*. Panicle long, *hardly narrowed to the abrupt apex; rachis pilose, prickles, especially those below the panicle, long, slender, declining*, with some glands and pricklets. *Flower-buds large, flowers 2·5–3 cm.; petals 5, 6, villous, roundish, clawed, deep rose-pink; filaments deep rose-pink, long*; styles greenish. *Anthers usually pilose.* Carpels glabrous or pilose. *Fruit large, drupels large*, sweet at first, then bitterish. *Tolerates chalk and clay.*

var. **albiflorus** Boul. 1900, *Fl. Cent. Fr.* **6**, 89; *R. leucanthemus* P. J. Muell. 1859, *Pollichia*, 122.

Petals pinkish, filaments white. By far the commoner form here and on the Continent.

var. **parvifolius** Gel.

Plant small in all its parts. 6 (Ashton Park, J. W. White).

E., W., (? S.), I. Common in Wales and the southern and midland counties of England, rare or absent in N. England, Scotland and Cornwall. General throughout Central Europe to Silesia.

> The names *R. leucostachys* Schleich., *R. leucostachys* Schleich. ex Sm. and *R. leucostachys* Sm. have been used by home and foreign authors for this species. Schleicher, however, did not publish *R. leucostachys*, nor did

Smith use any *R. vestitus* material for his description, or for his figure of *R. leucostachys*. His herbarium contains only specimens of *R. leucostachys*, none of *R. vestitus*. Schleicher's specimen of his '*leucostachys*' is in the British Museum; it consists of a panicle only: of *R. vestitus*. See below for *R. leucostachys* Sm.

169. **R. podophyllus** (*podophyllos*) P. J. Muell. 1861, *Bonpl.* 281; Sudre, 1908–13, 108, t. 106.

4x. Procumbent. Stem angled, striate, hardly furrowed, glaucescent, with simple and tufted hairs, a few pricklets and very short glands; *prickles rather numerous, unequally distributed, subequal and straight. Petiole long,* flat above Leaflets 3–5, subglabrous above, pubescent and felted beneath, *finely serrate-dentate; terminal leaflet* base emarginate or entire, *round to elliptic or obovate. Panicle* lax, pyramidal, not leafy, with 1–5-flowered branches above, and longer stalked, 3–12-flowered branches below, *terminal leaflets very cuneate; rachis* felted and hirsute, *prickles* numerous, subequal, *not much declining, subulate on the pedicels,* mixed on strong panicles with acicles and numerous very short glands. *Flowers 1·5–2·3 cm.*; sepals loosely reflexed; *petals roundish ovate, obovate-cuneate or narrow elliptic, pointed or apiculate, remote when expanded,* usually faint pink to white; filaments usually white, hardly longer than the pinkish-based styles. *Carpels glabrous or with a few long hairs. Receptacle glabrous.* Fruit very good and sweet, not large. On Wimbledon Common the petals and filaments are pink, as described by Sudre, against Mueller.

England: locally frequent; 17 (frequent from Selsdon to W. Wickham, etc.), 35, 49, 67. France, Switzerland, Bavaria.

170. **R. conspicuus** P. J. Muell. 1858, *Flora,* **41**, et 1859, *Pollichia,* 122; non auct. brit.

Robust. Stem angled, *red brown,* with long to *short patent and divergent hairs* and a few pricklets, some minute; *prickles red,* finely pointed, *subulate to acicular.* Leaves moderate to very large; leaflets 5, glabrous or glabrescent above, hairy and green to ± grey-white felted beneath, incise, coarsely and bluntly serrate, *with fine, long mucros,* sometimes jagged; terminal leaflet subcordate, roundish ovate to broadly obovate. Panicle large, leafy, lax, pyramidal, *branches* oblique, 3–5-flowered, *with rather short hairs, numerous prickles,* and a good many *unequal glands and acicles,* the pedicels and calyces felted and villous. *Flowers* up to 2·5 cm., *showy; petals* pilose on the margin, broad, elliptic-obovate, *red; filaments deep pink,* sometimes white, longer than the *reddish-based styles.* Carpels strongly pilose. *Fruit composed of a moderate number of large drupels.*

England: local; 16, 24, 29, 35, 36. France, W. Germany, Switzerland.

171. **R. andegavensis** Bouv. 1903, *Rub. Anj. in Assoc. Fr. Av. Sc.* 682; *R. umbrosus* Boreau, 1857, *Fl. Cent. Fr.* ed. 3, 200; *R. gymnostachys* Genev. pro parte, *Mon.* 201; non Rogers, 1900, 51.

Kent. Surrey. Stem angled, sides flat, *densely hairy,* pruinose, *pricklets and stalked glands rare; prickles long, strong, straight, hairy. Petiole prickles hooked. Leaflets* 5, *glabrous above,* felted and *velvety, shortly hairy beneath, red-edged, closely and sharply serrate,* or more openly serrate-dentate with rather long mucros; *terminal leaflet sometimes long-stalked, broad elliptic-obovate, or rather quadrate. Panicle* long, diminished slightly to the often *congested apex,* the *terminal flowers of the* short cymose *branches nearly sessile* and *the lateral flowers shortly stalked;* rachis hirsute, slightly glandular, prickles numerous, long, declining or hooked. *Sepals with short glands and pricklets.* Petals roundish, cuneate or abruptly clawed, pink. *Filaments* white or pinkish, *about equalling the rosy-based styles.* Carpels pilose. Fruit moderate. Stem and panicle sprinkled with red, sessile glands.

England: rare; 15 (Stockbury to Bimbury), 16 (Hosey Common), 17 (Holmbury Hill, abundant, and Newlands Corner). N.W. France.

172. **R. salisburgensis** Focke, 1877, *Syn. Rub. Germ.* 280; *R. hebecaulis* ssp. *salisburgensis* (Focke) Sudre, 1908–13, 125, t. 123.

High Weald. Small. Stem red-brown, blunt-angled, finely pubescent with scattered glands and pricklets; prickles bright red-brown, hard, straight. *Leaves* 3 (5)-*nate, rather small,* short-stalked; *leaflets* softly and shortly hairy beneath, *nearly evenly crenate-serrate; terminal leaflet* short-stalked, subcordate, *very round,* shortly cuspidate. *Panicle rather small, narrow, dense* at apex, sub-racemose or pyramidal with 3–5-flowered semi-erect branches; rachis and branches hirsute, prickles acicular, nearly patent. *Flowers up to 1·5 cm.;* sepals felted, glandular, loosely reflexed to subpatent; *petals bright pink, roundish,* fimbriate; *filaments pink, equalling the greenish styles.* Carpels glabrous. *Fruit perfect, globose.*

England: 14 (moor at Hindleap near Wych Cross). Central Europe.

†† Terminal leaflet long-pointed, not roundish.

173. **R. adscitus** Genev. 1860, *Mem. Soc. Acad. M. & L.* **8**, 88; Sudre, 1908–13, 110, t. 109; *R. hypoleucus* Lef. & Muell. 1859, *Pollichia,* 143; *R. micans* sensu Rogers pro parte, 1900, 48; non Gren. & Godr. 1848.

Stem angled, sides flat to furrowed, *pilose and pubescent with buried glands and acicles;* prickles short, straight, yellowish. Petiole prickles straight or slightly bent. *Leaflets* 3–5, glabrescent above, grey-white *felted and softly pilose beneath, sublobate, coarsely and deeply serrate; terminal leaflet* base indented or entire, *oblong-obovate-cuspidate* to elliptic-acuminate. *Panicle pyramidal, branches perpendicular to the rachis, numerous flowered; rachis* crooked below, hirsute, with numerous hidden glands, prickles weak, declining. Sepals long-pointed. Petals elliptic-ovate, narrowed below, pink. Filaments white, or pink-based, equalling or exceeding the greenish or reddish styles. Carpels glabrous. Receptacle hirsute.

E., W., I. Frequent in the west, rather scattered elsewhere; 1–3, 5, 10, 15–17, 20, 30, 35, 48, 58. H. 1, 38–40. Jersey. N. and N.W. France.

174. **R. leucostachys** Smith, 1824, *Eng. Fl.* **2**, 403; Syme, 1864, *Eng. Bot.* ed. 3, 172, t. 448; non Rogers, 1900, 50; Sudre, 1908–13, t. 102, figs. 6–10 (as *R. leucotrichus*). **Fig. 26.**

4x. S.E. England. Very shaggy in the shade. Stem angled, sides flat, *very hairy*, pruinose, stalked glands and pricklets rare; *prickles* subequal, *powerful and gashing. Leaflets* 5, nearly glabrous above, greyish felted and thickly villous beneath, *shallowly sometimes compound dentate-serrate to deeply incise serrate*, or subiaciniate when luxuriant; *terminal leaflet* short-stalked, base ± entire, ovate to *obovate or oblong-elliptic, long-pointed. Panicle long*, sub-pyramidal, *with usually small ovate and 3-nate leaves* above; rachis crooked below, densely villous, prickles firm, slightly declining, pricklets and glands few. *Flowers c. 2 cm.* Sepals white felted. *Petals rosy pink*, broad elliptic. *Filaments deep pink*, longer than the greenish styles. Carpels pilose. Receptacle hirsute. *Fruit subglobose.*

England: frequent in the south-east; 12–18, 20–22, 24, 30. France, Sarthe.
(*R. lasioclados* Focke. A plant in Paddlesworth Wood, E. Kent, was so determined by Focke, and is referred to under this name in Roger's *Handbook*. It is, however, *R. leucostachys* Sm. not *R. lasioclados*, a bramble which I have studied in Germany. Other specimens determined as *R. lasioclados* by Rogers are *R. ulmifolius × vestitus*: I have seen no *R. lasioclados* in Britain.)

175. **R. scutulifolius** Sudre, 1906, *Diagn.* 26; *R. adscitus* microg. *scutulifolius* (Sudre) Sudre, 1908–13, 112, t. 110.

Surrey. Robust. Stem procumbent, sharp-angled, purple, glaucescent, very hairy, with glands and small pricklets; *prickles* numerous, rather unequal, *narrow-based, patent. Leaves large or very large, bright green*; leaflets 5, glabrescent above, densely and softly felted and pilose, *the veins yellowish beneath*, broadly and irregularly crenate-serrate; *terminal leaflet elliptic-rhomboid, acuminate, base entire*. Panicle leafy, with long, ± spreading lower branches, narrowed above, the uppermost branches patent; rachis hirsute, glandular, prickles long subulate, pedicels felted and glandular. Flowers *c.* 2·5 cm.; sepals white felted, glandular, loosely reflexed; petals broad elliptic or obovate, clawed, pink or pinkish; filaments white, longer than the greenish styles. Carpels pilose. *The leaves grow very large in woods.*

England: apparently rare; 17 (Gee Wood, Selsdon; Spring Park, Addington). France (*Seine-Inférieure, Aumale*).

2. Sepals patent to erect.
† Leaflets moderately large, acuminate, coarsely toothed.

176. **R. Boraeanus** Genev. 1860, *Mem. Soc. Ac. M. & L.* **8**, 87; Sudre, 1908–13, 110, t. 108. **Fig. 27.**

4x. *Axes purple. Stem with thin tufted hair, scattered stout pricklets, some* minute, and a few acicles and short glands; prickles moderate, stout-based, straight or slightly bent. *Leaves 5-nate, pedate; leaflets thick,* grey felted and softly pilose beneath, slightly compound-serrate; *basal leaflets subsessile; terminal leaflet cordate, roundish ovate,* acuminate to obovate-cuspidate. *Panicle broad, equal,* obtuse, *branches semi-erect, the peduncles long, often unarmed, pedicels long*; rachis prickles few, weak, slanting or slightly falcate. Flowers 1·5–2·5 cm.; *petals broad obovate, pink or deep pink,* rarely white; *filaments* deep pink or only pink-based, *equalling the pink-based styles.* Anthers often pilose. *Carpels pilose.* Receptacle hirsute. Fruit small, subglobose.

E., W., I. Frequent; 2, 3, 5, 11, 13–18, 20, 23, 24, 27, 28, 34–37, 40, 48, 49, 66, 67. H. 1, 2, 23, 38. Guernsey, Jersey. W. France, Switzerland (one locality).

177. R. conspersus W. Wats. 1935, *Journ. Bot.* **23**, 255.

Axes mottled to wholly red. Stem obtuse-angled, slightly furrowed, *with tufted hair* concealing numerous very short glands; prickles yellowish at first, strong, subequal. Leaves yellowish green, *digitate, rather large, petiole thick, flat above*; leaflets 5, softly yellowish, pilose to felted as well beneath; *terminal leaflet elliptic-obovate, acuminate,* base nearly entire. *Panicle* sub-pyramidal, rather long, *dense at the rounded,* narrow, leafy *top,* middle branches long-peduncled, patent; *rachis stout,* pilose, *shortly glandular, weakly and sparsely armed.* Flowers *c.* 2·5 cm., incurved; *sepals acicular and gland-dotted; petals* broad elliptic *rhomboid,* undulate, bright pink; *filaments pinkish, hardly longer than the greenish styles. Carpels bearded.* Fruit rather large, ovoid, pleasantly flavoured.

E., I. Locally frequent; 12–14, 17, 69. H. 39 (Ballymena). Guernsey.

†† Terminal leaflet large, round or roundish, cuspidate.

178. **R. magnificus** P. J. Muell. ex Genev. 1880, *Mon.* 171; *R. vestitus* ssp. *conspicuus* var. *magnificus* (P. J. Muell. ex Genev.) Sudre, 1908–13, 104.

6x (3x, a seedling). *Very strong.* Stem obtuse-angled, reddish purple, glaucescent, felted and hairy, with scattered pricklets, some gland-tipped; prickles long, gradually narrowed. *Leaves large,* 3, 4, 5-nate, pedate, *petiole thick, flat above; leaflets subimbricate,* plicate, softly pilose and felted beneath, unequally serrate with long mucros; *terminal leaflet subcordate, short-stalked. Panicle very broad, lax, nearly equal*; rachis densely hairy, with numerous short glands and a few acicles, prickles crowded, slender, unequal. Flowers showy, *c.* 2·7 cm.; sepals aculeolate and glandular; *petals roundish, cuneate, retuse,* light pink; *filaments* white or pinkish, equalling the *yellowish or pinkish-based styles.* Carpels thinly pilose or glabrous. *Fruit large, oblong, drupels few, ripening slowly.* Young shoots an intense crimson. Sudre found this growing with *R. conspicuus,* together with intermediates, in woods in Marne. No *R. conspicuus* has been found with it in England.

England: rare; 17 (Winterfold Heath). France (*Marne, Sarthe, Eure, Seine-Inférieure,* Rouen), Switzerland (Vaud, Bonmont).

179. **R. acutidens** Boul. & Gill. 1892, *Rev. Rub. Spect.* 7; *R. vestitus* ssp. *acutidens* (Boul. & Gill.) Sudre, 1908–13, 104, t. 102.

Ireland. Stem angled, purple, *glaucescent, hirsute,* with scattered pricklets; *prickles* hairy, *purple, straight,* declining, moderately long. *Leaves* 3, 4, 5-nate, *petiole short; leaflets imbricate, glabrous above,* greyish, softly pilose beneath, *very sharply and unequally serrate, the teeth rather connivent, crisped.* Panicle short, *broad, always pyramidal,* truncate, *with* 1–2 simple leaves and 3-flowered, rather *long-peduncled, middle branches, and longer more compound branches below*; terminal flowers of the panicle shortly stalked; *prickles numerous, purple,* unequal, together with numerous short glands. Flowers 2–2·3 cm.; *petals pure white,* broad obovate; filaments white, longer than the greenish styles. *Carpels pilose.* Very fertile. Recalls *R. vestitus.*

Ireland: H. 1 (frequent and often luxuriantly developed beside the stone walls around Dingle). France (*Corrèze, Puy-de-Dôme, Saône-et-Loire*).

††† Terminal leaflet ovate, obovate or elliptical; acuminate or cuspidate.

180. **R. macrothyrsus** (*macrothyrsos*) Lange, 1870, *Fl. Dan.* 6, t. 2832; *R. adscitus* ssp. *macrothyrsus* (Lange) Sudre, 1908–13, 113, t. 110.

4x (**2x**, a seedling). *Stem* brownish purple, angled, *villous,* with a very few glands, acicles and pricklets; *prickles moderate, slender,* declining. *Leaves usually rather small,* 5-nate, pedate; leaflets subglabrous above, densely pilose on the veins and felted beneath, *rather deeply and sharply subcompound-serrate, or shallow and jagged; terminal leaflet obovate or elliptic, subtruncate, cuspidate. Panicle long or very long, narrowed,* usually with several simple leaves, branches spreading, cymose, like the peduncles felted and villose, glandular and acicular, prickles slender. Flowers *c.* 2 cm., cupped; *sepals armed with deep-purple prickles,* patent; *petals* elliptic-obovate, *fimbriate, bright pink;* filaments white, slightly longer than the yellowish styles. Carpels pilose. *Fruit rather small, drupels large, unattractive.*

E., I. Frequent in S.E. England, rare elsewhere; 9, 13, 15–20, 22–24, 33, 35–37, 69. H. 1, 3. Denmark, N.W. Germany.

181. **R. criniger** (E. F. Linton) Rogers, 1900, 52; *R. Gelertii* var. *criniger* E. F. Linton, 1894, *Journ. Bot.* **32**, 108.

Robust. Hirsute. Stem angled, slightly furrowed, *very hairy, pruinose,* with numerous sessile and some short-stalked glands and pricklets; *prickles* rather short, *much declining. Leaves large; leaflets* 3–5, plicate, undulate, greyish green above, densely pubescent and *greyish felted beneath,* sharply, closely, ±*compound or lobate-serrate; terminal leaflet elliptic-obovate or ovate, gradually acuminate. Panicle* long with long ascending lower branches, almost pointed at the top, *densely pilose, felted,* glandular and acicular; *prickles* long-based, numerous, weak, *much declining and retrorse-falcate.* Flowers moderate. *Sepals triangular ovate, attenuate,* patent. *Petals* glabrous on the margin, *ovate subrhomboid,* pink, pinkish or white. Filaments white, equalling or much exceeding the yellowish, reddish-based styles. Anthers sometimes pilose. Carpels usually pilose.

Endemic. England: 4, 5, 13, 15, 16, 18–25, 27–30, 36, 38, 39, 55, 57, 58, 60.

182. **R. eifeliensis** Wirtg. 1858, *Herb. Rub. Rh.* ed. 1, no. 94; *R. Schmidelyanus* var. *eifeliensis* (Wirtg.) Sudre, 1908–13, 120, t. 118, figs. 15, 16.

Robust. Hirsute. Stem angled, sides flat, *pruinose, very hairy,* with scattered glands, acicles and pricklets; *prickles* subequal, *moderately subulate.* Petiole long, flat above. *Leaflets* 4, 5, soft, *greyish felted and densely pilose on the veins beneath, undulate,* unequally to *slightly lobate-serrate; terminal leaflet* broad, *roundish-ovate to rhomboid-ovate, acuminate. Panicle dense, usually floriferous,* broad, equal to pyramidal with patent, cymose branches; rachis and sepals hirsute, prickles declining or hooked. *Flower-buds large;* sepals grey felted and hirsute, aculeate; *petals often 6,* pink or pinkish, *obovate, cuneate,* margin glabrous; filaments white, slightly longer than the greenish styles. Carpels glabrous or slightly pilose. Receptacle hirsute especially below the carpels. Fruit good, ovoid. A dwarf form occurs in S. Devon.

England: local; 2, 3, 22–24, 39, 55. Germany (Eifel).

183. **R. melanocladus** (Sudre) Riddels. 1948, *Fl. Gloucester,* 156; emend. W. Wats.; *R. pyramidalis* ssp. *melanocladus* Sudre, 1908–13, 48, t. 51; *R. hirtifolius* sensu Rogers pro parte, 1900, 48; non Muell. & Wirtg.

Mitcheldean Meend, W. Gloucester. Rather slender, axes red or deep purple, pruinose. Stem angled, striate, thinly hairy, with unequal pricklets and stalked glands; prickles hairy, stout, short, ± declining. Leaves subdigitate, rather tough; *petiole and petiolules short,* prickles numerous, unequal, falcate or hooked, some gland-tipped, together with a few pricklets and short glands; leaflets (3) 5, closely pubescent and felted beneath, serrate-dentate, undulate; *terminal leaflet base rounded or indented, ovate to elliptic-obovate, acuminate. Panicle long* with usually 1–2 simple leaves, *narrow and racemose above* with several, sharply ascending, 3–4-flowered branches below; *rachis nearly straight,* angled, felted and thinly hairy, glandular and acicular, prickles small, hooked; *leaflets mucronate or acute, becoming convex. Flowers, c. 2 cm.;* sepals ovate, leafy-tipped, glandular; *petals narrow elliptic, tapered below, rose pink;* filaments white, long; styles greenish. Carpels pilose. Recalls *R. polyanthemus.*

The foregoing description is made from notes and a specimen taken of a bush growing in a bleak situation between 800 and 900 ft. at Mitcheldean Meend, W. Gloucester in 1932. It is possible that Ley's record of *R. hirtifolius* for Harechurch Wood, Hope Mansel, a few miles away, may apply to this species: I have not seen his specimen; but his other records of *R. hirtifolius* in the Welsh border and Wales apparently apply to *R. ludensis*, which was confused with the Mitcheldean *R. melanocladus*.

Endemic. England: 34 (Mitcheldean Meend), 36 (Harechurch Wood?).

Ser. ii. MUCRONATI W. Wats.

(Type *R. mucronifer* Sudre)

I. Stem furrowed.

184. **R. mucronatiformis** (Sudre) W. Wats. 1946, *Journ. Ecol.* **33**, 340; *R. hypomalacus* ssp. *mucronatiformis* Sudre, 1908–13, 30, t. 24; *R. mucronatus* var. *nudicaulis* Rogers, 1900, 56.

Stem nearly glabrous, glaucescent, *becoming orange-scarlet; prickles* mostly patent, *long and subulate*, central petiolule with a few, long, gland-tipped acicles. *Terminal leaflet* broad-cordate, *oblong-obovate, with a long, twisted, cuspidate point*, incise-crenate to compound serrate, undulate, rough above, *thick, softly pilose and greyish felted beneath*. Panicle equal, branches 1–3-flowered, rachis angled and furrowed, glabrous below, felted above, stalked glands numerous, mostly very short, *terminal leaflets obovate-cuneate, mucronate*. Flower-buds large, flowers showy, incurved; sepals gland-dotted, appendiculate, patent; petals obovate, faintly lilac (rarely bright deep pink; ap. Rogers); *filaments white, equalling the greenish styles*; anthers glabrous. Carpels glabrous. Receptacle hirsute. Fruit ovoid-oblong, moderately large.

Endemic. South England, local; 6, 8–11, 22.

2. Stem not furrowed. Terminal leaflet nearly round, mostly short-pointed. Petals pink or deep pink.

† Anthers pilose. Filaments pink.

185. **R. mucronifer** Sudre, 1902, *Rub. Herb. Bor.* 56; 1908–13, 112, t. 111; *R. mucronatus* Bloxam in Kirby, 1850, *Fl. Leicester*, 43; non Seringe in D.C. 1825; *R. mucronatus* var. *drejeriformis* Frid. 1877, *Bot. Cent.* **70**, 407; Syme, 1864, *Eng. Bot.* ed. 3, 178, t. 451 (as *R. mucronulatus*).

Northern. Robust. Stem roundish, striate, hairy at first, becoming deep purple, *prickles* slender, *rather few. Leaflets* 3, 5, *imbricate*, pubescent or thinly hairy and greenish felted beneath, serrate or serrate-dentate; terminal leaflet base subcordate to nearly entire, *broad, roundish oblong-quadrate or obovate, mucronate to long-cuspidate*. Panicle leafy, lax, few-flowered, peduncles and pedicels long; rachis striate, felted and shortly pilose, with acicular prickles, numerous short and some long glands and acicles, bracts gland-fringed. *Flowers up to 3 cm.*; sepals greenish with a white margin, pilose, patent; petals obovate, rarely white; filaments pink, far longer than the pink styles. Carpels pilose.

E., S., I. Locally frequent in the north; 27, 38–40, 54, 55, 57, 62, 67, 68, 86, 90, 110. H. 38. N.W. Germany.

186. **R. Drejeri** G. Jensen, 1883, *Fl. Dan.* f. 51, 7, t. 3023; *R. infestus* ssp. *Drejeri* (G. Jensen) Sudre, 1908–13, 151, 145; *R. Purchasii* Bloxam, manscr. **Fig. 28.**

4x. Plant moderate to rather small. *Stem angled*, hairy, becoming *blackish-purple*, not pruinose, prickles numerous, strong-based, unequal, from patent to recurved. *Leaflets* 3, 4, 5, *tough, becoming plicate, rugose and convex*, subglabrous above, thinly hairy beneath, serrate-dentate; *terminal leaflet* subcordate, roundish to oblong-obovate, cuspidate; *basal leaflets subsessile.* Panicle rather short, pyramidal, racemose above, with patent mid branches; *rachis prickles mostly curved. Flowers c. 2 cm.*, incurved; sepals greenish with a white border, glandular and aculeate, almost clasping; *petals roundish-oblong*; filaments pink, rarely white, much longer than the yellowish styles. Carpels strongly pilose. Fruit ovoid, large, sweet.

E., S., I. Scattered and local; 14, 17, 24, 39, 55, 60, 63, 65, 87. H. 33, 38, 39. Denmark, N.W. Germany.

187. **R. badius** Focke, 1877, *Syn. Rub. Germ.* 276; *R. axillaris* ssp. *badius* (Focke) Sudre, 1908–13, 258, t. 215; *R. rubeolus* Weihe (specimen).

Leaves yellowish green. Stem red-brown, arching, procumbent, *glabrescent; prickles yellow*, subequal, *broad-based, subulate. Leaf pedate, petular prickles nearly straight; leaflets* ± *acute, imbricate, short-stalked, velvety* pilose *beneath*, grey when young, but not felted except on branch and upper leaves, serrate or serrate-dentate; *terminal leaflet subcordate, elliptic-ovate, short pointed; basal leaflets subsessile. Panicle* broad, lax, lower branches spreading, panicled, middle ones cymose, upper ones simple and patent, *villous*, glandular and acicular. Flowers *c.* 2·5 cm.; sepals patent-erect. *Fruit large, ovoid, soon crimson.*

E., I. Scattered; 7, 8, 15, 17, 24, 38, 54, 55. H. 23. N.W. Germany.
†† Anthers glabrous.

(*a*) Filaments pink or red. (Cf. *R. cinerosus*).

188. **R. Muelleri** Lef. in Muell. 1859, *Pollichia*, 180; Sudre, 1908–13, 114, t. 113.

E. Sussex. Stem angled, *deep purple, hairy, prickles subulate. Leaves large, digitate; leaflets all long-stalked*, glabrous above, velvety grey-white felted and pubescent beneath, *serrulate*; terminal leaflet cordate or entire, roundish-ovate to oblong-obovate, acuminate. *Panicle large and long, lax*, with a few simple leaves, branches spreading, *pedicels long*; rachis crooked, pilose, moderately glandular and acicular, *prickles acicular-subulate. Flowers up to* 2 cm., showy; sepals greenish within, long-tipped, loosely reflexed or patent;

petals elliptic; filaments longer than the red-based styles. Carpels and receptacle pilose.

England. 14 (by Newick railway station). France, Belgium, Germany (*Westphalia* and *Bavaria*), Switzerland.

189. **R. Briggsii** Bloxam, 1869, *Journ. Bot.* **7**, 33 (plate).

Devon. Cornwall. Stem low, blunt-angled, intensely villous, with numerous sunken, *very short, red glands* and acicles; prickles brown, unequal, slanting. Leaves digitate; *leaflets 3, 5, imbricate*, closely strigose above, softly pilose and felted beneath, *serrulate-denticulate; terminal leaflet roundish, or very broadly roundish-ovate, mucronate. Panicle* with 1–2 ovate leaves, short pointed, dense above, lax, with several semi-erect branches below, felted and *intensely villous, prickles rather short, acicular*, subpatent, stalked glands very short and numerous, acicles rarer. Flowers showy, 2–2·5 cm.; *sepals and petals 7 in some flowers; sepals* felted, gland-dotted, *clasping; petals broad, obovate-cuneate*, fringed at the apex, *deep rose; filaments about equalling the red styles.* Carpels long-pilose. Receptacle pilose. *The lowest leaflets are subsessile on all leaves.*

Endemic. England: very local and rare; 1 (a bush by a stone wall by the Truro road west of Par in 1951), 3 (several bushes formerly in Bickleigh Vale).

(*b*) Filaments white.

190. **R. cinerosus** Rogers, 1898 (for 1896), *Bot. Exch. Cl. Rep.* **1**, 513; 1900, 54.

A moderately small bush. Stem glabrescent, subpruinose, *blunt-angled*, with very small pricklets and glands, and *numerous short acicles; prickles* long, straight. *Leaflets 3, 4, 5* (*6, 7*), short-stalked, *imbricate*, pilose and grey felted beneath, *the felt often patchy*, finely serrate or serrate-dentate. *Panicle* long, subpyramidal, truncate, *mid branches patent, cymose*, villous, *prickles numerous, acicular, patent; rachis furnished below with numerous, short, pale acicles*, the *terminal leaflets obovate-cuneate-mucronate. Flowers* 1·5–2 cm.; sepals subpatent; *petals 5–7*, subglabrous on the margin, *roundish-obovate*, pink to pinkish above with a greenish claw; filaments sometimes white, longer than the *short, compact, greenish styles.* Carpels glabrous or pilose. Receptacle hirsute. Fruit small, subglobose. The young shoots and leaves are bronze. *Petioles and petiolules are usually dusky at their base.*

Endemic. E., W., S., I. Rather frequent; 4, 5, 12, 14–17, 20–24, 34–37, 40, 42, 43, 55, 58, 60, 62, 98, 106. H. 9, 16, 26, 38.

191. **R. atrichantherus** E. H. L. Krause in Prahl. 1890, *Krit. Fl. Prov. Schlesw.-Holst.* **2**, 61.

S. Bucks. Middlesex. Stem nearly *glabrous, blunt*, subpruinose, striate, *with scattered minute glands and pricklets*; prickles rather small, patent. Leaves 3-nate with the lateral leaflets bilobed, and 5-nate; *leaflets often large*, dull and glabrous above, glabrescent beneath, shallowly dentate-serrate; *terminal leaflet very broad, obovate-truncate, mucronate. Panicle* pyramidal-subcorymbose *with long-peduncled, numerous flowered lower branches; rachis* glabrous to felted and pubescent, armed *with weak pale prickles and minute acicles and glands. Flowers* 1·5–2 *cm.*; sepals greenish inside, loosely reflexed to patent; *petals elliptic, glabrous; filaments hardly equalling the reddish styles.* Carpels glabrous. *Fruit oblong, of numerous smallish drupels.*

E., W. Very local; 21 (Bayhurst Wood), 24 (Black Park and neighbourhood, abundant). Belgium, Denmark, N.W. Germany (especially *Schleswig-Holstein and Silesia*).

3. Stem not furrowed. Terminal leaflet acuminate, not roundish. Anthers glabrous.

† Sepals loosely reflexed to patent.

192. **R. Lettii** Rogers, 1901, *Journ. Bot.* **39**, 381.

Ireland. Robust. Stem tall, blunt, striate, thinly pilose, *becoming purple-speckled and pruinose,* with few glands or acicles; *prickles very long, subulate.* Leaflets (4) 5, glabrous above, softly pubescent and felted beneath, shallowly angular subcompound-serrate-dentate; *terminal leaflet ovate to obovate-acuminate, much narrowed to the entire base,* the *prickles large, sub-patent and spreading-falcate. Panicle* large, leafy and dense at the corymbose apex, upper branches deeply divided, lower branches semi-erect; rachis striate, felted and pilose, armed like the stem. Flowers moderate; sepals short-pointed; *petals narrow elliptic-obovate, pinkish*; filaments rather long. Carpels pilose. Fruit subglobose. All reports of *R. Lettii* in Britain rest on wrong identifications.

Endemic. Ireland: 1, 30, 37–39.

193. **R. Gelertii** Frid. 1886, *Bot. Tidsskr.* **15**, 237; Sudre, 1908–13, 55, t. 62; non Rogers, 1900, 56.

*4x. Very robust. Stem sharp-angled, *brownish red, glabrescent,* with numerous sessile and a few stalked glands and pricklets; *prickles numerous, slender, straight or nearly straight. Leaflets 5, very large,* glabrescent above, felted and pilose beneath, *sharply* and unequally to *compound-serrate; terminal leaflet short-stalked, long, narrow ovate-elliptic or oblong, gradually acuminate, sometimes narrowed below; basal leaflets often longer than the petiole. Panicle* broad, lax, pyramidal, with cymose, numerous flowered middle branches, upper ones deeply divided, corymbiform; rachis furnished with long subulate and falcate prickles, long acicles and intermediate prickles; *terminal leaflets long elliptic.* Flowers *c.* 2–2·5 cm.; sepals produced into a leafy tip; *petals narrow ovate-rhomboid,* pinkish or white; filaments long; styles yellowish. Carpels glabrous or pilose. Fruit large.

England: rare; 2 (near Launcells), 15 (Bigbury to Chartham), 22 (Wytham Wood). Denmark, Germany (*Schleswig-Holstein*).

†† Sepals patent to clasping.

(*a*) Armature, when strong, almost hystrican. Basal leaflets stalked, 1–2 mm.

194. **R. Leyanus** Rogers, 1895, *Journ. Bot.* **33**, 81; *R. Schmidelyanus* ssp. *Leyanus* (Rogers) Sudre, 1908–13, 120, t. 119. **Fig. 29.**

Robust. Stem angled, hairy or glabrescent, *pruinose, becoming brownish red, with a variable quantity of acicles and stalked glands; prickles some long, at first yellow. Leaves yellowish green, rather large,* (3, 4) 5-nate, pedate; *leaflets* glabrous above, glabrescent or velvety grey felted beneath, *finely serrate to dentate-sinuate, a little jagged,* mostly entire-based; terminal leaflet roundish, broad ovate or elliptic, acuminate. *Panicle ± long, pointed, pyramidal,* or racemose, truncate, or with compound patent branches; rachis angled, prickles slender, some long, recurved, with *sometimes very many unequal acicles and glands*; terminal leaflets obovate-cuneate, sharply compound serrate. *Flowers 1·5–2·2 cm.; calyx base discoid, sepals narrow, appendiculate,* aculeate, *soon erect; petals narrow, obovate,* pink or pinkish, glabrous or nearly glabrous on the margin; *filaments very long,* sometimes pinkish. Carpels glabrous or nearly glabrous. Receptacle pilose.

Endemic. E., W., I. Rather frequent in the west; 1–8, 11, 13, 20, 22, 23, 34, 36, 38–44, 49, 51. H. 'Pontydun'.

195. **R. hibernicus** (Rogers) Druce, 1908, *Pl. List*, 21; *R. Drejeri* var. *hibernicus* Rogers, 1897, *Journ. Bot.* **35**, 48.

N.v.v. Axes brown. Leaf pale green, not yellowish. *Stem moderately hairy,* glaucescent; prickles straight, unequal, ± numerous, distributed together with unequal acicles and *a few short glands* over the sides of the stem. *Leaflets 5, finely, shallowly and doubly dentate,* often jagged; *terminal leaflet rather narrow oblong-obovate,* rarely broad, cordate, ovate, long-pointed. *Panicle lax,* upper and middle branches nearly patent, the 2–4 lower ones when present long, oblique, numerous flowered; *rachis grey felted* at the summit, prickles straight; *the upper simple leaves not glandular above.* Petals obovate. *Carpels pubescent or pilose.* Flowering a fortnight after *R. dunensis.*

This species is much nearer to *R. Leyanus* than to *R. Drejeri*; and with *R. Leyanus* and *R. dunensis* would perhaps stand as well in HYSTRICES, near *R. spinulifer,* as in MUCRONATI.

Endemic. Ireland: H. 33, 37–39.

196. **R. dunensis** Rogers, 1901, *Journ. Bot.* **39**, 382.

N.v.v. Axes brown. Leaf yellowish. Stem a good deal hairy, prickles mostly slender, short, numerous, very unequal, mixed with *numerous, unequal, long glands* and glandular acicles. *Leaflets very broad,* concave, softly hairy beneath, *compound or slightly lobate, divaricate-serrate-dentate*; terminal leaflet cordate, roundish ovate or elliptic, long acuminate; *intermediate leaflets very long. Flowering branch pruinose and pallid below. Panicle dense,* narrowed and pyramidal above, *with a few, short, erect branches below, brown felted* and patent-pilose; prickles long, subulate, patent or declining, merging into acicles, glands numerous, short. *Several simple leaves conspicuously glandular above and on the base*, and whitish felted beneath. Flowers incurved; *sepals brownish*, densely glandular, erect; petals pinkish, narrow obovate. *Carpels glabrous or nearly so.* Fruit abundant, excellent. 'Armature often quite Koehlerian', ap. Rogers. Endemic. Ireland: H. 37, 38.

(b) Armature never at all hystrican.

197. **R. hypomalacus** Focke, 1877, *Syn. Rub. Germ.* 274; Sudre, 1908–13, 30, t. 24; *R. macrophyllus* var. *velutinus* Weihe & Nees, 1825, *Rub. Germ.* 35; *R. mucronatoides* A. Ley ex Rogers, 1900, 55; *R. Hansenii* E. H. L. Krause in Prahl. 1890, *Krit. Fl. Prov. Schlesw.-Holst.* 1, 60.

Rufescent. Stem angled, thinly pilose, sometimes with a few pricklets, stalked glands and acicles; *prickles yellow, long subulate*, not numerous. *Leaves large*, 3–5-nate, *leaflets* softly pilose to glabrescent beneath, *coarsely serrate, jagged; terminal leaflet* subcordate, *elliptic or oblong*, cuspidate-acuminate; *basal and lateral leaflets shortly stalked. Panicle subracemose* above, lax below, *often with long, ascending, panicled lower branches; rachis* crooked, intricately hairy, *with glands and acicles, some long-stalked, usually pallid; pedicels and calyces usually with numerous long acicles and glands;* terminal leaflets narrow obovate, cuneate, mucronate. Flowers *c.* 2·5 cm.; *sepals patent-erect; petals elliptic, glabrous* on the margin, *pink*, rarely pure white; filaments pink or white, equalling or much exceeding the styles. Anthers sometimes subpilose. Carpels pilose. *Receptacle hirsute, especially below the lowest carpels.* A. Ley describes *R. mucronatoides* as having sepals reflexed, but a fruiting specimen of his from New Radnor has them erect.

E., W., S., I. Thinly distributed; 13, 15, 20, 21, 23, 24, 30, 36, 38, 43, 55, 96. H. 27. Denmark, N.W. Germany.

198. **R. chaerophyllus** Sag. & Schultze, 1894, *Deutsche Bot. Monat.* 12; Sudre, 1908–13, 27, t. 20.

Robust. Stem purple, angled, slightly hairy, with numerous sessile and a few stalked glands; *prickles* rather numerous, *some very strong, broad-based, falcate.* Leaves large, stipules nearly filiform; *leaflets* short-stalked, imbricate, glabrescent above, green, at first pubescent beneath, *coarsely and unequally to compound serrate*, margin crisped; terminal leaflet cordate, broad roundish ovate to elliptic. *Panicle pyramidal, broad and lax with long spreading subcymose or panicled few-flowered branches; rachis crooked*, prickles very unequal and few, or with numerous stalked glands and acicles, some long. *Flowers up to 3 cm.*; sepals green felted with a white margin, patent or clasping; *petals elliptic-oblanceolate*, pinkish; filaments white, longer than the greenish styles. Carpels glabrous or pilose. *Fruit rather large, perfect.*

E., S., I. Rare; 8, 15, 16, 21, 24, 96. Belgium, Denmark (a variety), Germany, Austria.

199. **R. bracteosus** Weihe ap. Lej. & Court. 1831, *Comp. Fl. Belg.* **2**, 162; *R. orthoclados* A. Ley, 1896, *Journ. Bot.* **34**, 159; Sudre, 1908–13, 30, t. 25.

Stem suberect but not tall, arching, blunt-angled, thinly hairy, *with a few yellow glands and acicles;* prickles unequal, slender, declining. *Leaflets 3–5, glabrescent on both sides,* irregularly and partly patent serrate; *basal leaflets subsessile;* terminal leaflet short-stalked, roundish ovate to elliptic, long-pointed. *Panicle equal, subracemose* to compound, branches semi-erect; rachis flexuose, shortly hairy, with a few, pale, rather long glands and subulate prickles. Sepals narrow, appendiculate, green, white margined, pubescent, glandular and acicular. Petals obovate, white. Filaments long. Carpels pilose. Receptacle pilose. Fruit globose.

England: rare; 35 (Beacon Hill and district), 37 (Hedges near Bewdley, E. G. Gilbert, 1858), 39. Belgian Ardennes and N. Germany.

Ser. **iii.** **DISPARES** W. Wats.

(Type *R. Gremlii* Focke)

1. Leaves green beneath.

200. **R. dentatifolius** (Briggs) W. Wats. 1931 (for 1930), *Lond. Nat.* 73 quod synon. exclud. descript.; *R. Borreri* var. *dentatifolius* Briggs, 1880, *Fl. Plym.* 121.

Mainly western. Stem long, prostrate, furrowed, thinly hairy, purple and pruinose; *prickles long and slender. Leaves yellowish green,* 5-nate, *pedate; leaflets imbricate, plicate, concave,* glabrescent above, pubescent beneath, incise, sharply and *coarsely to subcompound serrate; basal leaflets subsessile; terminal leaflet narrow, cordate, ovate to elliptic, gradually acuminate.* Panicle rounded at the somewhat corymbose apex, narrower and lax below; rachis hirsute, prickles unequal, falcate. Flowers *c.* 2·5 cm.; *sepals greenish, patent to erect; petals narrow elliptic-obovate,* pinkish; filaments long. *Carpels long-pilose.* Fruit ovoid-oblong.

Endemic. E., W., I. Mostly in Devon; 2–4, 6, 9, 22 (Greenham common), 35, 46 (Dol y bont, Llanfihangel). H. 23.

201. **R. ahenifolius** W. Wats. 1956, *Watsonia*, **3**, 288; 1946, *Journ. Ecol.* **33**, 341 (nomen nudum); *R. dentatifolius* sensu W. Wats. 1931 (for 1930), *Lond. Nat.* 73 quod descript. exclud. synon.; non *R. Borreri* var. *dentatifolius* Briggs, 1880.

Young shoots copper-leaved. Stem blunt, furrowed, slightly hairy; *prickles numerous, yellow, stout-based,* short or moderately long, *sharply declining or falcate.* Leaves 5-nate, digitate, *prickles small, hooked; leaflets* with entire base, short-stalked, *long, narrow,* glabrous above, glabrescent beneath, from rather coarsely serrate to finely subcompound dentate-serrate; *terminal leaflet elliptic, long-pointed; basal leaflets short-stalked. Panicle equal, narrow, middle branches deeply divided,* armed like the stem. Sepals linear-tipped, aciculate,

127

patent or clasping. *Petals obovate, spatulate or oblanceolate*, glabrous on the margin, pink, unfolding widely. Filaments long. *Anthers pilose.* Carpels pilose. Fruit small.

Endemic. Chiefly in West Surrey; 14 (Copyhold), 17, 22 (Binfield).

202. **R. Daltrii** Edees & Rilst. 1945, *N. West Nat.* 161 (plate).

Staffs. Stem erect then bending, rooting? *Prickles yellowish, numerous,* broad-based, straight to falcate. Leaves 5-nate, prickles hooked; *leaflets glabrescent on both sides*, unequal to subcompound and partly patent-serrate; *basal leaflets very shortly stalked; terminal leaflet* long-stalked, *entire-based, obovate-elliptic*, sometimes *subcuneate*, rather shortly pointed. *Panicle* sub-racemose or more compound, the *terminal flowers very shortly stalked, peduncles and pedicels long*, semi-erect, *thinly pilose, with fine short prickles.* Flowers *c.* 2·5 cm.; *sepals green with a white margin*, appendiculate, patent; *petals white, glabrous, ovate-oblong (c. 10 × 6 mm.); filaments long, not connivent.* Fruit perfect. Remarkable for the *withered stamens hanging down on the reflexed sepals*, as in R. nessensis. General habit of *R. Gremlii.*

Endemic. N.W. Staffs. (Chortley Moss, Whitmore and Madeley).

203. **R. Gremlii** Focke, 1877, *Syn. Rub. Germ.* 266; *R. Colemannii* ssp. *Gremlii* (Focke) Sudre, 1908–13, 117, t. 116.

Middlesex. Stem low-arching, procumbent, or high climbing, ± hairy, *prickles yellow*, moderate, straight. *Leaves deep green*, (3) 5-nate, pedate, prickles hooked, *basal leaflets short-stalked or long-stalked; leaflets glabrescent on both sides*, or sometimes a little greyish felted beneath, *coarsely* and unequally incise *serrate; terminal leaflet long-stalked, ovate or ovate-oblong, gradually narrowed into a long point*, base rounded or emarginate. Panicle rather narrow and long, often leafy, lax below, with long ascending branches; *rachis* and the felted pedicels *densely patent-pilose*, prickles small, acicular, pricklets few and minute, glands various. Sepals felted, reflexed. *Petals narrow obovate-cuneate, pinkish or white, pubescent.* Filaments slightly or else far exceeding the styles. Carpels glabrous. Fruit perfect. *Leaflets especially on the panicle often ± cuneate.*

England: found only in Copse Wood, Ruislip, Middlesex. Mountain woods of South-central Europe, Vosges, Switzerland, Austria.

2. Leaves grey or white felted beneath. (Cf. *R. Reichenbachii* and *R. longifrons*).

204. **R. taeniarum** Lindeb. 1858, *Novit. Fl. Suec.* **5**, 1; *R. infestus* sensu Focke, Rogers pro parte, Sudre *et al.*; non Weihe, 1824; Sudre, 1908–13, t. 144 (as *R. infestus*).

Stem angled, red, glaucescent, furrowed, moderately hairy; *prickles deep red, broad-based, coalescent, unequal.* Leaves rather small, 5-nate, pedate, prickles hooked; *leaflets bright green and glabrescent above*, felted and *velvety pubescent beneath, crenate*-serrate-dentate; *terminal leaflet long-stalked, roundish obovate* to subrhomboid, subcordate or nearly entire; *basal leaflets subsessile. Panicle* equal *with several simple leaves,* mid branches patent; *rachis crooked,* pilose, *prickles unequal, red-based, decurrent, slanting and hooked.* Flowers showy, *c.* 2 cm.; *sepals erect; petals* remote, *roundish, notched, ciliate, pink* or white; *filaments deep pink, hardly longer than the styles. Carpels bearded. Fruit violet-black,* subglobose, *late,* 'not tasty; sweetish, slimy': Lindeberg.

E., W., S. Chiefly in the midlands and the north; 5, 6, 9, 11, 12, 18, 22, 23, 34–36, 38–40, 42, 43, 49, 55, 57, 59, 60, 62, 64, 65, 67–69, 74, 76, 86, 88, 97. Belgium, Netherlands, W. Germany, Denmark, Sweden.

205. **R. iodnephes** W. Wats. 1952 (for 1951), *Lond. Nat., Suppl.* 99.

Stem deep red to *violet,* glabrescent, glaucescent; *prickles* red-based, *rather short.* Leaves 5-nate, *petiole flat,. prickles small, geniculate;* leaflets acuminate. *Panicle* usually leafy, *subpyramidal, upper branches nearly equal, deeply divided;* rachis pubescent and felted, *prickles short, stout, recurved.* Flowers *c.* 2 cm.; *sepals* greyish green, leafy-tipped, *glandular punctate, aculeate, patent to loosely reflexed; petals* greenish white or pinkish, obovate, *glabrous* on the margin; filaments connivent. *Carpels glabrous.* Receptacle pilose. *Fruit* moderate, ovoid, *early,* sour.

Endemic. E., S. Very local; 17 (Barnes Common, Sheen Common, Palewell Common), 87 (Allan Water to Doune, abundant, *W. H. Mills,* 1953).

206. **R. Mercieri** Genev. 1869, *Mon. Rub.* 271; Sudre, 1908–13, 66, t. 73.

Robust. Axes purple. West Kent. Stem arched, angled, *furrowed, glabrescent;* prickles strong, patent, declining and ± recurved. *Leaflets glabrous above,* pilose as well as white felted beneath, *with deep unequal or subcompound angular, sharp teeth; terminal leaflet* moderately long-stalked, broad *ovate, gradually acuminate to rhomboid-obovate.* Panicle lax, subequal, leafy above, upper pedicels patent, ascending; *rachis* hairy *with sharply declining and falcate prickles. Flowers large,* 2·5-3 cm.; sepals felted, reflexed; *petals broad ovate or roundish, pink in bud; filaments* white, *a little longer than the styles.* Carpels slightly pilose. Fruit roundish, good.

England: rare; 16 (French Street, Hosey Common, Squerryes Park, Westerham). E. and S. France, S.W. Switzerland.

Ser. **iv.** RADULAE Focke

(Type *R. Radula* Weihe)

1. Sepals reflexed (to loosely patent in *R. echinatoides* and *R. aspericaulis*).
† Glands, acicles and pricklets on the stem all short and subequal.

207. **R. radula** Weihe ap. Boenn. 1824, *Prod. Fl. Monast.* 152; Sudre, 1908–13, 127, t. 124; *R. decipiens* P. J. Muell. 1859, *Pollichia,* 158.

4x, 5x. *Robust. Stem* arching, angled, *pilose; prickles subulate, mostly long. Leaves deep green, rather large,* 5-nate, digitate, *prickles straight and falcate;*

leaflets flat, glabrescent above, felted and pubescent, *with hard salient veins* beneath, unequally and *finely serrate-dentate; terminal leaflet base rounded or truncate, entire, ovate, gradually acuminate*, often long-stalked. *Panicle pyramidal, branches issuing at a low angle*, cymose; rachis hairy, *prickles subulate, declining, some very long. Flowers rather small, c.* 2 cm.; *petals pink, entire* at the apex; filaments white, rarely pinkish, much longer than the greenish or rosy-based styles. *Carpels usually glabrous.* Fruit roundish ovoid.

E., W., S., (I.). Frequent north of the Thames, rare south of it; 3, 5, 7, 12–14, 16, 17, 19–24, 27, 29, 30–40, 42, 49, 52, 55–58, 61, 62, 65, 67–69, 74, 80, 84–88, 90, 95, 99, 103–107, 110. Not Jersey. (? Ireland). N. and E. France, Belgium, Netherlands, Denmark, Sweden, N.W. Germany, Switzerland, Bavaria, Austria, Silesia, Hungary.

var. **microphyllus** Lindeb. *Herb. Rub. Scand.* no. 22.

Dwarf, in all parts much smaller, the terminal leaflet long, gradually acuminate; sepals patent. Sometimes occurs with *R. Radula.* It has been confused with *R. Powellii.* 1, 17, 20, 22–24, 36, 39, 55, 62.

R. radula × ulmifolius; *R. cotteswoldensis* Barton & Riddels, ap. Riddels. 1948, *Fl. Gloucester,* 160.

Frequent in 33 and 34.

(**R. pustulatus** P. J. Muell. ap. Sudre, 1903, *Bot. Eur.* 11; *R. Radula* microg. *pustulatus* (P. J. Muell.) Sudre, 1908–13, 128, t. 125.

Extinct. In most respects very like *R. radula* but with white petals. Panicle broader with longer, perpendicularly spreading branches. Terminal leaflet cordate with sharper more compound teeth, with long mucros. Carpels densely pubescent. Farm Bog, Wimbledon Common, Surrey. S. France.)

208. **R. sectiramus** W. Wats. 1933 (for 1932), *Lond. Nat.* 60. **Fig. 30.**

Elegant. Stem angled, nearly glabrous, becoming *dark brown, pruinose; prickles subulate, often in twos.* Leaves 5-nate, digitate, *petiole flat above; leaflets* pubescent and felted beneath, *serrate, partly patent-denticulate; terminal leaflet elliptic-obovate, cuspidate, base nearly entire. Panicle* equal, truncate, long, the *middle branches fasciculate,* and with the rachis felted and thinly villous, *armed with numerous, slender, straight prickles and very short glands. Flowers up to 3 cm.;* sepals loosely reflexed; *petals 5–8, not contiguous,* obovate, tapered below, *pilose, pinkish;* filaments white, much longer than the pallid styles. *Carpels pilose.* Receptacle pilose. Fruit moderate, rather oblong, dryish.

Endemic. England: 5, 14–18, 20, 21, 24, 34, 35, 39, 63, 65.

209. **R. macrostachys** P. J. Muell. 1858, *Flora,* **41,** 150; Sudre, 1908–13, 105, t. 103; non sensu Focke, nec sensu Rogers.

Berks., Oxon., Beds. Stem fuscous, blunt, *hairy,* with rather few, fine, short glands and minute pricklets; *prickles red-brown, some long subulate. Leaflets 3–5,* glabrous above, *pilose to velvety pubescent and white felted beneath, undulate,* rather coarsely and unequally serrate-dentate, *a little jagged; terminal leaflet* broad ovate or *obovate-oblong, acuminate. Panicle large, broad pyramidal, middle branches spreading at a low angle, long-peduncled, 5–7-flowered, cymose; rachis intricately villous, prickles long subulate, mixed with shorter prickles and long acicles;* pedicels felted. Flowers *c.* 2·5 cm.; sepals gland-dotted, hardly armed; *petals white,* ovate-oblong. Carpels pubescent. Receptacle pubescent. Fruit subglobose to slightly oblong.

England: local; 16 (Ryarsh Wood), 22 (south of Boars Hill), 23 (Crowell Hill), 30 (frequent). France (*Valois*), Germany (*Alsace*, Wissembourg), Switzerland, Austria.

210. **R. adenanthus** Boul. & Gill. 1881, *Assoc. Rub.* no. 429; non sensu Rogers, 1900; *R. decipiens* var. *juratensis* Schmid. 1888, *Cat.* 131.

W. Kent. Stem arched, angled, *a good deal hairy,* with rather numerous, buried, short glands, acicles and a few pricklets; *prickles moderately* declining. *Leaflets* 5, *greenish grey,* pubescent to felted *beneath,* rather unequally and *shallowly crenate-serrate, with fine distinct mucros, partly patent; terminal leaflet* rather *narrow elliptic-obovate, cuspidate. Panicle broad, very large,* corymbose at the top, with some simple grey-felted leaves, *numerous middle branches which are long-peduncled, patent and cymose, obtended by ovate or lanceolate-acuminate leaves; rachis* hirsute, shortly glandular, *prickles nearly acicular,* slanting. *Flowers c.* 2 *cm.;* sepals 5–6, loosely reflexed or partly patent; *petals* broad obovate, distinctly clawed, *pale pink; filaments pinkish after drying,* longer than the greenish styles. Carpels pilose. Receptacle hirsute. Fruit small.

England: 16 (Crockham Hill Common, roadside at 600′). North, central and south France, Switzerland, Hungary.

211. **R. Genevieri** Boreau, 1857, *Fl. Cent. France,* ed. 3, **2**, 193; Sudre, 1908–13, 131, t. 127. **Fig. 31.**

Slender and elegant. Stem blunt with flat, striate sides, closely pubescent; prickles long, slender, firm. Leaves rather small, pedate; *leaflets* (3, 4) 5, *often subcuneate,* glabrescent above, *silkily felted beneath, subcompound-serrate* and *partly patent-dentate; terminal leaflet obovate-cuspidate. Panicle very long, leafy, narrowed,* lax, mid branches 3-flowered to cymose; rachis flexuose, felted and shortly hairy, pedicels long and prickly. Sepals long-tipped. *Petals obovate, tapering below,* or spatulate, *notched, pale pink.* Filaments white, or pink-based, far longer than the *reddish styles. Carpels densely pilose.* Receptacle pilose. Fruit roundish, dryish. *Late flowering* with us. 'Stirps insignis': Chaboisseau.

E., I. Rare; 3 (Honicknowle), 16 (Crockham Hill Common West, in two stations), 35 (Cross Keys and St Mellons), 36 (Banks and woods at Clifford etc.). H. 38. N.W. France, Portugal, Bavaria.

212. **R. crispus** Lef. & Muell. 1859, *Pollichia*, 147.

Caernarvon. Rufescent. Stem angled, slightly furrowed, *glabrescent, with some stout pricklets; prickles* unequally distributed, *strong-based, some curved. Leaves* 5-*nate,* pedate, *prickles hooked; leaflets glabrous above,* pubescent and felted beneath, *coarsely subcompound crenate-serrate, jagged, crisped; terminal leaflet entire-based, elliptic obovate-(subcuneate), acuminate. Panicle broad,* pyramidal, *truncate,* lax, upper branches 1–4-flowered, patent or ascending; *rachis* thinly pilose, felted, prickles short, patent to recurved. *Flowers showy, c.* 2 cm.; *petals broad obovate, deep rose; filaments deep rose, far longer than the reddish-based styles.* Carpels pilose. Fruit globose.

Wales: rare? 17, 48, 49 (Bangor, Bethesda by the R. Ogwen). France (*Oise*).

R. crispus × **vestitus**; *R. leucostachys* var. *gymnostachys* sensu Rogers, 1900, 51; *R. gymnostachys* sensu Griffith, 1895, *Fl. Ang. & Carn.* 46; *R. macrothyrsus* × *rusticanus* Focke.

49 (Bangor and neighbourhood. Found here by *W. Wilson* in 1928).

213. **R. uncinatiformis** Sudre, 1906, *Diagn.* 33; *R. radula* ssp. *ericetorum* microg. *uncinatiformis* (Sudre) Sudre, 1908–13, 129, t. 125.

Norfolk. Stem furrowed, pilose, with pricklets and *sparse glands;* prickles firm. Leaves 5-nate, *petiole and petiolules villous,* prickles hooked; *leaflets villous and felted beneath, serrate* or *serrulate, crisped;* terminal leaflet broad obovate-cuspidate, base subentire. *Panicle narrow, equal, lateral leaflets subsessile; rachis villous and felted,* with hidden glands. Flowers *c.* 2·5 cm.; petals broad obovate, notched, pinkish; filaments long. Carpels and receptacle glabrous.

England: 27 (Ringlands Hill, west of Norwich). France (rare).

(*R. uncinatus* P. J. Muell. 1858, *Flora*, **41**, 154.

Not British. The various plants formerly so identified are assigned as follows: 13, Midhurst, *R. squalidus.* 14, Copyhold, *R. ahenifolius.* 17, Tooting, *R. retrodentatus*; Woking, *R. euryanthemus.* 24, Woburn, *R. thyrsiflorus*; Mop End, *R. teretiusculus.* 34, Lea Bailey and 35, Troy, both *R. angusticuspis.*)

†† Glands, acicles and pricklets more unequal, especially on leafstalks and panicle.

(*a*) Axes becoming nearly black in the sun.

214. **R. discerptus** P. J. Muell. 1859, *Pollichia*, 146; 1929 (for 1928), *Rep. Bot. Exch. Cl.* **8**, 86 (plate); *R. Genevieri* ssp. *discerptus* (P. J. Muell.) Sudre, 1908–13, 132, t. 128; *R. echinatus* auct. plur.; *R. rudis* auct. plur.

4x. *Common. Robust. Stem rainbow-arched, furrowed, hairy, prickles numerous, patent, some very strong and long.* Leaves digitate; *leaflets with thick, yellowish hair and grey felt beneath, sharply, coarsely* ± *compound-serrate, jagged; terminal leaflet base entire, ovate to obovate-elliptic, acuminate. Panicle* broad, dense, *equal, long and leafy; rachis villous,* prickles straight or ± curved. Flowers *c.* 2·5 cm.; sepals acuminate-appendiculate; petals 5–6, pink, margin slightly pilose, broad ovate or elliptic, apex entire; filaments white, longer than the yellowish styles. Carpels glabrous. Receptacle pilose. Fruit oblong. Flowering July. On clay as well as sand and gravel; Rogers says on chalk also.

var. **microphyllus** (Bloxam) W. Wats. 1956, *Watsonia,* 3, 288; *R. rudis* var. *microphyllus* Bloxam in Kirby, 1850, *Fl. Leicester,* 41.

A dainty dwarf form.

E., W., S., I. Frequent; 3–14, 16–27, 29, 30, 32–40, 42, 43, 49, 50, 55–59, 62, 86, 90, 99. H. 21, 37, 39. France, Portugal, Switzerland, Bavaria.

215. **R. echinatoides** (Rogers) Druce, 1927, *Fl. Oxford,* 142; *R. Radula* ssp. *aspericaulis* microg. *echinatoides* (Rogers) Sudre, 1908–13, 129, t. 126; *R. Radula* var. *echinatoides* Rogers, *Rep. Bot. Exch. Cl.* 8, 860 (plate).

4x. *Slender. Stem glabrous,* prickles numerous. *Leaflets 3, 5, glabrous above,* felted beneath, *subcompound-serrate, undulate, jagged; terminal leaflet obovate, cuspidate, base entire. Panicle strict,* long, interrupted below, the *lower branches nearly erect; rachis sharply angled, glabrous and acicular below,* prickles unequal, numerous, strong, long-based, declining, falcate and hooked. Flowers moderate; sepals patent or loosely clasping; *petals remote, obovate, notched, tapered below,* pinkish; filaments white, longer than the greenish styles. Carpels pilose. Fruit large, oblong.

Almost endemic. E., W., S., I. Especially northern; 6, 7, 9, 13–24, 27, 30, 35–40, 49, 51–53, 57, 62–65, 67, 68, 73, 84, 86–88, 92, 93. H. 1 or 2, 21, 23, 33, 37–39. N. France (one locality).

(*b*) Axes not becoming blackish.

216. **R. aspericaulis** Lef. & Muell. 1859, *Pollichia,* 141; *R. Radula* ssp. *aspericaulis* (Lef. & Muell.) Sudre, 1908–13, 129, t. 126.

*4x. *Robust. Stem thick,* angled, *sides concave, subglabrous, glaucescent,* with thinly scattered short glands and acicles; prickles unequal, some very long subulate. Leaves large, 5-nate; *leaflets* glabrescent above, felted and pubescent beneath, *serrate-dentate; terminal leaflet long-stalked, roundish obovate, shortly cuspidate,* sometimes a little lobate. Panicle large, interrupted, oblong-pyramidal; rachis hairy, with submerged glands, *prickles subulate, unequal, some very long;* leaflets serrulate. Sepals reflexed to patent; petals oblong, obovate, often notched, crumpled, pink to white; filaments slightly longer than the yellowish, reddish-based styles. Carpels pilose. Fruit subglobose.

E., W. Rather frequent; 9, 13, 15–18, 20, 21, 39, 48. France.

217. **R. malacotrichus** (Sudre) W. Wats. 1956, *Watsonia,* 3, 288; *R. apiculatus* microg. *malacotrichus* Sudre, 1906, *Diagn.* 136; 1908–13, 134, t. 130.

W. Kent. A weak, low, woodland bramble finely armed and very softly villous throughout. Stem blunt, hairy, with glands and acicles mostly equalling, a few exceeding, the hair; prickles straight or falcate, yellowish. Leaves 3 (5)-nate, pedate, short-stalked, prickles very slender, spreading-falcate; *leaflets narrow, ± cuneate*, greyish, softly pilose beneath, with spreading yellow prickles on the midrib, rather broadly serrate-dentate; *terminal leaflet elliptic-rhomboid, narrowed to a rounded base*, falcate-cuspidate. *Panicle short, racemose*, pedicels long; upper leaves grey felted as well as pilose beneath. *Flowers, c. 1·5 cm.*; sepals and petals 5–6; sepals loosely patent; *petals* obovate, apex rounded, subglabrous, *faintly lilac*; filaments white, slightly longer than the greenish styles. Carpels pilose. I see no likeness to *R. apiculatus*.

England: 16; about an acre of it in deep shade in Spring Park, West Wickham. Discovered by *Questier* at Bourneville, *Valois*, France, about 1860. Not known elsewhere.

2. Sepals patent or erect.

† Leaves in the open, grey or white felted beneath; glands very short.

218. **R. rudis** Weihe ap. Bluff & Fingerh. 1825, *Comp. Fl. Germ.* **1**, 687; Sudre 1908–13, 166, t. 160.

4x. *Rather common. A rather small bush. Stem rainbow-arched, deep purple, furrowed, glabrous*; prickles moderate, slanting or falcate. Leaves 3–5-nate, pedate; *leaflets subcuneate based*, entire, glabrescent above, serrate-dentate, the principal teeth coarse, angular, patent; terminal leaflet subrhomboid-ovate or elliptic-acuminate. *Panicle short, subcorymbose with widely spreading, cymose, intricate branches*, felted throughout, *the very short glands longer than the felt on the pedicels. Flowers c. 1·5–2 cm.; sepals triangular-attenuate; petals 5–8, pink, nearly or quite glabrous*; filaments long. Carpels glabrous or nearly so. *Receptacle glabrous.* Fruit small. A dwarf form at Crowell Hill, Oxon.

England: frequent, especially on the N. Downs; 6, 12–17, 20–24, 30, 33–36, 40, 55, 57. Guernsey. France, Belgium, W. and Central Germany, Switzerland.

219. **R. radulicaulis** Sudre, 1904, *Observ. Set. Brit. Rub.* 26; *R. Timbal-Lagravei* ssp. *occitanicus* microg. *radulicaulis* (Sudre) Sudre, 1908–13, 141, t. 137; *R. ericetorum* ssp. *sertiflorus* var. *scoticus* Rogers & Ley, 1906, *Journ. Bot.* **44**, 60; *R. Radula* ssp. *sertiflorus* sensu Rogers, 1900, 64; *R. Radula* sensu Syme, 1864, *Eng. Bot.* ed. 3, **3**, 184, t. 452.

Axes slender. Leaflets narrow. Stem purple, blunt, furrowed, hairy; *prickles rather short, sharply declining and recurved.* Leaves yellowish green, 3–5-nate, pedate; *leaflets thick, long and narrow*, base ± cuneate, entire, felted and pubescent beneath, serrate-dentate; *terminal leaflet oblong or obovate-elliptic, short-pointed. Panicle long, narrow, strict*, interrupted, branches sharply ascending, the upper ones 1–2-flowered; rachis blunt, felted and pubescent above, villous below with *recurved prickles.* Flowers 2–3 cm., incurved; *petals* broad elliptic, fimbriate, *deep pink; filaments pink* or white, long, anthers sometimes pilose; *styles pink-based.* Carpels strongly pilose or glabrous. Fruit· roundish.

E., S., I. Well distributed but rather rare; 3, 16, 17, 20, 21, 35–37, 40, 55, 62, 76, 87, 98, 99. H. 20. N.W. and S. France.

220. **R. prionodontus** Lef. & Muell. 1859, *Pollichia*, 117; *R. fuscus* microg. *prionodontus* (Lef. & Muell.) Sudre, 1908–13, 143, t. 138.

Stem pruinose becoming ochreous, furrowed, thinly hairy, *with scanty glands and acicles*; prickles rather short and straight. Leaves 5-nate, decidedly pedate; *leaflets serrate-dentate; terminal leaflet long-stalked, roundish,* subobovate *or reniform, short-pointed; basal leaflets short-stalked. Panicle usually short, lax, broad, branches spreading, few-flowered*; rachis felted to shortly hirsute, pedicels long felted with numerous acicular prickles; bracteoles subfoliaceous. Petals roundish, rosy, or more often pinkish. Filaments white slightly longer than the *red styles.* Fruit ovoid-oblong.

England: only in the Thames basin; 7, 15, 16 (frequent), 20, 24. France, *Aisne.*

†† **Leaves all green beneath, or the upper and branch leaves greyish felted; glands not always very short.**

(*a*) Petals faint lilac or white (cf. *R. micans*).

221. **R. granulatus** Lef. & Muell. 1859, *Pollichia*, 154; Sudre, 1908–13, 139, t. 134; *R. Radula* var. *Bloxamianus* Coleman ex Purchas, 1887, *Journ. Bot.* **25**, 102.

Stem reddish brown, angled, slightly furrowed, striate, glaucescent, rather hairy, *glands and acicles brown, crowded*; prickles straight, broad-based. *Leaves* 3, 4, 5-nate, *deep green, glabrous above, base* most often entire, *slightly glandfringed, margin unequally serrate-dentate; terminal leaflet long-stalked, roundishobovate or -elliptic, cuspidate-acuminate. Panicle* leafy, the *uppermost leaves glandular above,* dense, nearly equal, prickles numerous, moderate, *glands crowded, mostly equalling the hair; terminal leaflets obovate-cuneate. Flower buds shortly green-cristate*; petals 5–6, broad obovate; filaments white, longer than the greenish styles. Carpels glabrous. *Fruit* ovoid; *at first pale green.*

E., W., I. Locally frequent; 4, 13, 15–17, 20, 21, 24, 27, 32, 35, 38, 39, 45, 55, 57, 58, 60, 63. H. 38. France, Belgium, Germany, Switzerland, Bavaria, Austria.

222. **R. regillus** A. Ley, 1896, *Journ. Bot.* **34**, 217.

Yellowish green. Stem angled, brown, *glaucous, hairy; prickles* unequal, *subulate, declining.* Leaves 3, 4, 5-nate, pedate; *leaflets oblong, thinly pilose beneath,* unequally patent-serrate; *terminal leaflet obovate-oblong, truncate, cuspidate.* Panicle pyramidal, lax, lower branches 4–5-flowered with *long pedicels,* subracemose above, the branches sharply ascending; rachis villous, prickles weak and short. Flowers *c.* 2·5 cm.; sepals dark green felted, *with long tips; petals* expanded, *very narrow obovate; filaments* white, *about equalling the yellowish styles. Anthers often pilose.* Carpels glabrous. Name (*regillus,* royal) from Queen's Wood, Ley's original station for the bramble.

Endemic. E., I. Rare; 3 (Cornwood) (not 23), 34, 36 (Queen's Wood). H. 1, 2.

(*b*) Petals bright pink.

223. **R. micans** Godr. ap. Gren. & Godr. 1848, *Fl. France,* **1**, 546; non sensu Rogers, 1900, 48; *R. Schummelii* Boul. ap. R. & C. *Fl. France,* **6**, 465; *R. macrostachys* microg. *Wolley-Dodii* Sudre, 1908–13, 106, t. 131; *R. criniger* Rogers pro parte (the Edge Park bramble), 1900, 52. **Fig. 32.**

Rufescent. Stem furrowed, glabrescent, glaucescent, *with scattered short-stalked glands and acicles and numerous unequal pricklets; prickles yellow to cerise and red-brown, unequal, subulate.* Leaves short-stalked, pedate; leaflets 3–5, softly pilose to felted beneath, rather coarsely and unequally serrate, jagged; *terminal leaflet roundish, ovate to elliptic-obovate, acuminate. Panicle leafy, floriferous, large, pyramidal,* with cymose middle branches and long, ascending, numerous axillary branches, or merely subracemose; *peduncles and pedicels prickly.* Flowers showy, *c.* 2–2·5 cm.; *calyx discoid-based,* aculeate; *petals* 5, 6, 7, *glabrous,* ovate or elliptic, sometimes white; filaments pink-based, longer than the greenish styles. Carpels glabrous. *Fruit broad ovoid, irregular,* formed of rather large drupels.

E., W. Locally frequent; 3, 8, 13–18, 20, 39, 41, 58. France and Belgium, rare.

224. **R. pulcher,** Muell. & Lef. 1859, *Pollichia,* 148; *R. micans* ssp. *pulcher* (Muell. & Lef.) Sudre, 1908–13, 136, t. 133.

Stem blunt, hairy; prickles long-based, *strongly declining and falcate.* Leaves rather large, prickles small, hooked; leaflets 3–5, pubescent to felted beneath, sharply, unequally, ± patent-serrate; *terminal leaflet broad elliptic, acuminate. Panicle nodding, mostly subracemose, the branches short and equal,* or with 1–3 racemose axillary branches; rachis villous, *prickles short, recurved.* Flowers 2–2·5 cm.; sepals leafy-tipped, felted, glandular and aculeate; *petals roundish or broad ovate; filaments bright pink,* longer than the styles. Carpels glabrous. Fruit roundish-ovoid, fertile. *Late to flower.* Distinguish *R. grypoacanthus.*

England: very rare; 16 (Bostall Heath; and by the footpath near Sussex Shaw, Tunbridge Wells). France (rare; *Valois* and *Ardennes*).

Appendiculati

Ser. **v. APICULATI** Focke

(Type *R. apiculatus* Weihe)

Subser. †**Foliosi** W. Wats.

(Type *R. foliosus* Weihe)

1. Leaves grey or white felted beneath.

225. **R. foliosus** Weihe ap. Bluff & Fingerh. 1825, *Comp. Fl. Germ.* **1**, 682; Sudre, 1908–13, 145, t. 140; C. G. Trower, 1929 (for 1928), *Rep. Bot. Exch. Cl.* **8**, 862, no. 120 (plate); *R. flexuosus* Lef. & Muell. 1859, *Pollichia*, 240; *R. foliosus* microg. *flexuosus* (P. J. Muell.) Sudre, 1908–13, 146, t. 141, figs. 7–10; *R. saltuum* Focke ap. Gremli, 1870, *Beitraege Fl. Schweiz*, 30; *R. insericatus* ssp. *hyposericeus* (Sudre) Sudre, 1908–13, 150, t. 143, figs. 18–20; *R. fuscus* var. *macrostachys* sensu Rogers pro parte, 1900, 74.

Stem blunt, deep purple, ± hairy, glaucescent, with crowded, deep purple glands, acicles and pricklets; some prickles very long, declining and falcate. Leaflets 3 (4, 5), *thick and tough, deep green and shining above*, felted and hairy beneath, *serrate-dentate; terminal leaflet ovate to elliptic or elliptic-rhomboid, gradually acuminate*, base cordate to entire. Panicle either pyramidal with straight rachis and patent branches, or narrow, the rachis ± zigzag, leafy, the branches simple, deeply divided, or bunched in the axils; pedicels felted and with crowded, short, deep purple glands. *Flowers 1·5–2 cm.; sepals long-pointed, some reflexed, some patent and some erect on the same fruit; petals 5–7, pink (rarely white), fimbriate, narrow rhomboid; filaments* usually white, *equalling the red or pinkish-based* (greenish when the petals are white) *styles. Carpels pubescent. Fruit perfect, oblong, sweet.* A giant form occurs in a wood at Paddlesworth, E. Kent. The white-flowered form (Weihe's type and Lef. & Muell.'s type) on Chislehurst Common, by the Overflow Pond.

E., S., (I.). Common in most of England; 3, 4, 6–24, 26, 29, 33–40, 55, 57, 62, 93, 94. Reported from H. 38, 40. N.W. France, Belgium, Holland, Denmark, Germany, eastwards to *Silesia*, Switzerland, Austria.

226. **R. rubristylus** W. Wats. 1937 (for 1936), *Rep. Bot. Exch. Cl.* **11**, 220; *R. oigocladus* var. *Newbouldii* sensu Rogers, 1900, 66; non Bab. 1886.

Young shoots bronze. Stem red-brown to deep purple, glaucescent, *furrowed;* prickles subequal, *small-based, slender, straight. Leaflets* 3, 5, glabrous or glabrescent above, pilose and felted beneath, *unequally and sharply incise serrate, undulate, jagged; terminal leaflet long-stalked, subcordate, broad pentagonal-obovate. Panicle* equal and *often racemose, not leafy above, or with patent, deeply divided, 3–7-flowered middle branches; rachis prickles subulate.* Flowers moderate;

sepals loosely reflexed; petals white, seldom pink, oblong; *filaments long; styles red.* Carpels pilose. Fruit large, rather oblong.

(*R. Newbouldii* Bab. cannot be maintained. It is founded on a panicle of *R. discerptus* and a stem-piece of *R. granulatus.*)

Endemic. E., W., I. Well distributed in the west, but local there and rare elsewhere; 18, 22, 23, 34–37, 39, 40–42, 49, 58. H. 38.

227. **R. pseudadenanthus** W. Wats. 1956, *Watsonia,* **3,** 288; *R. adenanthus* sensu Rogers, 1900, 53; non Boul. & Gill. 1881.

4x. *Young shoots bronze.* Stem *furrowed,* pilose, becoming fuscous, glaucescent; some *prickles long subulate.* Leaflets 3–5, becoming convex, glabrescent above, *compound, partly patent serrate, undulate; terminal leaflet shortstalked, rhomboid-ovate-elliptic, acuminate, or obovate, subcuneate, base entire;* stalks of basal leaflets 1–3 mm. Panicle long pyramidal, mid branches cymose-partite, *pedicels rather long; rachis villous, with long acicles in the lower part,* otherwise armed as the stem. Flowers moderate; sepals loosely reflexed or patent; petals narrow elliptic, pink, pinkish or white; styles sometimes reddish-based. *Carpels glabrous.* Fruit slightly oblong.

E., S., I. Mainly western, rare; 1, 2, 39, 58, 66, 67, 71, 110. H. 2, 33, 35. Jersey.

228. **R. subtercanens** W. Wats. 1956, *Watsonia,* **3,** 288; *R. thyrsiflorus* P. J. Muell. 1858, *Flora,* **41,** 165; non Weihe, 1825; *R. obscurus* P. J. Muell. 1859, *Flora,* **42,** 72; non Kalt. 1845; *R. fuscus* var. *canescens* Boul. in Rouy. 1900, *Fl. France,* **6,** 95; *R. fuscus* var. *macrostachys* sensu Rogers pro parte, 1900, 74.

Robust. Axes and arms deep purple. Stem *villous, blunt, furrowed, with fine short glands and acicles;* prickles deep red, unequal, slender. *Leaflets* 3, 5, deep green, pubescent to felted beneath, *serrate-dentate; terminal leaflet* subcordate to entire, *roundish ovate, long-acuminate.* Panicle broad, pyramidal below, the lower axillary branches obliquely ascending, *numerous flowered, the upper branches equal, dense, patent, 1–3-flowered,* with 1–3 (4) simple leaves, the lowest one usually bi- or trilobed; *rachis and pedicels villous,* armed as the stem but with some *long, yellowish acicles* in addition. *Flowers 1·5–2·2 cm.;* sepals green, villous, felted, acicular and gland dotted, subpatent; *petals 5–7, narrow obovate, pilose, faintly pink;* filaments white, equalling the yellowish or reddish-based styles. Carpels pubescent.

This is the bramble that was pointed out by Focke in 1894 at Belmont, Herefordshire, as *R. macrostachys* P. J. Muell. and according to Rogers was subsequently confirmed without qualification by Gelert. The Yatton Wood specimens of 'macrostachys' issued in Set no. 126 are *R. foliosus;* they were not determined by Focke. I have noticed that *R. foliosus* grows mixed with *R. subtercanens* at Belmont.

229. **R. sagittarius** Riddels. 1930, *Journ. Bot.* **68**, 24; *R. mutabilis* ssp. *nemorosus* sensu Rogers, 1900, 72.

South Devon. A short bush, known by its strongly pruinose, furrowed and ± hairy stem, bearing long straight prickles; the shortly hairy terminal leaflet cordate, ovate, coarsely toothed and running into a very long slender point. The panicle not large, few-flowered, on short, very erect branches, the rachis pruinose below; petals narrow, white; stamens long. Carpels densely pubescent. Fruit large, oblong, of good flavour.

Endemic. S. Devon; 3 (formerly frequent).

230. **R. teretiusculus** Kalt. 1845, *Fl. Aachen Beck*, 282; *R. Schmidelyanus* ssp. *teretiusculus* (Kalt.) Sudre, 1908–13, 121, t. 119.

Stem subterete below, red-brown, hairy; prickles moderate, declining or falcate from a broad base. *Leaflets* 3, 4, 5, *rather small and neat*, pubescent and felted beneath, *finely serrate; terminal leaflet elliptic-obovate*, cuspidate, *with a rounded subcuneate base. Panicle long, lax*, pyramidal, furnished *with a few simple leaves towards the top and on the lower branches, pedicels long, very prickly; rachis villous*, prickles numerous, very slender, some hooked. *Flowers shorter than 2 cm.*; sepals prickly, long-tipped; petals obovate, narrowed below, white or lilac; *filaments equalling the greenish styles*. Carpels strongly pilose. Fruit small, subglobose. *All prickles yellow, rufescent.*

England: very local; 21 (near Copse Wood, Ruislip), 23 (at Mop End and thence to Beamond End, near Penn, in some plenty), 39 (wood between Swindon and Highgate, *E. S. Edees*). Recognized by me after seeing it at Aachen. Belgium (*Spa*), Netherlands, W. Germany, Bavaria.

2. Leaves green beneath or, especially the upper and branch leaves, slightly greyish felted beneath.

† Filaments slightly shorter than the styles. Stem pilose. Sepals becoming exactly patent soon after the fall of the petals.

231. **R. Bloxamii** Ed. Lees ap. Steele, 1847, *Hand. Field Bot.* 55; *R. pallidus* ssp. *Bloxamii* (Ed. Lees) Sudre, 1908–13, 154, t. 150; *R. multifidus* Boul. & Malbr. 1873, *Assoc. Rub.*

4x. *Robust, axes hirsute*. Stem angled, glaucescent; prickles yellowish, rather short, unequal, patent to sharply declining. *Leaflets* 5 (–7), *coarsely and very compound serrate; terminal leaflet obovate-oblong*, acuminate or *truncate-cuspidate*; base entire. *Panicle large, lax, with a leafy corymbose apex and distant, long-stalked, numerous flowered branches*; rachis zigzag, densely villous and acicular, prickles weak, slanting. Flowers 1·5–2·5 cm.; sepals greenish grey, leafy-tipped; *petals* 5–8, oblong, *white or pinkish; styles greenish*. Carpels pilose. I have sometimes seen the sepals clasping about mid August.

E., W. Well distributed and rather frequent in wooded districts; 2–4, 8–12, 14, 16, 17, 21–23, 33, 34, 36–40, 44, 49, 55, 57, 63, 64, 66, 69. Guernsey. N. France, Belgium, Denmark, N. Germany, Switzerland, Bavaria.

232. **R. largificus** W. Wats. 1928 (for 1927), *Rep. Bot. Exch. Cl.* **8**, 507.

A low, *dense-leaved bramble with* a blunt *furrowed stem*, the *rugose leaflets obovate-oblong, short-stalked and short pointed. Flowers densely collected together* at the head of the panicle and in the rather remote axils, decidedly *small* (*c. 1·5 cm.*), with *5–6, small white petals*, the erect stamens hardly drawing level with the yellowish, reddish-based styles. *Fruits standing very close together, large,* sweet and abundant.

Endemic. Chiefly in the south-eastern counties; 14, 16 and 17 (abundant), 23 (Freeland).

†† Stem glabrous, subglabrous or glabrescent.

233. **R. cavatifolius** P. J. Muell. ap. Boul. 1867, *Ronces Vosg.* 67; *R. foliosus* ssp. *cavatifolius* (P. J. Muell.) Sudre, 1908–13, 147, t. 141, f. 25–27.

Western. Stem furrowed, pruinose, with more prickles and acicles than glands; prickles moderate, declining. *Leaves large, yellowish green,* 5-nate; *leaflets all long-stalked,* openly simply to subcompound serrate; *terminal leaflet very broadly cordate, ovate, narrowed gradually into a long point. Panicle broad,* long, nearly equal, dense above with narrow simple leaves and lax below, *all branches patent,* 1–4-flowered, pedicels rather long; *rachis furrowed,* villous, with numerous buried glands and acicles, *prickles acicular. Flowers* *2·5–3 cm.*; sepals green with a white border, reflexed, *petals narrow obovate,* notched, white; *filaments* white, light pink when dried, *equalling* the greenish styles. *Carpels glabrous.*

England: very local; 2 (Longcoombe), 3 (near Lyneham), 34–36. France, Germany (*Schleswig*).

234. **R. corymbosus** P. J. Muell. 1858, *Flora,* **41**, 151; *R. insericatus* Gremli, 1871, *Oesterr. Bot. Zeitschr.* 34; non P. J. Muell. 1858; *R. foliosus* microg. *corymbosus* (P. J. Muell.) Sudre, 1908–13, 145, t. 141, figs. 1–3.

E. Sussex and W. Kent. Vigorous. Stem blunt, red-brown, *glands and pricklets usually few*; prickles strong-based, usually short, slanting and falcate. Leaves often large; *leaflets* 3, 5, plicate, glabrous above, softly hairy to glabrescent, or pubescent and felted beneath, *coarsely, partly patent serrate*; terminal leaflet subcordate, roundish cuspidate to ovate-elliptic acuminate, base contracted, obtuse. *Panicle floriferous, dense* and equal above, interrupted below, *the upper and middle branches divided to the base*; terminal leaflets roundish-oblong or elliptic-rhomboid; rachis felted and pilose, prickles few, short, recurved, *pedicels with sparse subsessile glands*, upper bracts green, long linear. Sepals green and reddish felted, aculeate and inconspicuously glandular, reflexed to patent and dubiously erect; *petals* pink, pinkish or white, *narrow oblong-obovate, apex rounded; filaments* white, *equalling the* sometimes reddish *styles. Carpels densely pilose.* Fruit moderate, good. *Flowering early July.*

England: local; 14 (Saxonbury Hill), 16 (Tunbridge Wells and Westerham). France, Belgium, Germany, Switzerland, Austria.

235. **R. exsolutus** Lef. & Muell. 1859, *Pollichia*, 241.

Stem blunt, with scanty glands and pricklets; prickles few, moderate. *Leaflets* 3 (4, 5), pale beneath; *finely serrate-dentate; terminal leaflet* roundish ovate to *obovate or broadly elliptic, cuspidate, base nearly entire. Panicle broad, lax, pyramidal, truncate,* the lower branches long, rather numerous flowered, *middle and upper branches denser, spreading,* 3–4-flowered, the *pedicels long, slender and ± spreading,* felted and pubescent with deep purple glands and weak prickles. Flowers 1·5–2 cm.; sepals loosely reflexed; *petals rather narrow elliptic-obovate, remote, pink;* filaments white, slightly longer than the yellowish styles. Fruit nearly globose. Recalls *R. podophyllus.*

England: rare; 16 (frequent in and off Old Park Lane, Bostall Heath; Gravelpit Lane, Shooters Hill), 20 (near inn, Woodcock Hill, south of Rickmansworth). France (*Valois*), Germany (rare).

✝✝✝ Stem, and the whole plant, densely hairy; terminal leaflet round or roundish ovate.

236. **R. fuscus** Weihe ap. Bluff & Fingerh. 1825, *Comp. Fl. Germ.* I, 682; *R. fusciformis* Sudre, 1906, *Diagn.* 38; *R. pallidus* ssp. *chlorocaulon* var. *fusciformis* (Sudre) Sudre; 1908–13, 154, t. 138.

Axes fuscous. Stem slightly furrowed; *some prickles very long, straight. Leaves deep green,* 5-nate; *leaflets* all rather *long-stalked,* softly hairy beneath, *coarsely subcompound-serrate. Panicle* rather long, equal or pyramidal, *with long linear-lanceolate bracts,* peduncles nearly patent, pedicels long, felted. Flowers large, *c.* 2·5 cm.; incurved; sepals green glandular and aculeolate, reflexed to patent, partly or wholly erect; *petals* not contiguous, *broad elliptic-ovate,* white or pinkish; filaments longer than the usually greenish styles. Carpels usually glabrous. *Receptacle hirsute, especially below the carpels.* Fruit rather large. Weihe and Nees describe the *glands* as *grey.* Normally they are deep red, but under some conditions of age and weather they burst and exude a grey matter.

E., W., S., I. Rather frequent; 3, 6, 7, 9, 11, 13–20, 22–24, 26, 30, 33–38, 40, 49, 55, 110. H. 33. N. France, Belgium, Denmark, Germany, Switzerland, Austria.

237. **R. acutipetalus** Lef. & Muell. 1859, *Pollichia*, 174; *R. fuscus* ssp. *acutipetalus* (Lef. & Muell.) Sudre, 1908–13, 143, t. 138.

4x. *Axes reddish to blackish purple.* Stem angled, slightly furrowed; *prickles numerous, subulate.* Leaves 5-nate, prickles hooked; *leaflets strongly ciliate, ± evenly serrate.* Panicle not leafy, rather broad, a little diminished to the apex, the upper branches 1–3-flowered, spreading, short, the lower ones short and semi-erect; rachis villous, prickles slender, slanting; the *terminal leaflets roundish, acute and red-edged.* Flowers *c.* 2·5 cm.; *sepals* ovate, shortly pointed, *with dark red glands and small prickles,* patent; petals elliptic, pink or pure white, *fimbriate on the apex*; filaments white, slightly longer than the greenish or reddish styles. Carpels pilose. A dwarf form is found.

E., I. Moderately frequent; 7, 14–17, 20–22, 24, 30, 35, 36, 40. H. 16, 23, 38. France, Belgium, Germany.

238. **R. trichodes** W. Wats. 1956, *Watsonia*, **3**, 289; *R. hirtus* ssp. *rubiginosus* sensu Rogers pro parte, 1900, 89; non *R. rubiginosus* P. J. Muell. 1859; *R. foliosus* sensu W. Wats. 1949, *Watsonia*, **I**, 75; non Weihe, 1825. **Fig. 33.**

Leaf and flowers rather small. Stem angled, becoming deep red and purple, pilose, with numerous, rather short, unequal glands and acicles; *prickles strong, subulate, unequal.* Leaves (3, 4) 5-nate, pedate; *leaflets strigose above, softly pilose especially on the veins beneath, prickles on the midribs uncinate,* a little unequal, *crenate-serrate; terminal leaflet roundish, cordate-ovate, shortly acuminate, to obovate-oblong. Panicle long,* nearly equal or somewhat narrowed, *with several ovate leaves,* branches spreading, (I) 3–5-flowered, pedicels rather long, felted and prickly. Flowers *c.* 1·5 cm.; *sepals green, white-bordered, ending, in the terminal flowers, in long linear, green appendages,* partly reflexed, patent and erect on the same flower; *petals 5–6, obovate, long-clawed,* notched, retuse or apiculate, *glabrous on the margin,* pinkish; filaments white, slightly longer than the yellowish, or reddish-based styles. *Carpels bearded. Fruit small, subglobose.*

Endemic. E., W. Fairly frequent in the southern counties; 4, 8, 14–17, 19–21, 23–25, 30, 33, 35, 42.

239. **R. hirsutus** Wirtg. 1841, *Prod. Fl. Rheinl.* 413; *R. pallidus* ssp. *hirsutus* (Wirtg.) Sudre, 1908–13, 155, t. 151.

Kent. Surrey. Stem low, deep red, *blunt, with rather few glands and pricklets; prickles* moderately subequal, *very slender. Leaves rather large,* 4, 5-nate; *leaflets broad, short-stalked, imbricate,* pubescent, with *the veins minutely felted beneath,* sharply and unequally serrate, the principal teeth larger and salient. *Panicle* subpyramidal, truncate, lax and interrupted below, *the middle and upper branches semi-erect, deeply divided;* rachis hirsute, with numerous dark red glands, many longer hairs and *a few acicles longer still, prickles acicular.* Sepals greenish with linear leafy tips, loosely reflexed to patent or some rather erect; petals white, oblong; *filaments white, equalling the greenish styles.*

England: very rare; 15 (Hurst Wood, Mereworth Woods), 17 (Spring Park, Addington). Belgium, Germany (*Coblenz*), Switzerland.

240. **R. hirsutissimus** (Sudre & A. Ley) W. Wats. 1928 (for 1927), *Bot. Exch. Cl. Rep.* **8**, 502; *R. Schlechtendalii* microg. *hirsutissimus* Sudre & A. Ley, 1908–13, 51, t. 55.

Herefordshire. Stem *rather sparsely and minutely glandular and acicular, the larger prickles* moderately strong-based, *slanting. Leaflets long-stalked, rather deeply crenate-serrate.* Panicle with 1–2 simple leaves, broad, lax, rather short, *the peduncles* of the middle branches *and the pedicels long and patent,* intricately villous, *with rather few sunken glands;* prickles short, falcate. Sepals loosely reflexed to ± patent; petals elliptic-ovate, pinkish; filaments white, about equalling the greenish or reddish styles. Carpels pubescent. Fruit globose.

Endemic. Rare; 36 (Welsh Newton Common, etc.).

†††† Stem hairy; terminal leaflet obovate, or broad rhomboid-ovate.

241. **R. Adamsii** Sudre, 1904, *Obs. Set. Brit. Rub.* 28; *R. Babingtonii* var. *phyllothyrsus* sensu Rogers; non *phyllothyrsus* Frid. 1896.

Axes blackish purple. Stem villous, with some *rather long acicles and intermediate gland-tipped prickles; prickles subulate, long, unequal.* Leaves 3, 4, 5-nate, pedate; *leaflets long-stalked, glabrescent beneath, shallowly,* irregularly or slightly compound *serrate-dentate; terminal leaflet subcordate, obovate or elliptic, acute* to acuminate. *Panicle long, with 3–6(12) ovate leaves,* lax, with very long branches divided half-way; *lateral leaflets short-stalked;* rachis villous, *prickles* strong, *some retrorse-falcate,* mixed with unequal acicles and stalked glands. Sepals greenish, long-tipped, ± reflexed, sometimes at length somewhat erect. *Petals glabrous on the margin,* obovate, pinkish or pure white. Filaments white, barely longer than the greenish or reddish styles. *Carpels glabrous.*

Endemic. England: locally frequent in woods in the London basin, rare elsewhere; 8, 16–18, 20, 21, 23, 24, 26, 30, 38, 39.

242. **R. Watsonii** W. H. Mills, 1949, *Watsonia,* I, 136.

Robust. Stem angled to slightly furrowed, thrusting forward in a low arch or climbing, *the larger prickles strong and long.* Leaves 3 (4, 5)-nate, pedate; *leaflets short-stalked and short-pointed, becoming convex and rugose,* pilose to pubescent and slightly greyish felted beneath, *rather broadly, shallowly and slightly compound serrate-dentate; terminal leaflet very broad rhomboid-ovate* to obovate-subcuneate, *base nearly entire. Panicle equal, dense and rounded above* with 1–3 simple leaves, and 3–5-flowered, short, oblique, cymose branches, and more erect axillary branches below; rachis villous, *prickles subulate; terminal leaflets obovate-subcuneate.* Flowers 2–2·5 cm.; sepals ± patent with the tips ascending; petals pinkish or white, elliptic-obovate, margin glabrous; filaments pinkish or white, hardly longer than the reddish-based styles. Carpels glabrous. *Fruit large ovoid-oblong, of numerous drupels, ripening rather late.*

Endemic. England: rather frequent on and near the Chilterns, elsewhere very local; 20, 24, 30, 31, 39, 40 (Wyre Forest).

243. **R. apiculatiformis** (Sudre) Bouv. 1923, *Rub. Anjou,* 62; *R. fuscus* ssp. *apiculatiformis* Sudre, 1908–13, 144, t. 139.

Shooters Hill. Slender. Stem slightly furrowed, glaucescent; *prickles* unequal, *yellowish, slender, long subulate.* Leaflets 3, 5, greyish pubescent to almost glabrous beneath, *finely to slightly compound serrate-dentate, jagged; terminal leaflet long, narrow oblong-obovate-cuspidate. Panicle long, narrow, equal or slightly tapered, leafless above, branches intricate,* lateral pedicels long and patent; rachis hirsute, not felted, very shortly glandular, with longer acicles, some gland-tipped; *prickles very numerous, long and slender*, mostly slanting. Flowers 1·5–2·5 cm.; sepals greenish felted, linear-tipped, reflexed to patent; *petals* fimbriate, elliptic, pink, pinkish or white, *often 8* on the terminal flowers; filaments white, about equalling the greenish styles. Carpels glabrous. Fruit oblong, of rather numerous drupels.

England: 16 (Shooters Hill at about 350′). France, W. Germany, Belgium.

244. **R. racemiger** (*racemigerus*) Gremli, 1871, *Oesterr. Bot. Zeitschr.* 128; *R. angustifolius* Lef. & Muell. 1859, *Pollichia*, 119; non Kalt. 1845; *R. fuscus* ssp. *angustifolius* (Lef. & Muell.) Sudre, 1908–13, 144, t. 139.

W. Kent. Stem brown, blunt; prickles rather short, much declining. Leaves large, pedate, prickles falcate; *leaflets 3–5, narrow, long*, nearly glabrous above, thinly hairy beneath, rather broadly and unequally crenate-serrate; *terminal leaflet base entire, obovate, acuminate. Panicle leafy, long, lax, subracemose-pyramidal*, middle and upper *branches deeply divided, sharply ascending*, few-flowered; rachis hairy and glandular, with some long acicles, prickles few, weak, slanting or falcate. Flowers *c.* 2·5 cm.; sepals long-leafy-pointed, reflexed; *petals 5–6, white*, broad elliptic, *margin glabrous*; filaments white, about equalling the greenish styles. Carpels glabrous. Fruit subglobose.

England: 16 (Chalket, near Pembury). France (Valois), Switzerland.

Subser. ††**Pallidi** W. Wats.

(Type *R. pallidus* Weihe)

1. Robust plants with large leaves. (Cf. *R. fuscus, R. Bloxamii* and *R. corymbosus*).

† Stem hairy.

245. **R. pallidus** Weihe ap. Bluff & Fingerh. 1825, *Comp. Fl. Germ.* I, 682; Sudre, 1908–13, 153, t. 149; *R. cernuus* P. J. Muell. 1859, *Pollichia*, 194. **Fig. 34.**

4x. *Stem* becoming purple, shortly hairy *with very short blackish glands*, etc.; *prickles short*, broad-based, *mostly slanting. Leaves 3–5-nate, pedate, roughish above*, glabrescent beneath, rather *bluntly and unequally serrate; terminal leaflet cordate, ovate-elliptic, attenuate or falcate, cuspidate. Panicle arched*, with one or several simple leaves, broadly pyramidal-truncate, *middle branches long, patent*, simple, subcymose or panicled with long patent pedicels; rachis crooked below, *prickles weak, slanting.* Flowers 2–2·5 cm.; *sepals purplish-greenish grey, narrow attenuate*; petals elliptic, apex retuse, pinkish or white, rarely pink; *filaments* white, *not much longer than the purple styles. Carpels numerous, glabrous.* Fruit of moderate size but well produced and possessing aroma. A cherry flavour has been noticed.

E., W. Rather frequent in moist woods; 3, 4, 6, 11–17, 20–24, 27, 30, 34–36, 38–42, 49 (below Pont Aberglaslyn, *W. H. Mills*), 55, 57, 58, 62, 69 (south end of Lake Windermere). France (rare), Belgium, Denmark, Germany, Austria.

246. **R. drymophilus** Muell. & Lef. 1859, *Pollichia*, 223; *R. pallidus* ssp. *drymophilus* (Muell. & Lef.) Sudre, 1908–13, 154, t. 150.

*4x. Stem at first glossy, deep red, with *deep red glands, acicles and pricklets, some long* and gland-tipped; *prickles* hairy, *subulate, some very long and patent. Petiole long, prickles slanting* and subfalcate; leaflets (3) 5, pilose beneath, unequally and irregularly crenate; *terminal leaflet* subcordate, *roundish or broad oblong, ovate, short-pointed. Panicle long, lax, pyramidal, blunt,* with simple leaves and narrow leafy bracts, upper branches 1–3 (–7)-flowered; rachis densely hairy, with numerous, very unequal, slender glands and acicles, and *long, slender, subulate as well as much recurved prickles.* Flowers 2–3 cm.; *sepals green, white-bordered,* pilose; *petals* white or pink, *broad ovate;* filaments white, equalling or exceeding the green or reddish styles. Carpels glabrous or pilose. Fruit large, roundish-ovoid.*

England: frequent in woods on the Chilterns, rare elsewhere; 13, 17, 18, 20, 21, 24, 30, 38. France, Belgium, Switzerland.

247. **R. Loehrii** Wirtg. 1854, *Herb. Rub. Rhen.* ed. 1, no. 22; *R. pallidus* ssp. *Loehri* (Wirtg.) Sudre, 1908–13, 154, t. 151.

Stem brownish, angled, *with fine, very short, dark red glands and acicles,* etc.; *prickles acicular, short, declining,* yellowish. Leaves 3, 4, 5-nate, prickles hooked, small; *leaflets short-stalked, greyish but not felted beneath, coarsely sublobate crenate-serrate; terminal leaflet* subcordate, ovate or *obovate, acuminate. Panicle leafy,* often to the top, long, rather narrow pyramidal, or subcorymbose and dense above, lower branches oblique, middle branches sometimes patent cymose; *terminal leaflets obovate,* ± *cuneate, coarsely lobate-serrate.* Flowers up to 3 cm.; *sepals green, triangular, ovate, attenuate,* felted, *soon erect; petals 5–7, ovate, pure white; filaments white, hardly longer than the green or red styles.* Carpels thinly pilose. Fruit ovoid, drupels many.

England: rather frequent in damp woods; 3, 12, 15, 16, 20, 24, 36, 62. Belgium, N.W. and W. Germany.

†† Stem glabrescent.

248. **R. spadix** W. Wats. 1952 (for 1951), *Lond. Nat.*, Suppl. 99; *R. podophyllus* sensu Rogers pro parte, 1900, 67; non P. J. Muell. 1861.

Stem long, pruinose and purple; prickles rather short, declining *from a chestnut brown base. Leaves deep green;* leaflets 3–5, becoming rugose, pubescent at first, and in the case of the upper leaves rather grey felted beneath; shallowly serrate with the principal teeth patent; *terminal leaflet cordate, broad ovate, mucronate. Panicle subthyrsoid,* the *terminal leaflets long oblong;* rachis hirsute, densely glandular, *prickles acicular*. Flowers *c.* 2·5 cm.; *sepals deep green, ovate-lanceolate*; petals pale pink, elliptic-obovate; *filaments white, hardly equalling the pallid, reddish-based styles*. Carpels glabrous. Very fertile. The leaves are a little glandular; those of the stem, on the back and the margin; those of the panicle on the upper surface.

Endemic. England: scattered; probably it will prove to be frequent; 16, 54, 57, 60, 65.

2. Plants moderate in size, with moderate or small leaves.
† Upper leaves greyish or grey-white felted beneath, on dry soils.

249. **R. chlorocaulon** (Sudre) W. Wats. 1946, *Journ. Ecol.* **33**, 340; *R. pallidus* ssp. *chlorocaulon* Sudre, 1908–13, 154, t. 150; *R. coombensis* Rilst. 1950, *Journ. Linn. Soc.* **53**, 420.

Stem blunt below, glaucescent, ± hairy, with glands, acicles and pricklets mostly sunken but some longer and gland-tipped; prickles unequal, small-based, rather long, straight, with fragile points. Leaves 3, 4 (5)-nate, pedate, prickles much curved; *leaflets plicate*, softly pilose to greyish felted beneath, *simply, sharply, rather evenly to rather compound serrate or serrate-dentate; terminal leaflet base entire, oblong or elliptic-obovate, acuminate. Panicle narrow, rather long, racemose* with short upper and long, sharply ascending lower branches, or more compound and subpyramidal, the middle branches then deeply divided with long pedicels; *bracts lanceolate or linear, green, terminal leaflets obovate-cuneate, lateral leaflets subsessile*; rachis villous, armed like the stem. Sepals pilose, felted and red-glandular; *petals long, narrow elliptic or obovate, pale pink or white*; filaments white, about equalling the yellowish styles. Carpels glabrous or pilose. Fruit perfect but small. *Flowering, in Cornwall, from the third week in May.*

E., W., I. Locally frequent. Overlooked? 2–4, 16, 17, 33, 39, 49. H. 3 (P. D. Sell). France, Belgium, Switzerland.

This bramble was collected by Hort at Llanberis as long ago as 1847: Babington thought it was *R. Lejeunei* 'possibly'. Rogers named it, much later, as *R. hostilis*. I have seen it in Bickleigh Vale, S. Devon, where it was probably known to Briggs, and it is abundant in the Port Gaverne valley in N. Cornwall. In *Flora Devon* it is put, from Ashburton, under *R. longithyrsiger*. As a Cornish bramble it once passed as *R. curvidens*, Mr Rilstone said recently in describing and naming it as a new species. It has long been known to me in more than one part of West Kent and Surrey, and occurs in Staffs.; and I have received it from other parts of the country and from Ireland. It was included in my 1946 list of British Rubi.

250. **R. Menkei** Weihe ap. Bluff & Fingerh. 1825, *Comp. Fl. Germ.* **1**, 679; Coste, 1903, *Fl. France*, **2**, 44, no. 1197 (plate); *R. distractus* P. J. Muell.

1861, *Bonpl.* 296; *R. tereticaulis* sensu Rogers, 1900, 91; non P. J. Muell. 1858.

Stem angled, patent-pilose; *prickles yellowish, broad-based,* unequal, much *declining and recurved. Leaflets 3–5, coarsely, unequally serrate, ± jagged; terminal leaflet short-stalked, elliptic, obovate, cuspidate, base entire. Panicle equal, dense and subcorymbose at the apex;* rachis crooked, with long, spreading hair, numerous, short glands and unequal acicles, *prickles strong, declining and curved.* Flowers *c.* 2 cm.; *calyx base discoid; petals glabrous on the margin,* broad ovate-elliptic, some pointed. Carpels glabrous. Fruit fertile, but not large. Small, starved plants in shade may be very deceptive.

England: rather rare; woods; 14–16, 27, 30, 39. N.E. France, Belgium, W. Germany, Switzerland.

251. R. morganwgensis Barton & Riddels. 1933 (for 1932), *Proc. Cotteswold Field Cl.* **24,** 218.

N.v.v. Wales. *Stem blunt, fuscous, glabrescent, with numerous, ochreous pricklets and few very short glands;* prickles numerous, unequal, falcate from long and *often confluent* bases. Leaves 3–5-nate, pedate; *leaflets of a regular and formal outline,* glabrescent above, *finely dentate or serrate; terminal leaflet roundish, truncate, cuspidate, base entire.* Panicle subpyramidal, obtuse, *with large leaves* and often leafy bracts nearly to the top; rachis furrowed, with short hair and close felt, deep red glands and acicles, prickles slender. Flowers moderate; *sepals triangular-acuminate, long-pointed, felted,* loosely erect; petals pink, broad, pubescent; filaments white, styles greenish.] *Carpels bearded, numerous. Fruit ovoid, well developed,* delicious.

Endemic. W., I. Rare; 41, 42, 49 (not 55). H. 38.

†† All leaves green beneath; petals pink.

252. R. argutifolius Muell. & Lef. 1859, *Pollichia,* 150; *R. glareosus* E. S. Marshall ex Rogers, 1912, *Journ. Bot.* **50,** 309; C. G. Trowers, 1929 (for 1928), *Rep. Bot. Exch. Club,* **8,** 861, no. 113 (plate).

Rufescent. Stem slender, angled, striate, *pruinose,* ± hairy, *with numerous, very short, dark glands,* and few, rarely numerous, long, gland-tipped acicles; *prickles very numerous, unequal,* long, slender, declining from a large base. Leaves 3, 4 (5)-nate, pedate, petiole prickles numerous, hooked, *glabrescent on both sides,* somewhat compound serrate; *undulate and jagged; terminal leaflet often twisted,* base emarginate, truncate or rounded, *narrow rhomboid or obovate-oblong, long-acuminate.* Panicle with 1–2 simple leaves, broad, nearly equal, lax, straggling, or more elongate, pyramidal and very floriferous; *rachis crooked,* hairy and felted, *with numerous, very short glands and few or rather numerous, very long acicles,* some gland-tipped; prickles numerous, weak. Flowers up to 3 cm.; *sepals green,* patent to erect, *leafy-pointed; petals twisted, narrow obovate-rhomboid,* some notched, usually glabrous; filaments white, longer than the *red styles*. Carpels glabrous or pilose. Fruit rather elongated. Occasionally found very luxuriant and panicle greatly developed, the armature then virtually hystrican.

South England, rather frequent on heaths and in woods (Surrey heaths and N. Downs woods); 7, 8, 11–17, 21–24. France, Belgium, Germany (*Bavaria*, rare).

253. **R. acutifrons** A. Ley, 1893, *Journ. Bot.* **31**, 13.

Western. Stem dark red, glaucescent, shallowly furrowed, striate, thinly hairy, *with stout pricklets and short, garnet glands*; prickles numerous, sharply declining or falcate. Leaves (3) 4, 5-nate; *leaflets subglabrous, dull above, thinly pilose and very finely and thinly felted beneath, crenate-serrate or serrate-dentate; terminal leaflet subcordate, roundish ovate, elliptic or obovate, obsoletely lobate, long-acuminate*. Panicle pyramidal with a rounded top and strongly ascending lower branches; rachis patent-pilose, felted and glandular, prickles slender, declining; the terminal leaflets obovate-cuneate. *Sepals* dark green, long linear-pointed, *soon clasping*; petals rather small, narrow obovate; *filaments long; styles red*. Carpels pilose. Fruit globose, acid. The epithet ʻ*acutifrons*ʼ is a misnomer; the terminal leaflet has a beautiful long ʻdrip-tipʼ. The base of the top panicle leaf is gland-fringed.

Endemic. E., W. Chiefly on the Welsh border, rare elsewhere; 3 (St Budeaux), 6 (Castle Orchard), (not 14), 35–37, 42, 52 (east of Menai Bridge, W. H. Mills).

†‡† All leaves green beneath; petals white. (Cf. *R. fuscus.*)

(*a*) Styles greenish.

254. **R. euryanthemus** W. Wats. 1946, *Journ. Ecol.* **33**, 340; *R. pallidus* var. *leptopetalus* Frid. ex Rogers, 1900, 75; *R. Schleicheri* ssp. *fissurarum* microg. *chloroxylon* Sudre, 1908–13, 203, t. 196.

Rufous in exposure. Very densely glandular and aciculate throughout. Stem rainbow-arched, slightly furrowed, subpruinose, *nearly glabrous, glands pallid; prickles numerous, unequal, subulate, mostly patent*. Leaves 3, 4, 5-nate, pedate; *leaflets long-stalked*, thin, *subpruinose above*, greyish and hairy beneath, *subcompound incise-serrate; terminal leaflet* subcordate, *obovate, long-acuminate*. Panicle leafy, *short, broad, the middle branches patent with interlacing pedicels*; rachis hirsute, with numerous, *weak, yellow prickles and crowded glands*; sepals at length loosely clasping; *petals remote, narrow, obovate-lanceolate*, nearly glabrous, *milk white*. Carpels glabrous. Receptacle thinly pilose. *Fruit* rather long ovoid, pale green, then *fuscous before changing to black*.

E., S. Widely distributed in open places and in woods, in some parts abundant; 11–21, 23–26, 34, 36, 38, 39, 55, 58, 110. France, Belgium, Germany (*Schleswig* and *Rhineland* to *Silesia*), Switzerland.

(Cf. *R. erubescens*, which sometimes has greenish styles.)

(*b*) Styles red.

255. **R. insectifolius** Lef. & Muell. 1859, *Pollichia,* 176; *R. fuscus* var. *nutans* Rogers, 1892, *Journ. Bot.* **30**, 304; *R. nuticeps* Barton & Riddels, 1933 (for 1932), *Proc. Cotteswold Field Cl.* **24**, 217.

4x. *Rufescent. Stem* slender, *much branched, the thin branches trailing over the top of the bush everywhere,* angled, *intricately hairy;* prickles hairy, stout-based, patent to recurved. *Leaves 3–6-nate,* pedate, *yellowish-green; leaflets rather small,* pilose on the veins beneath, *coarsely, sharply incise, somewhat compound crenate-serrate; terminal leaflet* with base nearly entire, *roundish ovate, acuminate attenuate. Panicle nodding, lax,* leafy throughout, pyramidal or nearly equal, the upper branches 1–3-flowered, pedicels long, patent, *bracteoles purplish, long and slender; rachis* flexuous, *furrowed, densely pilose,* prickles small, ± curved. Flowers *c.* 2 cm.; *sepals greyish green and purplish* with a white margin; petals narrow, white; filaments white, about equalling the *red styles;* carpels glabrous. Fruit subglobose.

South England, frequent, usually in the open; 3, 4, 6, 9–11, 13–22, 24, 29–31, 33, 34, 36, 38, 39, 40 (the Ercall, Leighton, 1841), 55. France.

256. **R. brachyadenes** P. J. Muell. 1866, *Boul. Ronc. Vosg.* 37; *R. Menkei* ssp. *brachyadenes* (P. J. Muell.) Sudre, 1908–13, 161, t. 155.

Stem blunt, *slender,* thinly hairy, *with crowded mostly short glands;* prickles broad-based, *falcate or slanting.* Leaves yellowish green, 4, 5-nate, pedate; *leaflets very short-stalked, glabrous above, glabrescent beneath,* coarsely serrate; terminal leaflet ovate-elliptic or obovate, long-pointed. *Panicle narrow,* equal, branches short, semi-erect, simple or divided to the base. Flowers 1·5–2 cm.; *sepals attenuate, clasping;* petals elliptic, white; *filaments shorter than the styles;* carpels pilose. Fruit long, ovoid.

England: rare, but locally plentiful; 3 (Ermington, Briggs, 1868), 14 (Rushlye Downs, near Tunbridge Wells), 16 (near Tunbridge Wells), 39 (Handsworth Wood, Bagnall, 1871), 40 (the Wrekin). France (*Seine-Inférieure* and *Vosges*), Switzerland.

257. **R. foliolatus** Lef. & Muell. 1859, *Pollichia,* 212.

Stem furrowed, *densely hairy;* prickles yellowish, short, sharply declining. Leaves yellowish green, 3, 4, 5-nate; *leaflets rather small,* pilose, rarely greyish felted beneath, *finely compound-serrate-dentate, undulate, jagged;* terminal leaflet elliptic, finely acuminate-cuspidate; *basal leaflets* broad ovate, *long-stalked. Panicle* nodding, narrow, tapered, *with simple leaves to the apex,* branches oblique, divided half-way, about 3-flowered; rachis felted and pilose, prickles weak. Flowers *c.* 1·5–2·5 cm.; *sepals greenish inside and out;* petals at first pale pink, glabrous on the margin, ovate, notched with a slender claw. Carpels pubescent. *Fruit soon becoming deep blood red.* Late young branches often pruinose.

South England: very local; 16 and 17 (plentiful on the Lower Greensand south of Westerham; Horsell Common, Woking). France, Belgium, Bavaria.

258. **R. acidophyllus** (*aciophyllus*, then *aciphyllus*) (Sudre) W. Wats. 1946, *Journ. Ecol.* **33**, 340; *R. pallidus* ssp. *hirsutus* microg. *aciphyllus* Sudre, 1908–13, 156, t. 151.

Stem brownish purple, blunt, villous; prickles numerous, ± subulate, patent, or falcate. *Leaflets* 3–5, nearly glabrous above, densely and softly pilose beneath, to greyish felted on upper leaves, unequally, *rather coarsely and deeply serrate* undulate; *terminal leaflet roundish, ovate, acuminate, the point twisted*, base truncate, entire. *Panicle with several simple leaves*, dense above, narrow, ± diminished to the top, branches oblique, partly fasciculate when strong; *rachis hirsute below, felted above, prickles nearly acicular*, slanting. Flowers 2–2·5 cm.; sepals green, aculeate, semi-erect; *petals 5–8, pilose, narrow obovate or oblanceolate*, mucronate, white or pinkish; filaments white, equalling or exceeding the *red-based styles.* Carpels shortly pilose.

England: rare; 14 (Benhall Mill Lane, near Tunbridge Wells), 20 (Sherrardswood and Mardley Heath). France (Valois), Germany and Switzerland (rare).

259. **R. erubescens** Wirtg. 1858, *Herb. Rub. rhen.* ed. 1, no. 93.

Stem angled, green to fuscous, hairy, with *very numerous*, mostly very short *glands, pricklets, acicles and unequal prickles.* Leaves 5-nate, pedate, pale green, prickles strongly curved, stipules linear-filiform; *leaflets* pubescent beneath, *unequal or subcompound serrate-dentate*, slightly jagged; terminal leaflet base entire or a little indented, oblong-obovate, cuspidate. Panicle long, nearly equal, interrupted, the lower two axillary branches nearly erect, with 1–2 simple leaves, slightly grey felted and glandular on the base; rachis hirsute, prickles numerous, fine, falcate. *Sepals* felted, hairy and aculeate, patent to erect and, *with the disc, turning purple at the base inside. Petals long, narrow elliptic, notched, veiny*, milk white or rarely pink (*R. morganwgensis* var. *Devoniae* Barton & Riddels.). *Stamens long, like the styles turning red in the sun.* Receptacle long-pilose, especially below the carpels. Fruit perfect, moderate in size.

England: very rare; 4 (Clawton), 5, 16 (Shooters Hill). Belgium, France, W. Germany.

Subser. †††**Scabri** W. Wats.

(Type *R. scaber* Weihe)

1. Stem ± pruinose and hairy.

† Leaves large, or leaflets long.

260. **R. microdontus** Muell. & Lef. 1859, *Pollichia*, 245.

Stem roundish below, *strongly pruinose over purple, hairy*; prickles numerous, acicular, short. Leaves yellowish green, 3, 4 (5)-nate, pedate; *leaflets* pubescent beneath, *finely crenate-serrate* to serrate-dentate; *terminal leaflet very broad,* subcordate, roundish or ovate, short-pointed. *Panicle with some ovate leaves,* rather long, equal, *with 2–3-flowered, semi-erect and panicled, deeply divided, spreading branches; rachis villous,* glandular and finely aculeate with a few, long acicles. Flowers 1–2 cm.; sepals triangular-ovate, prolonged, clasping; petals pinkish white, narrow obovate; filaments white, slightly longer than the greenish to reddish styles. Carpels glabrous. Fruit subglobose. *Readily detected by the greyish white stems.*

E., I. England, locally plentiful in the south-east and midlands; 14–16, 19, 21–24, 30. H. 23 (Crooked Wood, Mullingar). France (*Aisne, Oise*).

261. **R. thyrsiflorus** Weihe ap. Bluff & Fingerh. 1825, *Comp. Fl. Germ.* I, 684; Sudre, 1908–13, 152, t. 147; *R. hirtus* ssp. *flaccidifolius* sensu Rogers, 1900, 89; non P. J. Muell. 1861.

4x. *Robust. Stem thick, roundish,* purplish, *glabrescent, with numerous, very short, slanting and recurved prickles. Leaves* large, firm, *shortly stalked; leaflets* 3, 4, 5, *very broad, shortly stalked,* imbricate, glabrescent above, shortly hairy beneath, crenate or serrate-dentate to incise subcompound-serrate; terminal leaflet broad ovate or roundish from a subcordate base, short-pointed. *Panicle* with several simple leaves, *long thyrsoid* with, in the open, up to about 24, semi-erect, 1–3-flowered, deeply divided branches, with 2–3 panicled branches below; *or, in shade, laxer with a crooked rachis, fewer branches, all panicled, and very large leaves;* rachis felted and hirsute, with mostly buried glands and acicles, and short prickles. Flowers *c.* 2 cm.; sepals and petals often 6; *sepals large,* felted, *with linear tips; petals white,* elliptic, notched, tapered below, glabrous on the tip; filaments white, about equalling the greenish or reddish styles. Carpels usually glabrous. Fruit ovoid, drupels many.

England: fairly well distributed, occasionally in the open, but most often in woods; 2 (Grampound, at roadside), 3 (Woodlands), 15, 21, 22, 24, 30, 33, 34. France (very rare), N.W. Germany and Bavaria, Switzerland.

262. **R. derasifolius** (Sudre) W. Wats. 1952 (for 1951), *Lond. Nat., Suppl.* 95; *R. tereticaulis* ssp. *derasifolius* (Sudre) Sudre, 1908–13, 196, t. 192; *R. minutiflorus* sensu Rogers, 1900, 89; non P. J. Muell. 1859.

A low bramble with big leaves, usually dominant in dark woods. Stem roundish, densely hairy, with sunken glands and acicles, together with longer tuberculate-based acicles; prickles short, weak falcate. *Leaves large,* 3 (5)-nate, short-stalked, petiole prickles fine, hooked; *leaflets convex,* softly patent-pilose on the veins beneath, rather equally crenate-serrate; *terminal leaflet* subcordate, *broad, obovate to oblong-elliptic* with a very short point. Panicle lax, pyramidal-corymbose with several, distant, few-flowered branches below, and a large trilobed and other simple leaves above; *lateral leaflets nearly sessile; rachis villous,* pedicels long, densely glandular, felted and *with weak, yellowish prickles.* Flowers up to 3 cm., *sepals and petals often 6, 7;* sepals leafy-pointed, loosely erect; *petals remote, long, narrow elliptic,* white or pink; filaments white, equalling the greenish styles. Carpels glabrous or pilose. Fruit globose.

England: rather frequent in W. Kent, rare elsewhere, but ± abundant where it occurs; 15–17, 20. France, Belgium, Germany, Switzerland.

263. **R. putneiensis** W. Wats. 1956, *Watsonia,* 3, 289.

Robust. Stem low arching, blunt, *pruinose, hairy below, glabrescent higher up, with numerous short* and a few longer *red glands and acicles; prickles* unequal but *mostly short, numerous, slanting.* Leaves rather large, yellowish green, becoming rugose; leaflets 3, 4, 5, glabrescent beneath, irregularly crenate-serrate; *terminal leaflet roundish or obovate-cuneate, rounded cuspidate.* Panicle equal and subracemose above, interrupted, with longer, panicled branches below, finely pubescent and felted with *crowded, very short glands and pricklets;* prickles acicular, some recurved; terminal leaflets obovate-cuneate. *Flowers 2–3 cm., starry; sepals* deep green, *narrow-triangular,* slenderly appendiculate, reflexed to patent; *petals narrow,* elliptic, tapered below, often notched, glabrous, *pale pink;* filaments white, slightly longer than the greenish to reddish styles. Carpels glabrous. Fruit roundish ovoid. Top leaves of the panicle greyish felted beneath in the open.

Endemic. England: 3 (near Cornwood Railway Station, 1951), 17 (Putney Heath (type specimen *Herb. Watson*) 12 July 1943), 30 (Aspley Wood, Woburn, Beds.).

264. **R. curtiglandulosus** (Sudre) W. Wats. 1946, *Journ. Ecol.* 33, 341; *R. tereticaulis* ssp. *curtiglandulosus* (Sudre) Sudre, 1908–13, 196, t. 192.

Western. Stem long, slender, roundish below, *glabrescent;* prickles weak, short. *Leaves* 3–5-nate, pedate, *prickles straight; leaflets glaucescent* above, at first hairy beneath, *ciliate, simple serrate; terminal leaflet, short-stalked, narrow, long* ovate-elliptic, acuminate, *base* subcordate, *lop-sided.* Panicle with a pyramidal top of 1–3-flowered branches with long ascending pedicels, the terminal leaflets broad, acute, obovate-rhomboid; rachis villous with sunken glands and few, weak, falcate prickles. Flowers 1·5–2 cm.; *sepals narrow ovate-lanceolate, long-pointed, clasping;* petals white; filaments longer than the styles. *Carpels very pilose.* Fruit ovoid-oblong, small.

E., I. Woods. Rare; 36 (Clifton Wood, near Upper Sapey), 37 (Southstone Rock, Upper Sapey), 46 (Monachty Dingle). S. France, Belgium, Germany, Switzerland, Austria, Hungary.

†† Leaves moderate. Petals often very small.

265. **R. scaber** Weihe ap. Bluff. & Fingerh. 1825, *Comp. Fl. Germ.* 1, 683; Sudre, 1908–13, 193, t. 190; *R. dentatus* (Bab.) Bloxam in Kirby, 1850, *Fl. Leicest.* 39; *R. Bellardii* sensu Genev. 1880, *Mon.* 75; non Weihe & Nees, 1825. **Fig. 35.**

Frequent. Stem round to blunt angled; lower *prickles* weak and straight, *upper ones recurved from a stout base.* Leaves 3, 4, 5-nate, pedate; *leaflets* plicate-rugose, pilose and *greyish blue-green beneath*, sharply, *rather evenly*, sometimes deeply *serrate*; terminal leaflet base cordate to entire, broad ovate, elliptic or obovate acuminate; *lateral and basal leaflets short-stalked.* Panicle pyramidal, dense at the top, usually hanging down, sometimes large, pyramidal, lax with widely spreading branches; rachis flexuose, prickles yellowish, weak, *pedicels* felted and hairy, *the glands short but longer than the felt and conspicuous*, sometimes with much longer ones intermixed. Flowers 1–2 cm.; *sepals triangular-lanceolate, acuminate, soon clasping; petals* narrow, *glabrous on the margin*, flesh pink or white; filaments about equalling the greenish or red styles. *Carpels glabrous.* Receptacle pilose. Fruit roundish, drupels small and many. Winter-green. On the open heath it makes a small bush with small leaves and very small flowers, with upstanding minute petals like those of the raspberry; in woods it throws up strong arching stems and produces large, widely branching panicles.

E., I. Rather thinly distributed, especially in northern and western England; 3, 6, 13, 15–17, 20, 21, 23, 24, 26, 27, 30, 35, 37–39, 55, 68 (above Eglingham). H. 2, 23, 25. France, Belgium, Germany to Silesia, Switzerland, Austria.

266. **R. tereticaulis** P. J. Muell. 1858, *Flora*, **41**, 173; Sudre, 1908–13, 194, t. 191; Coste, 1903, *Fl. France*, **2**, 45, no. 1202; non sensu Rogers, 1900, 91.

Stem slender, trailing, round to angled, *becoming fuscous; prickles acicular, small-based.* Leaflets 3 (5), slightly hairy on the veins, the upper leaves slightly greyish felted beneath, sharply serrulate; *terminal leaflet* short-stalked, elliptic, obovate, *long and finely acuminate*, base nearly entire. Panicle dense and equal at the top, the branches deeply divided, 3-flowered, often with another stalked flower at the base; *rachis* crooked, pubescent and felted, *with very short glands and acicular prickles.* Flowers starry, 1–2 cm.; sepals greyish green, felted, long-tipped, erect; petals greenish white, narrow-oblong; *filaments white, slightly shorter than the yellowish, reddish-based styles*, anthers greenish. Carpels glabrous. Fruit globose-ovoid. Flowering about 1 July.

England: very rare; 3 (by Ivybridge Station), 15 (Bigbury), 30 (Aspley Wood, Woburn). France, Belgium, Germany to *Silesia*, Switzerland, Austria.

††† Leaves moderate. Petals larger.

(*a*) Stamens conspicuously shorter than the styles at first.

267. **R. praetextus** Sudre, 1900, *Rub. Pyr.* 78; *R. glaucellus* ssp. *luteistylus* microg. *praetextus* (Sudre) Sudre, 1908–13, 171, t. 165.

Kent. Surrey. Stem slender, climbing, *round* to angled, *thinly hairy; prickles moderately strong, short, slanting and falcate. Petiole prickles straight. Leaflets 3*, slightly hairy and green beneath, nearly evenly serrate; terminal leaflet elliptic, acuminate. *Panicle nodding*, narrow, *not very long*, lax, the branches 2–3-flowered, spreading or ascending, prickles numerous, short, falcate, glands very numerous and very short. *Flowers up to 1·5 cm.*; *sepals green with a contrasting white border*, narrow ovate, acuminate, *clasping*; petals white or pink, rhomboid-elliptic, notched; *filaments* white or pinkish, *much shorter than the red styles. Carpels glabrous.* Fruit oblong.

England: very local; 15 (Kings Wood, Sutton Valence; lane west of St Michaels, Tenterden), 16 and 17 (plentiful on the Lower Greensand south of Westerham, especially on Hosey Common). S. France, Pyrenees.

268. **R. derasus** Lef. & Muell. 1859, *Pollichia*, 239; *R. foliosus* microg. *derasus* (Lef. & Muell.) Sudre, 1908–13, 146, t. 141.

Essex. Stem angled, *becoming nearly glabrous; prickles subulate, subequal. Leaflets 3, 4, 5, tough*, glabrescent above, pubescent and grey felted beneath, *serrulate-denticulate; terminal leaflet* base ± entire, *ovate to elliptic, acuminate. Panicle very long, stiff, pyramidal-oblong, floriferous*, the branches mostly patent, all *dense-flowered; rachis stout, round*, felted, hirsute, with numerous, mostly small, bright red glands and longer acicles, *prickles subulate.* Sepals long-pointed, felted, aculeate, glandular, spreading to erect. Petals glabrous on margin, pink, pinkish or pure white, narrow elliptic, notched or erose; filaments white; styles reddish. Carpels glabrous or pilose. Fruit subglobose, freely and perfectly produced.

England: very rare; 18 (Danbury Common, where it was shown to me in 1935 by Mr G. C. Brown), 19 (Middlewick, Donyland, G. C. Brown). France (*Valois*).

(*b*) Stamens at least as long as the styles.

269. **R. vallisparsus** Sudre, 1906, *Diagn.* 42; 1908–13, 168, t. 162.

4x. Kent. Luxuriant. Stem long, roundish, *becoming white with the pruina*, thinly pilose; prickles not long, declining or falcate from a broad base. *Leaflets a pleasant green, rather large, 3 (4)*, green and glabrescent to greyish felted beneath, deeply serrate to shallowly serrate-dentate, *jagged; terminal leaflet* rather narrowed below, *elliptic-obovate, cuspidate-acuminate.* Panicle lax, pyramidal-truncate, the middle and upper branches 1–4-flowered, deeply divided; rachis crooked, felted and thinly pilose with crowded, mostly short, deep red glands, glaucescent below. Flowers *c.* 2·5 cm.; *sepals* greenish felted, *patent to half erect; petals rhomboid-elliptic, notched, remote, pinkish*; filaments white going reddish, *not much longer than the styles; styles* yellowish, *usually becoming intense red.* Carpels pilose. *Fruit rather large*, ovoid, sweet and aromatic. Easily detected by the white stems.

England: very local; 16 (distributed over much of Shooters Hill). France, Switzerland, Bavaria.

Appendiculati

270. R. frondicomus Foerst. 1878, *Fl. Aachen; R. scaber* b. *foliosus* Kalt.

Kent. Stem brownish red, round to blunt angled, glabrescent, with short, scattered glands, acicles and pricklets; *prickles very short, subulate, declining.* Leaves 3, 4-nate, pedate, *petiole prickles weak, straight; leaflets tough, plicate-rugose, twisted,* green and glabrescent beneath, shallowly crenate-serrate with outward-pointing mucros; *terminal leaflet subcordate, narrow oblong-elliptic, acuminate.* Panicle with numerous, linear-oblong leaves, long, branches semi-erect, the middle ones mostly divided to the base; rachis felted and hirsute, with numerous unequal, deep red glands, prickles weak, pedicels felted. Flowers 1·5–2 cm.; *sepals clasping;* petals narrow oblong, white, glabrous on margin; filaments white; styles reddish. Carpels glabrous. Fruit ovoid, rather large, of numerous smallish drupels. The leaves are nearly always buckled at the point when growing, giving rise to a fold when they are pressed.

England: very rare; 16 (Hayes Common, in three places, first noticed in 1948). Germany (in two places near Aachen, one beyond Burtscheid, W. Watson, 1937).

2. Stem hairy, not pruinose.

271. R. longithyrsiger Ed. Lees ex Bab. 1878, *Journ. Bot.* **16**, 176–7; *R. foliosus* microg. *longithyrsiger* (Ed. Lees ex Bab.) Sudre, 1908–13, 146, t. 141; *R. pyramidalis* (Bab.) Bab. 1849, *Bot. Gaz.* **1**, 121; non Kalt. 1845; *R. Menkei* sensu Ed. Lees, 1853, *Phytologist*, **4**, 920.

Stem long, procumbent, becoming purple; prickles purple, short, slanting from a stout base. *Leaves 3 (4, 5)-nate, pedate, petiole flat above; leaflets* greyish green and thinly hairy, not felted beneath, *serrate-dentate; terminal leaflet twisted, obovate, the sides rather straight below the middle.* Panicle from racemose, the pedicels spreading, *to very large pyramidal with the axillary branches nearly erect; rachis and branches strict,* prickles subulate. *Calyx flat-based,* sepals greenish, ovate-lanceolate, attenuate, hairy, leafy-tipped, *loosely clasping; petals often 7–8 in the terminal flowers, ± erect,* narrow obovate, retuse, faintly lilac in bud, greenish-based, sometimes tinged red outside; filaments about equalling the greenish, or reddish-based, or wholly red styles. Carpels glabrous. Floral disc broad. Receptacle cylindrical. Fruit oblong.

Endemic. E., W., I. Frequent in the west, local but sometimes locally abundant elsewhere; 1–5, 14–17, 23, 24, 34–37, 39, 43, 44, 49, 68 (Alnwick, Massey). H. 2 (Muckross).

272. R. homalodontus (*omalodontos*) Muell. & Wirtg. ap. Wirtg. 1860, *Herb. Rub. Rhen.* ed. 1, no. 146; *R. foliosus* ssp. *homalodontus* (Muell. & Wirtg.) Sudre, 1908–13, 147, t. 141.

Stem blunt, striate, *becoming blackish purple, with minute pricklets* and small, equal glands and acicles; prickles rather short. Leaves 3, 5-nate; *leaflets cordate-based*, pubescent beneath, *serrate-dentate; terminal leaflet roundish ovate, acuminate*. Panicle with 1–4, large, simple leaves, broad and nearly equal, the branches compound, the pedicels long and spreading, or all the branches nearly simple and ± erect; rachis clothed with dense, short and long hair and very numerous, short glands and acicles and some longer ones, prickles few, weak, oblique. Flowers 2–2·5 cm.; *sepals* with greenish, leafy tips, *loosely erect; petals obovate, white*; filaments equalling styles. Carpels glabrous.

England: rare, woods; 17 (Newlands Corner), 18 (Epping Forest), 23 (Combe Wood, Cuddesdon), 30. France (*Valois* and *Nord*), Germany (*Eifel*).

273. **R. truncifolius** Muell. & Lef. 1859, *Pollichia*, 139; *R. insericatus* ssp. *truncifolius* (Muell. & Lef.) Sudre, 1908–13, 119, t. 143; *R. peninsulae* Rilst. 1950, *Journ. Linn. Soc.* **53**, 418.

Axes deep purple. Flower-buds, pedicels and top leaves closely grey-white felted. Stem angled *with flat sides*, glands and acicles very short; prickles moderate, subulate, declining. *Leaves 3* (4, 5)-*nate*, pedate; leaflets glabrescent or greyish felted beneath, finely or rather coarsely serrate-dentate, the principal teeth sometimes larger and patent; *terminal leaflet* roundish or more often ± broadly *obovate-subcuneate, with a ± abrupt, short point. Panicle* long, narrow pyramidal, *terminating upwards in as many as 12–24, closely placed, 1–2 (3)-flowered, spreading branches*; rachis densely pilose below with added felt above and numerous, mostly short, deep purple glands, and fine, small, slanting prickles. Flowers *c.* 2 cm.; *sepals* glandular-punctate, awned, not armed, *loosely reflexed or patent; petals 5, 6 or 7*, narrow obovate-cuneate or elliptic, tapered below, rosy pink, pink or pinkish; filaments white, sometimes pinkish in drying, or pink-based, equalling or slightly exceeding the red styles. Carpels glabrous or pilose. The structure of the fully developed panicle strongly recalls that of *R. longithyrsiger* and *R. thyrsiflorus* in this group.

England: local; 1 (Hendra, Truro), 2 (frequent, Haye Lane, Callington), 3 (Cornwood railway station; near Shaugh Bridge), 16 (south of Shooters Hill in several spots). France (*Valois*), Belgium, Germany.

274. **R. concolor** W. Ley, 1846, *Bot. Centralblatt*, **1**, 434.

4x. *Stem round*, green, becoming red-brown, with very short acicles, glands and pricklets; *prickles very short, recurved. Leaves rather small*, 3, 4, 5-nate, pedate, *petiole short, prickles small, curved; leaflets green on both sides*, glabrescent beneath, *finely serrate*; terminal leaflet short-stalked, subcordate-ovate, acuminate. Panicle long, lax, leafy, *branches short, deeply divided or bunched in the axils of the 3-nate and simple leaves*; rachis wavy, hairy and with crowded, very short glands and *tiny, hooked prickles. Flowers 1–2 cm.; sepals green*, ovate-lanceolate, attenuate, *soon clasping*; petals glabrous, narrow elliptic-cuneate, pinkish; filaments white, equalling the greenish or reddish styles. Carpels glabrous. Fruit roundish, perfect.

Appendiculati

England: very rare; 15 (Bigbury), 16 (Shooters Hill, Oxleas Wood). Extinct at Eupen, the only known station for it on the Continent. Cf. W. Watson, 1951, *London Naturalist, Suppl.* 96.

3. Stem glabrous or glabrescent, ± pruinose.

275. **R. cyclophorus** (Sudre) W. Wats. 1946, *Journ. Ecol.* **33**, 340; *R. melanoxylon* microg. *cyclophorus* Sudre, 1908–13, 165, t. 158.

Kent and Surrey. Robust. Stem roundish, purple, shining at the growing point; *prickles* unequal, fragile-pointed, declining, *purple. Leaves large*, 5-nate, prickles straight or nearly straight; leaflets broad, imbricate, plicate, thinly hairy beneath; serrate-dentate, undulate; *terminal leaflet subcordate, roundish, cuspidate. Panicle* with several, long-stalked, ovate leaves which are glandular above, *large and very compound, the middle branches 3–20-flowered, cymose; rachis stout,* felted and pubescent, prickles short, glands some short and some long. *Flower-buds large, depressed,* pubescent; flowers *c.* 2·5 cm.; sepals ending in very long, linear tips, erect; petals remote, pink, pinkish or white, elliptic, short-clawed; filaments white, equalling or exceeding the red-based styles; *disc broad.* Carpels thinly pilose. *Fruit soon turning crimson, large, ovoid-oblong,* aromatic.

England: local; 15 (Bigbury Hill), 16 (Shooters Hill; Lesness Wood; Old Park Wood, Plumstead; Bexley Wood), 17 (Walton Heath, Mrs M. L. Wedgwood, 1923). France (*Valois* and *Puy-de-Dôme*).

276. **R. pseudo-Bellardii** (Sudre) W. Wats. 1952 (for 1951), *Lond. Nat.* Suppl., 97; *R. tereticaulis* var. *pseudo-Bellardii* Sudre, 1904, *Ronces Bretonnes,* 20.

Very strong. Very large-leaved. Stem round, with numerous short glands and acicles, and some longer acicles and intermediate prickles; prickles short, subulate. Leaves 3–5-nate; *leaflets* all *elliptic-cuspidate, entire-based, serrate. Panicle* long, narrow, *branches* rather short, 1–3-flowered, *some divided to the base*; rachis very shortly hairy, prickles few, weak, but crowded on the pedicels. Flowers *c.* 2·5 cm.; *sepals* green, *loosely erect*; petals elliptic, notched, glabrous, pinkish, slightly longer than the styles. Carpels glabrous. Receptacle pilose. *Fruit* moderate, ovoid, *soon blood-red.* The 5-nate leaves at once distinguish it from *R. Bellardii.*

England: very rare; 14 (Buckhurst Park), 17 (Coombe Wood, Kingston). Brittany, Belgium, Bavaria, Switzerland.

277. **R. luteistylus** Sudre, 1900, *Rub. Pyr.* 76; *R. glaucellus* ssp. *luteistylus* (Sudre) Sudre, 1908–13, 171, t. 165.

Bedfordshire. A large-leaved, large-panicled bramble. Stem angled, brownish red, with *scintillating* particles and *sparse, short glands and acicles*; prickles short, slanting and falcate. *Leaves large* or very large, 3, 4, 5-nate; leaflets thin, subimbricate, thinly hairy beneath, *evenly and shallowly serrate; terminal leaflet cordate, broad or very broad ovate or obovate, acuminate. Panicle broad, pyramidal, very floriferous*, the upper part narrowed, dense with the entangled pedicels of panicled branches, springing often from the axils of felted, simple leaves, with more and more remote, long-peduncled, 20–30-flowered branches; rachis pilose with numerous, fine prickles and crowded, short, red glands and acicles. *Flowers 1–2 cm.; sepals deep green, conspicuously bordered with white felt*, long, slender tipped, *semi-erect*; petals elliptic, short-clawed, apiculate, fimbriate, pinkish; *filaments* white, *distinctly shorter than the reddish-based, yellowish styles*. Carpels glabrous. Fruit small, abundant. *Flowering late in July.*

England: 30 (Aspley Wood, Woburn). France *(Pyrenees, Tarn)*, Germany, *Silesia*, Switzerland.

278. **R. scaberrimus** Sudre, 1898, *Rub. Pyr.* 19; *R. glaucellus* ssp. *scaberrimus* (Sudre) Sudre, 1908–13, 170, t. 165.

Tunbridge Wells. Robust, large-leaved. Stem stout, roundish, brownish *to deep violet, slightly glandular*; prickles rather clustered, short, unequal, broad-based. *Leaflets 3–5, short-stalked, broad, cordate, imbricate*, glabrescent to greyish felted beneath, sharply serrate; *terminal leaflet roundish or very broadly ovate or obovate*, shortly pointed. *Panicle* leafy, long and broad, *with 3–7-flowered branches divided cymosely above the middle; rachis round*, felted and pubescent, densely and shortly glandular, aciculate, prickles small; pedicels felted, with subsessile glands and numerous tiny prickles. Flowers 2–3 cm.; sepals felted and pubescent, gland-dotted and prickly, patent; petals broadly ovate, narrowed below, retuse, pinkish; filaments longer than the styles. Carpels glabrous. Fruit globose-ovoid.

England: 14 (hedge east of Tunbridge Wells cemetery). France, Belgium, Germany.

279. **R. laxatifrons** W. Wats. 1946, *Journ. Ecol.* **33**, 340; *R. acutifrons* var. *amplifrons* A. Ley, 1902, *Journ. Bot.* 1902, **40**, 69.

Herefordshire. Stem slender, *red-brown, furrowed, with rather numerous pricklets but fewer glands and acicles*, all short; prickles short, slender, declining and falcate. *Leaves 3–4-nate*; leaflets thinly pilose beneath, evenly, rather angularly crenate; *terminal leaflet cordate, roundish-ovate, cuspidate. Panicle* long, a little tapered, *with ovate to small linear-lanceolate leaves to the apex*, the upper branches 3–5-flowered, the lower ones up to 7-flowered, all divided above the middle, or the whole panicle merely subracemose; *rachis and branches villous*, prickles fine, falcate, with numerous sunken and rarely longer glands and acicles. *Flowers small, 1–1·5 cm.; sepals narrow triangular-ovate, green, villous, soon clasping; petals narrow elliptic*, pinkish to white; filaments white, longer than the greenish styles. Carpels glabrous.

Endemic. England: 36 (Big Wood, Whitfield; A. Ley's original station: I know of no other).

Appendiculati

Subser. ††††**Obscuri** W. Wats.

(Type *R. obscurus* Kalt.)

1. Stem glabrous or glabrescent, rarely remaining very slightly hairy.
(Cf. *R. aggregatus* Kalt.)

280. **R. obcuneatus** Lef. & Muell. 1859, *Pollichia*, 196; *R. oigocladus*
sensu Rogers, 1900, 65; non Lef. & Muell. 1859; *R. longithyrsiger* var.
botryeros Rogers, 1900, 77; *R. cenomanensis* Sudre, 1904, *Observ. Set. Brit.*
Rub. 127.

4x, 5x. *Stem blunt,* striate, purplish *glaucescent, a little hairy* at first; prickles
red, moderate to short, slanting and falcate. *Leaves* 3 (4, 5)-nate, *rather large,*
prickles mostly straight; leaflets pilose on the veins at first, sometimes
glabrescent, or slightly felted beneath, finely and sharply serrate or serrate-
dentate; terminal leaflet base emarginate or entire, *roundish or broad obovate,*
narrowed downwards, rather abruptly *cuspidate.* Panicle ± equal or subpyramidal, *dense, domed and subracemose above,* interrupted below with erect,
panicled, axillary branches; *rachis stout and stiff,* felted and hairy with mostly
sunken glands and acicles and a few subulate prickles. *Flowers up to 3·25 cm.*;
sepals greenish with a white edge, patent to loosely erect; *petals pink to*
pinkish, elliptic-obovate, often notched; *filaments white equalling or slightly*
exceeding the red, or reddish-based, *styles.* Carpels pilose, rarely glabrous.
Receptacle pilose. *Fruit large, oblong.*

E., W., S., I. Well distributed, but local; 1–5, 11, 12, 14, 16, 22, 23, 27,
30, 33, 35, 36, 38, 43, 51, 53, 90, 94. H. 33. Jersey, France (*Aisne, Oise*),
Belgium, Switzerland, Bavaria.

281. **R. obscurissimus** (Sudre) W. Wats. 1946, *Journ. Ecol.* **33**, 341; *R.*
obscurus ssp. *obscurissimus* Sudre, 1899, *Bull. Soc. Bot. Fr.* **46**, 91; 1908–13,
159, t. 153.

4x. *West Kent and Surrey. Robust. Axes reddening. Stem blunt, pruinose*;
prickles unequal, some long subulate. *Leaves large,* 3 or 5-nate; leaflets green
beneath, serrate or serrate-dentate; *terminal leaflet subcordate, roundish to broad*
oblong-obovate, shortly pointed. Panicle short pyramidal with usually one
simple leaf, subracemose above; the terminal leaflets roundish, cuspidate, the
upper leaves greyish felted beneath. *Flowers 3 cm.*; sepals patent; petals 5–6,
ovate, margin glabrous; filaments and styles pink-based. Carpels glabrous.
Fruit moderate. It occurs with white petals, white stamens and red styles in
Spring Park, Addington, 17.

England: very rare; 16 (Barnet Wood, Hayes), 17 (Littleheath Wood,
Selsdon), 20. France (*Tarn,* frequent), Germany (one station in *Rhine*
Province).

2. Stem ± hairy.

† Sepals patent or erect. Filaments long.

(*a*) Panicle prickles straight and mostly patent.

282. **R. entomodontus** (*entomodontos*) P. J. Muell. 1862, *Ann. Fl. Fr. et All.;*
R. obscurus ssp. *entomodontus* (P. J. Muell.) Sudre, 1908–13, 157, t. 153.

Robust. Stem purplish red, blunt-angled, *intricately hairy; prickles* unequal, *patent, some long subulate.* Leaves 3, 4, 5-nate, pedate, *petiole prickles spreading-falcate; leaflets* imbricate, *thickly hairy on the veins to white felted beneath; terminal leaflet roundish ovate, acuminate-cuspidate, coarsely incise serrate*-undulate, to subcompound-serrate. Panicle equal, broad, dense, leafy above, middle branches deeply and irregularly divided, with long, remote, axillary branches below; *rachis hirsute.* Flowers *c.* 2·5 cm.; *petals broad* ovate or elliptic with a slender claw, usually *pink;* filaments white or pinkish; styles often red. Carpels glabrous. Fruit rather large.

England: rather rare; 2 (Polperro valley), 3 (Plympton (as *R. macranthelos* sensu Rogers non Marss.)), 16, 17, 30, 62. France (*Oise*).

283. **R. erraticus** Sudre, 1899, *Bull. Soc. Bot. Fr.* **46**, 91; *R. obscurus* ssp. *erraticus* (Sudre) Sudre, 1908–13, 158, t. 153.

Axes slender, villous. Stem deep red, round below to angled with flat sides, *pruinose; prickles* deep red, *acicular. Leaves yellowish green,* 3 (5)-nate, *prickles fine, slanting;* leaflets glabrescent above, softly hairy and pubescent to greyish felted beneath, *finely, shallowly, rather evenly serrate; terminal leaflet elliptic-obovate, acuminate.* Panicle long, leafy throughout, the upper branches usually 3-flowered, the lower ones 5–7-flowered, deeply divided; *rachis* flexuose, *villose, prickles numerous, weak, slanting.* Flowers *c.* 2 cm.; sepals green, short-pointed, erect; petals elliptic, pinkish, margin glabrous; filaments white; styles red. Carpels glabrous. *Fruit* roundish ovoid, *deep crimson before ripe.* Very fertile and with very good pollen.

Another form of *R. erraticus* occurs in 36 (Belmont, near Hereford) and 49 (Llanberis Falls), distinguished by its smaller size, roundish to obovate terminal leaflets, rosy petals, short stamens and pilose carpels. It is *R. fuscus* ssp. *obscurus* sensu Rogers 1900, 74; non *R. obscurus* Kalt. It is possibly *R. erraticus* var. *obscurus* Sudre.

England (Wales): rare; 13 (Cissbury), 17 (Wareham Hill, Witley), 23 (Aston Wood, Aston Rowant). France, Belgium, Germany, Switzerland.

284. **R. adornatiformis** (Sudre) Bouv. 1903; *R. insericatus* ssp. *adornatiformis* Sudre, 1908–13, 150, t. 143; *R. obscurus* sensu Genev.

Stem slender, sides flat, reddish purple; *prickles* patent, *numerous, slender, subulate,* unequal, *some long. Petiole prickles straight,* numerous. Leaflets 3–5, thinly pilose on the veins and greyish felted beneath, *coarsely, unequally, and deeply crenate-serrate; terminal leaflet roundish or broad elliptic, shortly pointed; basal leaves 5-nate, very shortly stalked. Panicle lax, broad, short, pyramidal-*

truncate, the middle branches patent, 1–3 (5)-flowered and long, ascending lower branches; *rachis crooked, hirsute*, prickles and glands crowded, with a few, fairly long acicles and stalked glands intermixed. Flowers 1·5–2 cm.; sepals felted and pilose, densely glandular; petals remote, broad elliptic, rose-pink, fimbriate; filaments pink or white, equalling or slightly exceeding the greenish or red styles. Carpels slightly pilose. Fruit small, subglobose.

England: not frequent; 15 (Ileden Wood), 16 (Hosey Common, frequent), 17, 20 (Oxhey Wood), 30. France, S. Germany, Switzerland, Styria.

285. **R. obscurus** Kalt. 1845, *Fl. Aach. Beck.* 281; Sudre, 1908–13, 156, t. 152; non *R. fuscus* ssp. *obscurus* sensu Rogers, 1900, 74.

Robust. Stem blunt-angled, densely long-pilose, purple, glaucescent; prickles long, patent, bright red. Leaves 3, 4, 5-nate, pedate, *petiole short, prickles straight*; leaflets glabrescent above, becoming convex, at first softly hairy beneath, *the branch and upper leaves grey white felted*, sharply unequal or ± compound serrate; terminal leaflet elliptic-obovate-acuminate, subcuneate, base nearly entire. *Panicle large and compound*, narrowed to the rounded apex, *mid branches long-peduncled*, 3–5-flowered, cymose, *lower branches still longer, numerous flowered*; rachis flexuous, densely pilose, prickles short, yellowish, straight. *Sepals* felted and pilose *with leafy tips cresting the bud*, patent-erect; petals narrow, rose; filaments white or rose; styles pale. Carpels pilose. *Fruit large and good, ovoid-oblong.*

England: rare; 15 (Chartham Hatch to Bigbury), 16 (Tunbridge Wells Common; Shooters Hill), 17 (St Ann's Hill), 39. Belgium, Germany, Switzerland, Austria.

(*b*) Panicle prickles slanting, falcate or hooked.

286. **R. rufescens** Lef. & Muell. 1859, *Pollichia*, 152; *R. Hystrix* ssp. *rufescens* (Lef. & Muell.) Sudre, 1908–13, 181, t. 178; *R. velatus* Lef. ap. Genev. 1872, *Mem. Soc. Ac. M. et L.* **28**, 32; *R. rosaceus* ssp. *infecundus* (Rogers) Rogers, 1900, 80. **Fig. 36.**

4x. *Prickles and axes rufescent, leaves yellowish green.* Stem arching, blunt; *prickles* rather scattered, *subulate, fragile-pointed.* Leaves rather large, petiole prickles nearly straight; *leaflets harsh above and soft beneath, with a few pricklets and glands on the midrib and the basal margins*, simply serrate or serrate-dentate; *terminal leaflet roundish* or oblong-obovate or somewhat narrowed below, *shortly pointed*, branch leaves and upper leaves slightly felted. *Panicle pyramidal, lax, floriferous, the long pedicels entangled*; rachis villous below, felted above, densely glandular with some longer acicles, prickles acicular. Flowers *c.* 3 cm.; *calyx rather large with a flat base and shouldered sides, sepals* narrow ovate, attenuate, *clasping the blood-red carpels; petals 5–8*, pilose, rosy, *remote*, narrow elliptic, retuse; filaments white, rarely pink, *styles red.* Carpels glabrous or pilose. Fruit ovoid. A dominant bramble in many woods in the home counties, especially on pebbly and sandy soils.

E., W., S., I. Common in England in woods, in shade or sun; 3, 6–8, 10–13, 15–24, 26, 29, 30, 33–42, 49, 53, 55, 57–59, 67, 70, 97, 99. H. 38. N. France and Belgium.

287. **R. Purchasianus** (Rogers) Druce, 1908, *Brit. Pl. List*, ed. 1, 22, no. 851.

Whole bramble slender, very villous, leaflets subcuneate. Stem prostrate, round to flat-sided, fuscous; prickles numerous, unequal, confluent, declining and recurved. *Leaves rather small,* 3, 4, 5-nate, pedate, petiole prickles hooked; *leaflets irregularly subcompound-serrate, undulate*; terminal leaflet obovate, with a long point and an entire base. *Panicle not leafy, narrow, slightly tapered,* dense, with deeply divided 3-flowered branches and long, linear bracts above, and short, erect branches below; lower leaves often 4, 5-nate; rachis and branches villous and armed with numerous weak prickles. *Flowers c. 1·5 cm.; sepals* villous, aculeate, *narrow* triangular or oblong-attenuate, *soon erect; petals narrow* elliptic, tapered below, pale pink or bright rose, subglabrous, *often 7 on the terminal flowers*; filaments white or pink, longer than the red styles. Carpels glabrous. *Fruit small, subglobose, sweet.*

Endemic. England: rare except in a few western counties; 13 (Madehurst and Slindon), 23 (Crowell Hill), 34–37, 57.

288. **R. cruentatus** P. J. Muell. 1859, *Pollichia,* 294.

Stem angled becoming fuscous and sometimes pruinose, glands, acicles and pricklets rather scanty; *prickles short and broad-based in woods, long but unequal, straight and curved mixed in the open. Leaves deep green,* 3–5-nate; leaflets subimbricate, moderately pilose and *green beneath*, subcordate, *rather flatly crenate in shade* but deeply, sharply and coarsely serrate in the open; *terminal leaflet roundish ovate or a little rhomboid*, usually shortly acuminate. *Panicle* leafy nearly to the top, *moderately long, lax, with a crooked rachis*, short, deeply divided, few-flowered upper, and long, numerous flowered, lower branches, all hirsute, glandular and aciculate, *prickles* numerous, unequal, *slanting and curved. Flowers very showy, arresting, c. 2·5 cm.; sepals green, ovate-lanceolate or triangular-attenuate, patent-erect; petals 5, 6, 7, ovate or elliptic, deep rose-pink; filaments and styles also deep rose-pink*, or the styles sometimes whitish; filaments about equalling the styles. Fruit rather small, roundish.

England: local; 14, 16, 18, 20, 24, 39. France, Belgium, Denmark, W. Germany, Austria.

289. **R. grypoacanthus** Lef. & Muell. 1859, *Pollichia,* 133.

Stem roundish, becoming deep purple; *prickles rather short, much recurved. Petiole prickles numerous, hooked. Leaflets* 5, entire-based, glabrescent above, thinly pilose beneath and *finely grey felted, rather evenly serrate or serrate-dentate*; terminal leaflet elliptic or elliptic-oblong, sometimes contracted below; *basal leaflets large. Panicle* leafy to the top with the leaves grey felted, long and equal or subpyramidal, *most of the branches short, deeply divided,*

3–5-flowered with sometimes a further stalked flower at the base; rachis hirsute, *with* sunken glands and acicles and *numerous, small, hooked prickles.* Sepals hirsute, aculeate, at length completely clasping. Petals ovate, notched, rose-pink. Filaments pink, equalling the red styles. Carpels pilose. Fruit ± globose.

England: rare; 11 (Pike Hill), 17 (The Chart Quarry, near Limpsfield), 21 (Copse Wood). France, *Aisne.*

290. **R. aggregatus** Kalt. 1845, *Fl. Aach. Beck.* 275; *R. obscurus* ssp. *aggregatus* (Kalt.) Sudre, 1908–13, 157, t. 153.

Extremely vigorous. Stem green, *stout, blunt, glaucescent,* ± *hairy, glabrescent above,* with numerous to few very short glands, acicles and larger pricklets; *prickles yellow to ochreous, short, declining from an extended base. Leaves large or very large,* bright green, 3–5-nate, thinly hairy to greyish felted beneath, *shallowly, only slightly irregularly, crenate-serrate;* terminal leaflet roundish to broad ovate, shortly pointed, base cordate to entire. *Panicle very strong, large, the very dense subcorymbose top numerous flowered,* with distant, long-stalked, cymose, then, below, panicled axillary branches; rachis crooked, ± hirsute and felted above, glabrescent below, the upper prickles sometimes subulate on strong panicles and peduncles, *more often small, slanting and falcate;* pedicels and calyces felted, with red glands and prickles; *upper simple leaves grey felted beneath.* Flowers *c.* 2 cm.; sepals erect; *petals* 5–7, roundish cuneate to obovate, pilose, *rosy pink; filaments* white, turning carmine, *slightly longer than the red styles.* Carpels pilose. Receptacle hairy. Fruit roundish ovoid. 'Pollen perfect or with very rare deformed grains', Sudre. I collected a specimen of *R. aggregatus* at Neuhofen, near Aachen, Kaltenbach's station, in 1937.

England: very rare; 20 (Sherrards Park Wood, Welwyn Garden City). Germany (*Neuhofen and Marienhagen, in the Rhineland*), Belgium (*Vleurgat*).

291. **R. fuscoviridis** Rilst. 1940, *Journ. Bot.* **78**, 166.

Cornwall. Axes hirsute and fuscous, the young shoots bronze. Stem angled, with numerous sunken glands and acicles and some unequal pricklets; prickles purple and pilose below with glossy yellow tips. *Leaves* 5-nate, *petiole short, green and rough with hairs on the veins beneath,* equally incise crenate or sharply serrate; *terminal leaflet* cordate, roundish or *broad obovate, cuspidate.* Panicle equal, often short and dense, subracemose with 1–2 simple leaves which are greyish felted beneath; or longer with distant and erect racemose branches from the axils of 3-nate or 5-nate leaves below; rachis slightly wavy, clothed and armed like the stem. *Flowers c. 2–3 cm., incurved; sepals* ovate, acuminate with a short linear point, becoming *concave-patent under the fruit; petals broad obovate,* narrowed below, *bright pink;* filaments white or pinkish, longer than the *reddish styles,* anthers green, subpilose. Carpels pilose. *Fruit rather large,* sweet at first then soft and mawkish, recalling the fruit of *R. vestitus. Flowering from the first week of June,* in east Cornwall.

Endemic. Rare; 2 (Langreek, Longcombe, Newbridge, Port Gaverne).

†† Sepals patent or erect. Filaments short, equalling or shorter than the styles.

292. **R. sprengeliiflorus** (Sudre) W. Wats. 1956, *Watsonia*, **3**, 289; *R. erraticus* ssp. *sprengeliiflorus* Sudre, 1906, *Diagnoses*, 41, t. 153.

Stem angled, purple, *glabrescent; prickles* rufescent, numerous, rather long, unequal, declining and falcate. *Leaves yellowish green*, 3, 5-nate; *leaflets short-stalked, imbricate, entire-based*, pilose beneath, *finely and unequally crenate-dentate; terminal leaflet* broad elliptic-ovate and *like the intermediate leaflets finely attenuate. Panicle* broad, lax, obtuse, *with several small simple leaves, branches* cymose or corymbose, 3–6-flowered, *spreading*, or the lower branches panicled; rachis, peduncles and pedicels slender, villose with numerous sunken glands, acicles and *short, curved prickles*. Flowers *c.* 2 cm.; *sepals greenish, white-bordered, ovate, attenuate, leafy-tipped, becoming erect; petals narrow* elliptic-obovate, notched, *rose*; filaments white; styles greenish. Carpels densely pilose or subglabrous. Fruit moderate, globose.

England: rare; 14 (Hindleap, near Wych Cross, frequent; Heathfield), 35 (west of Tintern). France (*Aisne*).

293. **R. obscuriformis** (Sudre) W. Wats. 1956, *Watsonia*, **3**, 289; *R. obscurus* microg. *obscuriformis* Sudre, 1907, *Fl. Toul.* 75.

Stem blunt angled with hollow sides, with short glands and acicles; *prickles yellowish, small*, unequal, *falcate*. Leaves 5-nate, pedate; *leaflets* thinly pilose beneath, *sharply erect-serrate; terminal leaflet broad obovate, acuminate*, slightly narrowed below. *Panicle short, leafy, subracemose* and *dense, branches sharply ascending*, with usually two erect axillary branches; rachis hirsute with submerged glands and acicles, and weak, slanting prickles. Flowers *c.* 1·8 cm.; sepals ovate, greenish grey, patent; petals elliptic, tapered below, glabrous, notched, pink; *filaments* white, *slightly shorter than the red styles*. Carpels glabrous. Receptacle glabrous. Fruit subglobose.

England: 16 (Jack Wood, Shooters Hill, at *c.* 360′). E. and S. France, Bavaria.

††† Sepals reflexed.

294. **R. insericatus** P. J. Muell. 1858, *Flora*, 184; Sudre, 1908–13, 148, t. 142.

4x. *Axes slender, reddening.* Stem red, arched then prostrate, long, angled, striate; prickles red, unequal, some smaller ones standing on the sides. Leaves 3, 5-nate, pedate, *bright green; leaflets silkily hairy* on the veins beneath, to sparsely felted on the upper leaves, *incise serrulate-denticulate* with some teeth patent; *terminal leaflet* from a rounded base, *rather narrow (rhomboid) elliptic, gradually and finely acuminate. Panicle* pyramidal, not leafy, subracemose at the narrow and pointed apex, *the middle branches patent, the lowest ones long and spreading*; rachis villous, with numerous, mostly short glands and acicles, and fine prickles, *pedicels felted, the numerous glands almost sessile; terminal leaflets* serrulate. Flowers 1·5–2 (2·5) cm.; sepals triangular-ovate, with long, linear tip; *petals narrow* elliptic or ovate, narrowed to the base, *remote*, pink; filaments white, shorter or not much longer than the red styles. Carpels glabrous. Receptacle pilose. *Fruit small, roundish, perfect.*

England: 16 (Shooters Hill, frequent). France, Belgium, W. Germany, Switzerland.

ssp. **Newbouldianus** (Rilst.) W. Wats., 1956, *Watsonia*, **3**, 289; *R. Newbouldianus* Rilst. 1950, *Journ. Linn. Soc.* **53**, 417; ? *R. insericatus* microg. *silvigenus* Sudre, 1903, *Bat. Eur.* **13**, no. 36.

Differs from the main species in the stem being glaucescent and less hairy, the leaflets more sharply and deeply incise-serrate, in the *long stamens, pilose carpels*, and the *much larger, oblong fruit*. It is also a *more robust* bramble.

England: 1–4. ? S. France.

295. R. rhombophyllus Muell. & Lef. 1859, *Pollichia*, 175; *R. insericatus* ssp. *rhombophyllus* (Lef. & Muell.) Sudre, 1908–13, 148, t. 143.

Stem reddish purple with scattered, short glands and acicles; *prickles* often in twos and threes, *strong-based, patent and declining*. Petiole prickles hooked; *leaflets* 3, 4, 5, glabrous above, pilose and greenish to *grey-white felted* beneath, *shallowly*, rather *coarsely angled* and *unequally crenate-serrate, a little jagged; terminal leaflet obovate-rhomboid, acuminate. Panicle leafy, pyramidal, dense-flowered,* the *pedicels being short and the flowers numerous, the lowest branch usually very large;* rachis crooked, hirsute with a moderate number of glands and acicles, prickles numerous, strong, long, slightly falcate; leaves felted beneath. Flowers showy, *c.* 2·5 cm.; sepals grey felted and pilose, slightly glandular-punctate; *petals rosy pink; filaments rosy pink*, longer than the yellowish, *rosy-based* styles. Carpels glabrous. *Fruit large, rather long.*

England: rare; 16 (Eltham Common), 17 (Tilburstow Hill), 24 (Rassler Wood, Medmenham). France, Germany, Belgium.

296. R. Gravetii (Boul. ex Sudre) W. Wats. 1946, *Journ. Ecol.* **33**, 341; *R. insericatus* ssp. *Gravetii* Boul. ex Sudre, 1908–13, 149, t. 143.

Stem blunt, striate, densely villous with very small glands and acicles; prickles numerous, slanting and falcate. *Leaves large,* 3–5-nate, pedate, light green; *leaflets* nearly glabrous and glaucescent above, *very softly hairy and green beneath*, rather *regularly crenate-serrate; terminal leaflet emarginate, elliptic, acuminate.* Panicle pyramidal but not pointed, the middle and lower branches ± long and spreading but the pedicels short; *rachis villous, wavy*, prickles subulate, slanting, *terminal leaflets elliptic, evenly crenate-serrate*, finely acuminate. Sepals loosely reflexed. Petals elliptic or narrow obovate with a slender claw, margin glabrous, deep pink. Filaments pink, equalling or slightly exceeding the deep red-based styles. Carpels pilose or glabrous. Fruit moderate, deep crimson before turning black.

England: very rare; 15 (Bigbury to Chartham Hatch), 18 (Epping Forest, High Beach). Belgium, W. Germany (? Switzerland).

Subser. ††††† **Incompositi** W. Wats.

(Type *R. apiculatus* Weihe)

1. Stem either glabrous or, if slightly hairy at first, soon becoming glabrous.

† Leaves felted beneath.

297. **R. apiculatus** Weihe in Bluff & Fingerh. 1825, *Comp. Fl. Germ.* **1**, 680; Sudre, 1908–13, 132, t. 129; *R. anglosaxonicus* Gelert, 1888, *Bot. Tidssk.* 16, 81; *R. curvidens* A. Ley, 1894, *Journ. Bot.* **32**, 143.

4x. *Vigorous.* Stem at length red, angled, striate, glaucescent, with scattered short glands and *numerous acicles* and pricklets; prickles unequal, slender, finely pointed. *Leaves large,* (3) 5-nate, pedate; *leaflets long, twisted,* glabrescent above, *felted and roughish with short, erect hairs beneath,* unequal, in part patent serrate, *jagged; terminal leaflet usually elliptic obovate, shortly acuminate, often subcuneate below, base rounded; the intermediate leaflets narrower and more narrowed to the base.* Panicle leafy, *long,* interrupted, *narrowed to the top,* branches rather short, semi-erect; rachis angled, closely pubescent and felted with glands and acicles, *the acicles being particularly numerous below the panicle,* prickles yellowish, sharply declining. *Flowers* 2·5–3·5 (4·5) *cm.; sepals* felted, glandular and with whitish acicles, *loosely reflexed or patent;* petals 5–6, elliptic or ovate, subglabrous, *rose-pink or pinkish; filaments pink or white, about equalling the bone-coloured styles.* Carpels usually glabrous. Fruit moderately large, roundish ovoid.

E., W., I. Rather frequent in woods on clay; 2–4, 6, 8, 9, 11, 13–17, 19–24, 26, 35–38, 40, 41, 44, 49, 51, 55–57. H. 1, 2, 26, 38–40. France, Belgium, Germany, Switzerland, Austria, Hungary, Caucasus.

298. **R. heterobelus** Sudre, 1904, *Observ. Set. Brit. Rub.* 8; *R. granulatus* microg. *heterobelus* (Sudre) Sudre, 1908–13, 139, t. 135; *R. penhallowensis* Rilst. 1950, *Journ. Linn. Soc.* **53**, 419; *R. cognatus* Brown in Smith pro parte, 1892, *Eng. Bot.* ed. 3, *Suppl.* 101; *R. praeruptorum* sensu Rogers, 1892, *Journ. Bot.* **30**, 301; *Set. Brit. Rub.* no. 15; non Boul. 1868; *R. debilis* sensu Bab.; non Boul. 1868. **Fig. 37.**

Moderately strong. Axes crimson. Stem low-arching, blunt, striate, *shining* to glaucescent, with a few, or many hairs at first, *scattered short glands and acicles, and stout pricklets; prickles bright red, patent, small-based, often a short one against a very long one. Leaves* rather large; leaflets 5, *very softly hairy to felted beneath,* deeply, *coarsely,* often subcompound *crenate-serrate,* undulate; terminal leaflet subcordate, roundish ovate, obovoid-rhomboid or elliptic, acute to acuminate or somewhat cuspidate. *Panicle with 1–5 simple leaves, rather narrow, dense and equal, or lax pyramidal, the branches spreading at a low angle;* rachis crooked and glabrescent below, grooved, felted and moderately hairy above, with short glands and various intermediate gland-tipped structures, *upper prickles subulate, long, exactly patent*; lateral leaflets very shortly stalked. Flowers up to 3 cm.; sepals greenish, grey felted with a white margin, glandular and prickly, loosely reflexed to patent; petals 5–6 rhomboid-ovate, rose-pink; filaments white or pink, much longer than the yellowish styles. Fruit moderate. *In shade the plant becomes everywhere more villous.*

Almost endemic. E., W., I. Well distributed but not common; 1–3, 7–9, 13, 14, 16–20, 23, 24, 30, 36, 37, 39, 40, 43, 48, 49, 52, 57, 62. H. 22, 33. S. France (*Tarn, one station*), Belgium (*one station*).

299. R. raduloides Rogers, 1900, 57.

Axes dark red. Stem furrowed *with* few or numerous stalked glands and *unequal acicles, some gland-tipped; prickles* numerous, unequal, *some subulate, long and patent.* Petiole prickles straight or slightly falcate. Leaflets 3–5, plicate, glabrous above, *soft grey-white felted and pilose beneath, sharply incise, doubly serrate*, with long and in part patent mucros; terminal leaflet elliptic-obovate, acuminate, base entire. Panicle usually long and narrow with few simple leaves, middle branches 3–7-flowered, patent; *rachis pilose above, with crowded, rather long-stalked glands and acicles exceeding the hair, prickles numerous, long, subulate, nearly patent.* Flowers up to 3 cm.; sepals felted and pilose, prickly, patent to erect; *petals 5–6, ovate-elliptic; bright pink*, filaments white, slightly longer than the greenish styles. Fruit moderate.

Endemic. E., S., I. Rather infrequent except in the west; 3, 6–9, 19, 22, 23, 33–38, 43, 48, 49, 74, 76 (not in 16, 17).

300. R. melanoxylon Muell. & Wirtg. 1861, *Herb. Rub. rhen.* ed. 1, no. 181; Sudre, 1908–13, 164, t. 158, non Rogers, 1900, 59.

4x. *Young shoots bronze. Stem often flexuous*, angled, *deep fuscous*, glaucescent, *with longish acicles and pricklets*; prickles unequal, strong, subulate. Leaves 3, 4, 5-nate, petiole prickles declining or slightly falcate; leaflets pubescent, *and if not felted becoming glabrous beneath*, coarsely incise-serrate to subcompound serrate-dentate; *terminal leaflet* widely cordate or emarginate, roundish ovate, oblong *or slightly obovate, acuminate-cuspidate, basal leaflets rather long-stalked.* Panicle with several *simple leaves, glandular on the upper side*, ± equal, upper branches dense, semi-erect, long-peduncled, 3-flowered, often with a further stalked flower at the base, with longer branches below; rachis crooked below, glabrescent, prickles numerous, strong, slender, declining and often a little falcate at the tip. Flowers 1·5–2·5 cm.; sepals greenish felted, ovate-lanceolate, prolonged, patent or loosely erect; *petals narrow* ovate or elliptic, some acute, fimbriate, remote, spreading, pink, pinkish or white; filaments pinkish-based or white, much longer than the greenish or reddish-based styles. Carpels pilose or glabrous. Fruit oblong. Cf. *R. Adamsii.*

England: locally frequent; 7, 13, 14, 16, 17, 19, 21, 22, 24, 30, 34, 35, 39. France, Belgium, Germany, Bavaria, Hungary, Switzerland.

301. **R. pascuorum** W. Wats. 1946, *Journ. Ecol.* **33**, 341; *R. Borreri* var. *virgultorum* A. Ley, 1894, *Journ. Bot.* **32**, 143; *R. infestus* var. *virgultorum* (A. Ley) Rogers, 1900, 60.

Western. Always *a low bush* with rather *small leaves* and a *large panicle.* Stem angled, *greyish slate purple*, glands few, pricklets numerous; prickles unequal, yellow to brown, slanting. Leaves digitate, *petiole prickles hooked; leaflets glabrous above*, felted beneath, evenly serrate-dentate; *terminal leaflet rather long-stalked, roundish* ovate or *obovate-cuneate.* Panicle leafy, large, broad and compound, *floriferous, with longer branches* and some 5-nate leaves *below*; rachis moderately pilose with numerous yellow prickles. Flowers *c.* 1·5 cm.; *sepals patent; petals roundish obovate* with a slender claw, rose-pink to pinkish; *filaments long*, white; styles greenish. Carpels pilose or glabrous. Fruit roundish.

Endemic. E., W., I. Mainly in the Welsh border counties, frequent in Herefordshire 'leasowes'; 3, 23, 35–37, 39, 40, 48. H. 40.

302. **R. longifrons** W. Wats. 1937 (for 1936), *Rep. Bot. Exch. Cl.* **11**, 221; *R. longifolius* W. Wats. 1935 (for 1934), *London Nat.* 62; non Host, 1835.

4x. Stem at first erect then procumbent, angled, *green to reddish brown*, glaucescent, with few to many, yellowish, unequal glands, acicles and pricklets; prickles short, unequal, slanting. *Leaflets 5, long and long-pointed*, glabrous above, greyish green and glabrescent to *finely chalky white felted beneath, coarsely subcompound patent-serrate; terminal leaflet short-stalked, elliptic-obovate*, base nearly entire; *basal leaflets rather long-stalked.* Panicle long, lax, narrow with some leafy bracts, equal or subpyramidal, branches 1–5-flowered, sharply ascending, or more spreading and divided to the base, sometimes with several distant, long, but few-flowered, lower branches; rachis felted. Flowers 1·5–2 cm.; sepals felted, aciculate, reflexed to patent; *petals oblanceolate, pointed, white; filaments white, equalling the greenish styles.* Carpels pilose. *Recalls R. Reichenbachii.*

Endemic. England: local; 14 (Eridge and Withyham), 16 (in several stations around Tunbridge Wells).

†† Leaves green beneath, the upper ones sometimes slightly felted.

303. **R. melanodermis** Focke, 1890, *Journ. Bot.* **28**, 133; *R. melanoxylon* sensu Bab. 1887, *Journ. Bot.* **25**, 21.

4x. Stem round to blunt angled, furrowed, *becoming black and pruinose*, striate, thinly hairy at first, with *very short glands* and unequal acicles and pricklets, some gland-tipped; prickles strong, unequal. *Leaves 3, 4, 5-nate, petiole prickles hooked; leaflets* plicate, hairy and harsh on both sides, *coarsely and rather deeply unequal serrate; terminal leaflet short-stalked*, ± cordate, moderately broad elliptic, *obovate or oblong-obovate, with a* ± *abrupt, rather long point. Panicle narrow*, hardly diminished to the top, with some *simple leaves, glandular above* and mid branches 3–7-flowered, cymose, furnished with slender, falcate prickles and *densely crowded, short glands and acicles.* Flowers 1–2 cm.; *sepals* greenish grey felted, ovate, finely pointed, *becoming patent and semi-erect;* petals 5–6, elliptic, narrowed below, crumpled, glabrous, white with a greenish claw; filaments white, slightly longer than the styles. Carpels pilose or subglabrous. *Fruit rather large, oblong-ovoid.*

Endemic. E., W., I. Rather infrequent except in the south-west; 4 (Sticklepath), 9, 11, 14, 16, 17, 19, 23, 24, 34, 41, 49, 55. H. 2, 39.

304. **R. inopacatus** Muell. & Lef. 1859, *Pollichia,* 117; *R. Colemannii* ssp. *inopacatus* (Muell. & Lef.) Sudre, 1908–13, 117, t. 116.

Robust. Stem stout, thinly hairy at first, with small pricklets and acicles and a *few glands; prickles strong, mostly patent,* unequal. *Leaves rather large,* 5-nate, *petiole prickles straight or slightly falcate; leaflets* glabrescent above, at first velvety pilose then glabrescent beneath, unequally, *coarsely* and in part patent *serrate; terminal leaflet subcordate, roundish ovate, shortly acuminate.* Panicle broad, lax, short, mid branches spreading, 3–5-flowered, cymose; *rachis* moderately pilose and like the peduncles and pedicels, *armed with numerous, strong, unequal, ± patent prickles.* Flowers *c.* 2·5 cm.; sepals felted, reflexed; *petals roundish-ovate,* pilose; filaments white, longer than the styles. Carpels pilose. Fruit moderate.

England: rare; 17 (Fairmile Common and Netley Heath), 18, 39 (Hanchurch Hills), 57. France (*Aisne, Oise*).

305. **R. Reichenbachii** Koehl. ex Weihe in Bluff & Fingerh. 1825, *Comp. Fl. Germ.* **1**, 685; Wimm. & Grab. 1829, *Fl. Sil.* **1**, 685; W. Wats. 1949, *Watsonia,* **1**, 77.

Robust. Stem glabrous, round, becoming blackish in the sun, glaucescent, *with scattered small pricklets, acicles and very rare glands which are wanting on some internodes; prickles irregularly distributed,* moderately unequal, broad-based, slanting or falcate. *Leaves large,* 5-nate, *short-stalked; leaflets very long-pointed,* glabrous above, densely to thinly pubescent beneath, passing into greyish green felt on the topmost leaves, *unequally, deeply, sharply serrate; terminal leaflet nearly round or roundish-ovate, acuminate-cuspidate; basal leaflets longstalked* (up to 1 cm.). *Panicle* broad, pyramidal, *domed at the apex,* dense and leafy, all *branches* semi-erect, numerous flowered, subcorymbose, *often in pairs;* rachis crooked, finely felted and with numerous, weak, subulate prickles, acicles and glands. *Flowers showy, c.* 2·5–3 *cm.*; sepals linear-tipped, erect; *petals broad ovate, white, glabrous;* filaments white, longer than the greenish styles. *Carpels strongly pilose. Fruit large, ovoid, abundant.*

England: 14 and 16 (east and south-east of Tunbridge Wells, around Hawkenbury and thence to Pembury). Germany, *Saxony and Silesia.*

2. Stem considerably hairy at first, leaves ± white felted beneath. (Cf. *R. heterobelus.*)

306. **R. retrodentatus** Muell. & Lef. 1859, *Pollichia*, 168; *R. fuscus* ssp. *retrodentatus* (Muell. & Lef.) Sudre, 1908–13, 144, t. 139; *R. Schmidelyanus* var. *breviglandulosus* Sudre 1908–13, 119; *R. vectensis* W. Wats. 1937, *Journ. Bot.* **75**, 198; *R. Borreri* sensu Rogers, 1900, 61; *R. Schmidelyanus* ssp. *Borreri* sensu Sudre, 1908–13, 144, t. 139. **Fig. 38.**

4x. *Rufescent. Low growing.* Stem angled, hairy at first with rather few glands and acicles; prickles hairy, *numerous, unequal, mostly falcate. Leaves* 5-nate, usually digitate, *yellow-green, short-stalked,* petiole prickles numerous, some hooked; *leaflets short-stalked,* plicate, softly felted and pilose beneath, *unequally, sharply and deeply serrate* undulate, *jagged; basal leaflets longer than the petiole; terminal leaflet* broad or narrow *obovate, cuspidate, narrowed to the entire base.* Panicle leafy, equal, or narrowing to the top, mid branches up to 5–7-flowered cymose; rachis crooked, hirsute, with sunken glands and numerous, unequal, falcate prickles. Flowers *c.* 2·5 cm.; sepals long-pointed, densely pilose and felted, shortly glandular and aculeate, *patent or in part reflexed; petals 5–8,* pinkish, *elliptic-obovate, narrowed below, often notched,* fimbriate; filaments white, longer than the yellowish, reddish-based styles. Carpels subpilose. *Receptacle pilose, especially below the lowest carpels.* Fruit setting well, roundish.

E., W., I. Mainly western, locally frequent; 1–6, 8–14 (17 extinct), 22, 34–36, 39–41, 45, 48, 57, 58, 69, 72. H. 1, 2, 11, 12, 16, 21, 37, 38. Guernsey, Jersey, France, Belgium, Styria.

307. **R. phaeocarpus** W. Wats. 1946, *Journ. Ecol.* **33**, 341; 1937, *Journ. Bot.* **75**, 157 (nomen); *R. Babingtonii* sensu Rogers, 1900, 69; Sudre, 1908–13, 151, t. 146; non Bell Salt. 1845.

*4x. *Stem climbing, blunt, furrowed, fuscous,* with numerous short glands and acicles, *pricklets and smaller prickles from a swollen base;* prickles unequal, mostly rather short. Leaves 5-nate, petiole flat above; *leaflets thick,* dull and glabrescent above, *softly and densely hairy as well as felted beneath, the principal teeth prominent; terminal leaflet oblong, cuspidate,* base nearly entire. *Panicle very long, broad,* with several simple leaves above, grading into reddish bracts, *middle branches long-peduncled, patent,* 3–5-flowered, cymose; *rachis villous,* with sunken glands, acicles and numerous acicular prickles. Flowers *c.* 2 cm.; sepals yellowish, villous, prickly, *becoming patent-convex* and dark red at the base within; petals pinkish, oblong; filaments slightly longer than the styles. Carpels pilose. *Fruit* roundish, *fuscous before ripening, rather small.*

Endemic. England and Wales, chiefly southern England, where it is frequent; 5–7, 11–24, 27, 34–38, 43, 44, 57, 58. Not Ireland; the Co. Galway form referred to by Rogers is *R. Griffithianus.*

Appendiculati

308. **R. Griffithianus** Rogers in Griffiths, 1895, *Fl. Anglesey and Carn.* 48; *R. Genevieri* subsp. *discerptus* var. *griffithianus* (Rogers ap. Griffiths) Sudre, 1908–13, 132, t. 128, figs. 20, 21.

Western. Robust. Stem angled, purple and *pruinose*, with numerous unequal glands, acicles and pricklets; *prickles bright red-based*, very unequal, *a few, very long and patent.* Petiole prickles partly patent. *Leaflets* mostly 5, *grey-white felted beneath*, sharply incise, ± compound, *partly patent-serrate; terminal leaflet* base subcordate to entire, roundish (subrhomboid) ovate to *obovate*, rather long pointed; basal leaflets short-stalked. *Panicle leafy, broad, the middle branches spreading, 5–7-flowered, cymose*, and the lower axillary branches very long; *rachis* felted and pilose with numerous, rather unequal, mostly sunken glands and acicles, together *with some very long, subulate, patent prickles in the upper part*; pedicels felted. Flowers *c.* 2 cm.; *sepals closely felted, glandular, punctate and aciculate*, patent to erect; *petals* ovate, elliptic or obovate, fimbriate, *pinkish or pink*; filaments white, far longer than the styles. Carpels pilose. Fruit roundish, moderate.

E., W., S., I. Frequent in the west; 35, 40, 49, 50, 55, 67, 71, 104, 110. H. 1, 2 and Co. Galway. France, *Sarthe.*

309. **R. indusiatus** Focke, 1877, *Syn. Rub. Germ.* 284; *R. hebecarpus* microg. *indusiatus* (Focke) Sudre, 1908–13, 182, t. 180.

South-east England. Stem fuscous in the sun, *reddish yellow at the back*, with rather numerous glands, acicles and *minute pricklets; prickles subulate,* ± *patent.* Leaflets mostly 5, glabrescent above, velvety pilose and white felted beneath, equally and finely serrate, or the principal teeth much larger and salient, or ± compound-serrate; *terminal leaflet ovate, emarginate, with a long, fine point*, or obovate from an entire base. Panicle equal or with some longer, sometimes paired, oblique, axillary branches; *rachis, peduncles and pedicels hirsute, with patent subulate to acicular prickles* and mostly sunken glands and acicles. Flowers 1·5–2 (3) cm.; sepals felted, glandular and prickly, reflexed; *petals pinkish at first*, elliptic-obovate, fimbriate and downy; *filaments white, hardly longer than the greenish styles.* Carpels pilose, rarely glabrous. Fruit roundish, rather large.

England: rare; 14–18. Belgium (very rare), Bavaria (infrequent).

310. **R. euanthinus** W. Wats. 1946, *Journ. Ecol.* **33**, 341; *R. anglosaxonicus* ssp. *vestitiformis* A. Ley ex Rogers, 1900, 58.

Western. Robust, recalling R. vestitus. Stem deep purple, *very hairy, glaucous with scattered stout pricklets*, glands and acicles; *prickles* unequal, *stout-based, mostly patent. Leaf becoming subcoriaceous in the sun*, petiole prickles hooked; *leaflets all short-stalked*, pilose and grey felted beneath, crenate-serrate and *a little jagged; terminal leaflet roundish or obovate-subcuneate with a short abrupt point*, or elliptic-rhomboid acuminate. *Panicle long, rather broad*, hardly narrowed above, *mid branches patent, 3–7-flowered*, lower branches spreading; rachis and branches closely felted and villous with sunken, deep purple

171

glands and various prickles. *Flowers showy, 2–2·5 cm.; sepals* greenish white, felted and ± hirsute, with a white felted border, *gland-dotted and prickly, reflexed; petals broad, obovate-cuneate,* rose-pink or pinkish; filaments white or red, far longer than the greenish styles. Carpels pilose. A species probably derived from a cross of *R. apiculatus* with *R. vestitus.*

Endemic. E., W., I. Locally frequent in the west; 11, 33–37, 40, 44, 49. H. 2, 40.

311. **R. tardus** W. Wats. 1935 (for 1934), *Rep. Bot. Exch. Cl.* **10**, 794.

4x. *S.E. England. Rufescent. Strongly armed.* Stem angled, with a fair number of acicles and pricklets, some gland-tipped, also a *few stalked glands;* prickles unequal, *the longer ones subulate from a swollen base.* Leaf digitate, petiole long; leaflets plicate, glabrescent above, velvety pilose and felted beneath, unequal or subcompound-serrate or serrate-dentate, undulate; *terminal leaflet* oblong or *elliptic-obovate, cuneate, base entire.* Panicle equal, dense above, interrupted and lax below; rachis pilose to pubescent and felted, together with the peduncles and pedicels armed like the stem, but with *some longer stalked glands and acicles.* Flowers showy, 2·5–3 cm.; sepals bearing glands and red prickles, patent; petals 5–7, pink or pinkish, elliptic, ovate-rhomboid or obovate, margin fringed; filaments white or pinkish, far longer than the *red styles.* Carpels glabrous. *Fruit* rather small, *ripening slowly, and late. Flowering about 3 July.* The scanty minor armature of the stem contrasts with the sometimes almost hystrican armature of the panicle.

Endemic. England: locally frequent; 15, 16, 17.

Ser. **vi. GRANDIFOLII** Focke

As the two typical species on which he based the Series, Focke names the Madeiran *R. grandifolius* Lowe and the West Mediterranean *R. incanescens* Bertol. (**2x**): the former shows the pointed pyramidal type of panicle, the latter the equal and abrupt type. Both these panicle forms are represented in the British species, e.g. *R. Lejeunei* Weihe and *R. disjunctus* Lef. & Muell. respectively. In his last work Focke includes in GRANDIFOLII the putatively hybrid products of *R. incanescens* and *R. grandifolius* with APICULATI and HYSTRICES, that is to say, so many as were known to him. If the same were done here, *R. Reichenbachii, R. longifrons* and probably *R. melanoxylon* would have to be included. But Focke's views do not work out well in practice and have not been adopted.

I. Stem leaves greenish, greyish or white felted beneath.

† Stem hairy.

312. **R. Leightonii** Ed. Lees ex Leighton, 1841, *Fl. Shropshire*, 233; *R. ericetorum* Lef. ex Genev. 1880, *Mon. Rub.* 162; *R. Radula* ssp. *ericetorum* (Lef.

Appendiculati

ex Genev.) Sudre, 1908–13, 128, t. 125; *R. Radula* ssp. *anglicanus* (Rogers) Rogers, 1900, 63.

Stem becoming light red, with numerous *rather long glands and acicles; prickles* crimson-based, strong, subequal, patent. *Leaflets large,* 5, softly pubescent and grey felted beneath, *serrate-dentate; terminal leaflet long-stalked, elliptic-obovate, cuspidate, base rounded. Panicle* often nodding, not leafy above, *long and compound,* mid branches patent, 3–5-flowered, cymose, lower axillary branches panicled, oblique; *rachis* hirsute *with longish glands and acicles and very unequal, slender prickles. Flowers up to 3 cm.*; sepals reflexed; *petals bright pink,* elliptic, narrowed to the base; *filaments pink, far longer than the reddish styles.* Carpels pilose. Fruit roundish to oblong. (Submitted by Leighton to Nees in 1836, who thought it near to *R. Lejeunei* although distinct from it.)

E., S. Rather frequent; 3, 4, 9, 11, 16–18, 20–24, 30–32, 33, 35, 36, 38–40, 55, 57, 58, 90. Jersey. N.W. France.

313. **R. disjunctus** Muell. & Lef. 1859, *Pollichia,* 216; *R. Lejeunei* sensu Focke pro parte, 1877, *Syn. Rub. Germ.* 316; non Weihe in Bluff & Fingerh. 1825; *R. Lejeunei* var. *ericetorum* sensu Rogers, 1900, 71; non Lef.; *R. Moylei* Barton & Riddels. 1933 (for 1932), *Proc. Cotteswold Field Cl.* **24,** 218; *R. ericetorum* var. *cuneatus* Rogers & Ley, 1906, *Journ. Bot.* **44,** 59.

Robust. Stem climbing, angled, slightly furrowed, glaucescent, *with numerous stout pricklets,* acicles and glands; *prickles* unequal, *mostly slanting. Leaflets large, often long and narrow in shade,* 3, 4, 5, not contiguous, nearly glabrous above, green and softly hairy beneath to pubescent and felted, unequal to subcompound-crenate-serrate and jagged; *terminal leaflet oblong-obovate-acuminate, narrowed below* to the somewhat truncate base. *Panicle large and long with an abrupt apex, lax,* ± *patent,* upper and middle *branches* 1–3 (5)-flowered, lower ones panicled and rising from axils below; rachis shortly hirsute, with numerous slanting prickles and *very numerous, deep purple, acicles and glands,* some longer than the hair; pedicels and calyces finely glandular. Flowers 2·5–3 cm.; sepals felted, long-tipped, patent or partly erect; *petals* ovate or elliptic, notched, narrowed below, *usually pinkish; filaments long, usually white; styles usually reddish. Carpels pilose.* Fruit short, ovoid-oblong. (Around Cranbrook and Tenterden in Kent a form occurs with deep pink petals, stamens and styles.)

E., W., I. Frequent; 1, 6, 7, 9, 11–18, 21–24, 30, 33, 35, 36, 38, 39, 41–43, 58, 66. H. 8, 38. Guernsey. France (*Aisne*), Belgium.

314. **R. diversus** W. Wats. 1928 (for 1927), *Rep. Bot. Exch. Cl.* **8,** 508; *R. hirtus* ssp. *Kaltenbachii* sensu Rogers, 1900, 88; non *R. Kaltenbachii* Metsch, 1856; nec Focke, 1877.

*4x. Stem climbing, blunt, *glaucous,* lineolate; *prickles* slender, broad *crimson-based,* unequal. Leaves large, yellowish green; *leaflets* 3–5, pilose and

felted beneath, ciliate, *sharply subcompound-serrate*, rather narrow *obovate-cuneate*, elliptic-acuminate or rhomboid, *base entire. Panicle hanging in bud, pointed pyramidal, usually large, with crowded flowers on racemose branches and a racemose apex*; rachis felted and hirsute, glandular and acicular, prickles weak; the terminal leaflets cuneate. Flowers 2–2·5 cm.; *sepals greenish, felted within and without*, linear-tipped, patent; *petals* ovate-cuneate, often pointed, pale or deep *pink; filaments* white, *much longer* than the reddish-based *styles.* Carpels glabrous or subglabrous. Fruit subglobose, sweet. Goes deepest into the wood, and begins to flower after all other brambles there have finished.

Endemic. E., W. Woods, frequent; 2, 6, 8, 12, 13, 15–17, 21, 22, 33, 34, 36, 39, 41, 42, 55.

315. **R. hostilis** Muell. & Wirtg. ap. Wirtg. 1861, *Herb. Rub. rhen.* ed. 1, no. 139; non Rogers, 1900, 81; *R. adornatus* ssp. *hostilis* (Muell. & Wirtg.) Sudre, 1908–13, 174, t. 169.

Young shoots light bronze. Stem low-arching, procumbent, angled, glaucescent, with unequal glands and acicles; *prickles unequal, none very long, broad-based, some curved. Leaves rather small,* 3–5-*nate, prickles hooked; leaflets* pilose and felted beneath, serrate-dentate, *jagged; terminal leaflet elliptic-obovate or oblong* (narrow on old bushes), *cuspidate. Panicle* not, or not very, leafy, *long and equal,* subracemose above and subcorymbose, *or large, compound, dense-flowered, pyramidal,* the branches often in pairs; rachis and peduncles felted and villous, with mostly very short glands, prickles numerous, unequal, some gland-tipped. Flowers 2–2·5 cm.; sepals felted, narrow and attenuate, erect; *petals* narrow, remote, 5, 6, 7, notched, glabrous, *pink, often flushed red outside;* filaments white or pink, far longer than the greenish or red styles. Carpels glabrous. Receptacle pilose. Fruit large, ovoid-oblong.

S. England: common in Middlesex and S. Bucks., rare elsewhere; 8, 9 (extinct in 17), 20–22, 24, 31. France (*Finistère*), Germany (*Rhineland*), Switzerland.

316. **R. Wedgwoodiae** Barton & Riddels. 1933 (for 1932), *Proc. Cotteswold Field Club,* **24**, 212; including var. *Sabrinae* Barton & Riddels. *l.c.*; *R. mutabilis* var. *Regnorum* W. Wats. 1931 (for 1930), *Rep. Bot. Exch. Cl.* **9**, 437.

Stem angled, slightly furrowed, *hirsute, with* few glands but *numerous very unequal acicles and stout pricklets; prickles strong,* clustered, unequal. *Leaves large,* yellowish green; *leaflets* 5, *thick, pilose and felted beneath, long, rather narrow and cuneate, incise sublobate-serrate; terminal leaflet elliptic, cuneate, acuminate. Panicle large and compound,* broad to *subpyramidal,* the upper branches patent, 1–3 (5)-flowered, the lower branches ascending, panicled; rachis clothed and armed like the stem. *Flowers up to 3 cm.; sepals* hairy and felted with a narrow white edge, *reflexed; petals large,* narrow elliptic, tapered

below, *rosy in bud* then pinkish; *filaments white, very long; styles reddish-based*
Carpels and receptacle very hairy. *Fruit large.*

Endemic. E., W. Rather rare; 11, 13, 17, 34, 41, 42.

†† Stem glabrous or glabrescent. Petals pink.

(*a*) Stem becoming blackish purple.

317. **R. furvicolor** Focke, 1914, *Biblioth. Bot.* **83**, 216; *R. melanoxylon* sensu
Rogers, 1900, 59; non Muell. & Wirtg. 1861.

Stem furrowed, with few or many, mostly short, glands and acicles; *prickles
subulate, stout based, unequal. Leaflets* 3 (5), softly pubescent and felted beneath,
serrulate; *terminal leaflet subcordate, roundish, or roundish ovate, rather shortly
pointed. Panicle long,* ± equal or broader below, dense, with 1–3-flowered
branches, *patent above,* and *others* more or less remote, *cymose-umbellate below;*
rachis angled, striate, *glabrescent below,* felted and shortly pilose above, armed
like the stem; the *terminal leaflets roundish, serrulate.* Flowers 2–3 cm.; sepals
felted, cuspidate, patent or patent-erect; petals 5–8, elliptic; *filaments long,*
pink or white, longer than the bright pink or greenish styles. Carpels
glabrous or bearded. Fruit ovoid.

Endemic. E., W., S. Frequent in the north; 32, 36, 39, 44, 57, 66, 68, 86,
87, 95, 96, 98.

(*b*) Stem becoming light or deep red, or brown.

318. **R. angusticuspis** Sudre, 1904, *Observ. Set. Brit. Rub.* 35; *R. anglosaxonicus* ssp. *setulosus* Rogers, 1900, 58.

Slender and elegant. Young shoots bronze. Stem red, closely covered with very
unequal, slender, largely falcate prickles, very unequal and numerous
pricklets and acicles, and a few, short glands. *Leaflets 5, glabrous above,* all
rather *long-stalked and* ± *cuneate, finely incise-serrate, undulate; terminal leaflet
broad obovate-rhomboid, slenderly acuminate,* base entire or nearly so. *Panicle
often leafy above, large, broad, lax, pyramidal, with large, spreading, panicled, leafy
lower branches; leaflets serrulate.* Flowers 1·5–2 cm.; *sepals* long-tipped, *prickly,*
clasping; *petals narrow, obovate-cuneate,* notched, *bright pink;* filaments
white, longer than the *reddish-based styles.* Carpels pilose. Fruit roundish,
perfect.

Endemic. E., W., I. Fairly frequent in the west, rarer elsewhere; 3–5, 20,
22–24, 30, 33–36, 41, 42. H. 38.

319. **R. squalidus** Genev. 1869, *Essai Monogr.* 128; Syme, 1864, *Eng. Bot.*
t. 453 (as *R. Koehleri*); *R. Timbal-Lagravei* ssp. *occitanicus* var. *squalidus*
(Genev.) Sudre, 1908–13, 141, t. 137, fig. 9; *R. mutabilis* var. *Naldrettii*
J. W. White, 1907, *Watson Bot. Exch. Cl. Rep.* **2**, 86.

*S.E. England. Stem glabrescent, with very fine glands and numerous
pricklets; prickles yellowish brown, long-based, rather short, numerous, confluent,*

unequal, the smaller ones stout-based. *Leaves yellowish green*, petiole-prickles hooked; leaflets 5, glabrescent above, pubescent to grey-white felted beneath, serrate-dentate; *terminal leaflet* short-stalked, ovate to elliptic or oblong-obovate, *falcate-cuspidate, base nearly entire*; basal leaflets rather short-stalked. *Panicle very large and compound, dense*, with patent, numerous flowered branches; *rachis* felted and thinly pilose, *pricklets numerous, fine, some recurved*; lateral leaflets short-stalked. *Flowers 1–2 cm.*; sepals erect; *petals obovate, entire, rose-pink*; filaments white, long; styles greenish. *Carpels very pilose.* Fruit small, subglobose, good.

E., I. Local; 13–17. H. 2, 8. W. France, Austria.

R. **leucostachys** Sm. × R. **squalidus** Genev.; *R. carpinifolius* sensu Borrer; non Weihe & Nees; *R. Grabowskii* sensu Syme, 1864, *Eng. Bot.* ed. 3, 3, 173, t. 449; non Weihe, 1829.

Henfield, W. Sussex.

320. R. **pseudoplinthostylus** W. Wats. 1956, *Watsonia*, 3, 289; *R. plinthostylus* sensu Rogers, 1900, 83; *Set no. 72*; non Genev. 1868.

Western. Stem sharp-angled, *glabrous*, with few, short glands and acicles; *prickles short-based, subulate.* Leaves large, pedate, petiole prickles slanting and retrorse-falcate, stipules filiform; *leaflets 3, 4, 5*, glabrous above, pubescent to felted beneath, *irregularly serrate or serrate-dentate, base entire; terminal leaflet short-stalked, long-pointed, elliptic with a subcuneate base. Panicle* with various simple leaves above, *very long, lax, subpyramidal*, the 1–3-flowered upper branches patent, *the lower branches sharply ascending, leafy panicled; rachis pubescent and felted, with numerous short prickles*, pricklets and a few glands; pedicels felted, finely glandular. *Flowers 1·5–2 cm.* Sepals loosely reflexed to patent. *Petals narrow, obovate, pink. Filaments white, slightly longer than the pinkish styles. Carpels pilose. Receptacle pilose, especially below the carpels,* where the hairs are seen protruding in the young flowers. Fruit small.

Described from the living bush in Foxholes Wood, Bailie Gate, Dorset in 1931 and 1951. Wolley-Dod's specimen from Broxton Woods, Cheshire has flowers with bright pink petals.

Endemic. Apparently very rare; 9 (Foxholes Wood), 58 (Broxton).

321. R. **mutabilis** Genev. 1860, *Mem. Soc. Ac. M.-et-L.* 84; *R. obtruncatus* ssp. *mutabilis* (Genev.) Sudre, 1908–13, 176, t. 171.

N.v.v. Axes and glands purple-red. Stem angled, sides flat or slightly impressed, *glabrescent, glaucescent*, with numerous, mostly very small glands, acicles and pricklets; prickles moderate, straight, slanting, slightly unequal. *Leaflets 3–5, greyish white felted and pilose beneath, doubly and shallowly serrate; terminal leaflet broadly ovate-rhomboid* to elliptic, acuminate. Panicle pyramidal, narrow, truncate, subracemose above, the lower branches again racemose; *rachis* felted and pilose, *prickles numerous, very slender, slanting and retrorse-falcate*; terminal leaflets rhomboid. Flowers 1·5–2 cm.; *sepals very prickly*, loosely clasping; petals obovate, pale pink; filaments white, longer than the reddish-based styles. Carpels glabrous. Fruit ovoid-oblong.

England: rare; 17 (West End Common, near Bisley, *C. Avery*, 1936), 23 (Heythrop), 37 (Woods near Bewdley, *E. G. Gilbert*, 1857 as *R. podophyllus*). France, Germany.

2. Stem leaves green beneath. Top panicle leaves sometimes greyish felted.

† Petals white.

322. **R. Turneri** W. Wats. 1937, *Journ. Bot.* **75**, 159; *R. glandulosus* Sm. pro parte 1824, *Eng. Fl.* **2**, 404.

Western. Rufescent. Stem reddening, angled, glabrescent with small pricklets, acicles and glands; prickles unequal, the shorter ones often gland-tipped, the longer ones slanting and falcate. Leaves (3) 5-nate, pedate; *leaflets* glabrescent above, *pubescent and green felted beneath, sublobate*, crenate-serrate; *terminal leaflet* broad to rather narrow *obovate, subcuneate; lateral and basal leaflets short-stalked*. Panicle long, equal, with a few simple leaves, grey felted beneath, subracemose above, the branches nearly patent, the lower branches ± compound, nearly erect; rachis sharply angled, armed much like the stem. *Flowers starry, 1·5–2·5 cm.*; sepals felted, appendiculate, loosely reflexed to ± patent; *petals long, narrow, obovate*, glabrous on the margin; *filaments long, white; styles greenish*. Carpels glabrous or pilose. Fruit abundant, globose.

Endemic. E., W., S., I. Western; 45 (Pont-faen, *Alston*, 1931), 49 (Bangor Garth), 69 (Rydal Woods, *Dawson Turner*, ante 1824, *Borrer*, 1845), 72 (Jardine Hall, *Babington* as *R. Koehleri* b. *cuspidatus*). H. 2 (Killarney, *Babington*).

†† Petals pink. Sepals reflexed.

323. **R. Powellii** Rogers, 1894, *Journ. Bot.* **32**, 47; *R. rosaceus* ssp. *Powellii* (Rogers) Rogers, 1900, 81.

Epping Forest. A small, low plant with small leaves and flowers. Stem deep purple, angled, pilose, with *rather long, blackish glands and acicles*, some gland-tipped; prickles falcate. Petiole prickles hooked. *Leaflets deep green, subcoriaceous; glabrous and shining above*, pubescent beneath, *teeth* angular, incise, open and *compound*, the principal ones larger and patent; *terminal leaflet narrow oblong-elliptic, gradually acuminate*. Panicle equal, narrow, lax, subracemose above with ascending long pedicels, and a few, sometimes very long, panicled branches below; *rachis* hairy, *with longish glands and acicles* and long-based, slanting and falcate prickles. *Flowers 1–1·5 cm.*; sepals long, leafy pointed, from patent to *decidedly reflexed; petals narrow elliptic-obovate*, pilose; filaments white, much longer than the greenish styles. Carpels glabrous.

Fruit small. The top panicle leaves are slightly greyish felted beneath. All the leaves when young are a little glandular above, below and on the margins.

Of this small bramble a pretty, diminutive form occurs, as a large bush, half a mile from the Wake Arms, Epping Forest.

Endemic and confined to Epping Forest, 18.

It has been reported from other counties, but all such specimens that I have seen have been the dwarf form of *R. radula*, except Wolley-Dod's Shooters Hill bramble, which was *R. argutifolius*.

††† Petals pink. Sepals patent-erect.

(*a*) Terminal leaflet roundish (-ovate).

324. **R. rotundifolius** (Bab.) Bloxam in Kirby, 1850, *Fl. Leicester*, 39; *R. glandulosus* var. *rotundifolius* Bab. 1848, *Ann. Nat. Hist.* Ser. 2, **2**, 40; *R. hirtus* var. *rotundifolius* (Bab.) Rogers, 1900, 88; *R. Lejeunei* sensu Ed. Lees, 1853, *Phytologist*, **4**, 819; *R. Bellardii* var. *Lejeunei* sensu Ed. Lees in Steele 1847, *Handb. Field Bot.* 55; *R. Hiernii* Riddels. in Martin & Fraser, 1939, *Fl. Devon*, 281. **Fig. 39.**

Rufescent in the sun. Often luxuriant in woods. Stem thick, blunt, striate, hairy, *glaucous*, with numerous, mostly very short glands and unequal acicles and pricklets; *prickles very numerous below, short, confluent*, mostly slanting. Leaves 3, 4, 5-nate, pedate, *leaflets* pubescent to glabrescent or greenish grey felted beneath, *finely and shallowly crenate, or partly dentate; terminal leaflet roundish to oblong*, base subcordate to entire; *lateral leaflets short-stalked. Panicle often very compound*, pyramidal, but not pointed, with long-peduncled, patent, panicled lower branches, hanging in fruit; *rachis glaucous below*, villous, armed like the stem. Flowers 1·5–2 cm. incurved; sepal-tips leafy, felted and pilose, gland-speckled, prickly; *petals* 5, 6, *moderately broad, elliptic, clawed, nearly glabrous*; filaments white, sometimes pink-based, short or long; styles yellowish. Carpels glabrous or thinly pilose. Fruit subglobose.

Endemic. E., S., I. Locally rather frequent; 2–6, 8, 13, 15, 17, 20–22, 24, 37–39, 55, 56, 59, 60, 62, 63, 104. H. 1, 33.

325. **R. rosaceus** Weihe ap. Bluff & Fingerh. 1825, *Comp. Fl. Germ.* **1**, 685; W. Wats. 1929 (for 1928), *Rep. Bot. Exch. Cl.* **8**, 862 (and two plates by Trower); 1949, *Watsonia*, **1**, 78; non sensu Focke, 1877; nec Rogers, 1900; nor hardly any author after Weihe & Nees; *R. serpens* ssp. *leptadenes* var. *calliphylloides* (Sudre) Sudre, 1908–13, 220.

4x. *Axes deep purple. Stem blunt-angled, furrowed, hairy* with numerous short glands and acicles; *prickles short*, unequal, some very small. *Leaves large, deep green*, 3–5-nate; *leaflets* pilose beneath, *coarsely, rather incise-serrate, jagged; terminal leaflet* short-stalked, *roundish*, acuminate. *Panicle* often hanging, *very floriferous, large, not long, nor leafy, very broad, pointed, pyramidal*, racemose at the top, the pedicels being patent, followed below by a series of *nearly patent, 3-flowered, middle branches* and then by more and longer, racemose, *lower* ones, all approximately horizontal; pedicels short, making the *flowers crowded*. *Flowers 1–2 cm.; sepals* greenish, *long leafy-tipped, clasping the fruit; buds cristate* (rosaceous); *petals* 5–7, or up to 11 superposed, *roundish* or broad elliptic-ovate, *short-clawed, rosy pink;* filaments white or pink, mostly longer than the *red styles.* Carpels glabrous or pilose. Fruit globose-ovoid. Up to about 100 flowers may be open at the same time on a single panicle.

England: frequent in the south-east in dark hilly woods and in open places near them; 13, 15–21, 32. Belgium and W. Germany.

326. **R. formidabilis** Lef. & Muell. 1859, *Pollichia*, 128; *R. Lejeunei* ssp. *formidabilis* (Lef. & Muell.) Sudre, 1908–13, 178, t. 174.

4x, 5x. *Surrey and Bucks. Rufescent. Stem* climbing or prostrate, *angled, furrowed, thinly hairy with scattered strong-based, often gland-tipped acicles* and a few glands; *prickles numerous, strong-based, subequal.* Leaves yellowish green, *petiole prickles large, falcate; leaflets* 5, glabrescent beneath, *very coarsely and sharply,* sometimes slightly doubly *serrate; terminal leaflet* long-stalked, *roundish or broad cordate, ovate, cuspidate. Panicle* hanging, large, bluntly pyramidal, lax above *with long peduncles and long pedicels,* and long, numerous flowered branches ascending from below; rachis slightly villous, armed like the stem. *Flowers up to 3 cm.; sepals* greenish, prickly, loosely reflexed to *patent,* with the long *tips partly erect; petals* ovate or obovate, clawed, pilose, *deep pink; filaments deep pink, longer than the flesh-coloured styles.* Carpels glabrous. *Fruit large,* ovoid-oblong. *Flowering late.* A dangerously prickly bramble.

England: rare, except in Surrey; 17, 24 (Lee). N. France.

(b) Terminal leaflet obovate.

327. **R. hastiformis** W. Wats. 1946, *Journ. Ecol.* **33**, 341; *R. thyrsiger* Bab. 1886, *Journ. Bot.* **24**, 227; non Banning & Focke, 1877.

4x. *Western. Robust. Axes deep reddish purple.* Stem blunt, striate, glaucescent, with long and short hair, short glands, acicles and stout pricklets; prickles subequal, strong-based. *Leaves large,* (3) 5-nate, pedate; leaflets glabrescent beneath, unequal or subcompound-serrate; *terminal leaflet oblong-obovate, ± truncate-cuspidate. Panicle leafy, very long, tapering to the racemose top,* all branches semi-erect, 1–2 (3)-flowered; *rachis straight,* felted and hirsute with *numerous short, mostly equal, glands and acicles; prickles few, short and weak.* Flowers 2·5 cm.; sepals leafy-tipped, erect; petals 5, 6, 7, narrow, tapered down-wards, pilose; filaments pink or white, long, anthers sometimes pilose; *styles red.* Carpels pilose. *Receptacle pilose, especially below the carpels. Fruit large, thimble-shaped, well-flavoured.*

Endemic. E., W. Rare; 2–4, 13, 41.

328. **R. Lejeunei** Weihe ap. Bluff. & Fingerh. 1825, *Comp. Fl. Germ.* **1**, 683; Sudre, 1908–13, 177, t. 173; non sensu E. Lees, 1853, *Phytologist*, **4**, 819; nec sensu Rogers, 1900, 70.

5x. *Strong. Panicle very large, floriferous, hanging.* Stem climbing, long, blunt-angled, thinly hairy with very short glands, tubercular-based acicles and pricklets; prickles unequal, large-based, tapered. *Leaves large, 3–5-nate;* leaflets glabrescent above, pubescent, glabrescent, or slightly felted beneath, unequal or slightly compound-serrate; *terminal leaflet rather long, obovate-subcuneate, acuminate. Panicle pyramidal, with numerous, panicled, axillary branches, then cymose middle branches and a subracemose apex;* rachis felted and pubescent with rather short, subulate prickles, *unequal acicles and stalked glands, some very long. Flowers 2–3 cm.;* sepals greenish, felted with a narrow white border; *petals glabrous, broad elliptic* with a slender claw; filaments red, pink or white, slightly longer than the greenish, pinkish-based styles. Carpels subglabrous. Receptacle pilose. Fruit ovoid.

England: rare; 2 (near Cutmere Bridge, St Germans, *Briggs,* 1875), 3 (Bickleigh Vale, 1939, 1951), 15 (two miles west of Linton), 16 (Cobham Park). France, Belgium, W. Germany and Bavaria, N. Italy (*Piedmont*), N. Spain, *Pyrenees,* Portugal.

(*c*) Terminal leaflet ovate-acuminate, elliptic, or elliptic-rhomboid. Flowers small.

329. **R. festivus** Muell. & Wirtg. 1861, *Herb. Rub. Rhen.* ed. 1, no. 138; non 1925, *London Cat. Brit. Pl.* ed. 11, 16, no. 552; *R. floribundus* W. Ley ap. Rabenh. 1846, *Bot. Centralbl.* 1, 435; non Weihe, 1825.

Stem roundish, hairy, with sunken glands and acicles and longer unequal pricklets; prickles yellowish, rather unequal, rather *small. Leaflets 3, 4, 5, thinly or densely, shortly hairy beneath,* with or without greyish felt, *unequally serrate-dentate; terminal leaflet ovate-acuminate, a little contracted below* to a truncate base; the lower leaflets shortly stalked. *Panicle long pyramidal, or nearly equal and truncate; lower panicles much prolonged, with progressively smaller leaves to the apex;* rachis wavy, villous and armed like the stem, but with some prickles hooked. *Flowers 1·5–2 cm.;* sepals greenish, gland-dotted, aculeate, loosely erect; petals narrow, obovate, tapered below, glabrous; filaments white or pink, longer than the greenish or red-based styles. Carpels pilose. Fruit subglobose.

W. Ley collected this at Eupen and Aachen. I have seen his specimens and have collected it myself at Spa and Polleur.

England: rare; 18 (Danbury Common pit), 19 (W. Bergholt), 39 (Heathy-lee at 1000 ft.), 56 (Oxton). Belgium, W. Germany, Switzerland.

330. **R. breconensis** W. Wats. 1956, *Watsonia,* 3, 290; 1946, *Journ. Ecol.* 33, 341 (nomen nudum); *R. Lejeunei* sensu Rogers, 1900, 70; non Weihe & Nees, 1825.

Wales. *Stem fuscous, slender,* furrowed, hairy, with numerous glands and very numerous, short acicles and pricklets; *prickles yellow,* unequal, *rather* small and very slender. Leaves 3–5-nate, pedate; *leaflets rather long and narrow, softly and shortly hairy beneath, very finely pointed serrate; terminal leaflet elliptic, narrowed to both ends,* finely acuminate; *intermediate and basal leaflets very shortly stalked.* Panicle broad below, usually with one of the leaves simple, the rest 3-nate, *terminal leaflet narrow and long cuneate,* the lower and middle branches patent, long-peduncled, 3–6-flowered, narrowing upwards to a pointed, racemose apex; *rachis* crooked, *villous,* prickles weak, mostly falcate. *Flowers 1–1·5 cm.;* sepals narrow, ovate, attenuate, villous, prickly; *petals* obovate, tapered below, *rose pink; filaments long, deep pink.* Receptacle glabrous. Fruit small.

Endemic. Wales; 42 (Glanau and Llanwrtyd), 43 (near Llandrindod), 45 (Brynau, Blackpill; and Pont Faen).

331. **R. Rilstonei** Barton & Riddels. 1933 (for 1931–2), *Proc. Cotteswold Field Cl.* **24**, 213. **Fig. 41.**

S.W. England. Stem reddish purple, angled, ± hairy with some short glands and numerous, mostly small, intermediate pricklets and glandular acicles; *prickles yellow,* numerous, unequal, slender, slanting and falcate. Leaves 3–5-nate, prickles retrorse-falcate, stipules filiform; *leaflets all short-stalked and entire-based, softly hairy beneath,* sharply, deeply, ± compound, partly patent serrate; *terminal leaflet obovate or elliptic-rhomboid, gradually acuminate.* Panicle of moderate size, leafy above, broad, lax, pyramidal-truncate, the upper branches 1–3-flowered, semi-erect with longish pedicels, the lower ones longer, also with semi-erect pedicels, and leafy; rachis crooked, villous, armed like the stem. *Flowers 1·5–2 cm.; sepals* appendiculate, *green with a white edge,* prickly; *petals narrow rhomboid-ovate,* fimbriate, *remote;* filaments white, scarcely longer than the greenish or reddish-based styles. Fruit moderate.

Endemic. Frequent in Cornwall; 1, 2, 3 (Fingle Gorge), 4 (Colebrook, *Briggs,* 1868).

Sect. **VII. GLANDULOSI** P. J. Muell.

(Type *R. Bellardii* Weihe)

Ser. **i. HYSTRICES** Focke. Stem angled, *usually strong.* The longest prickles strong and broad-based, usually straight. Panicle with cymose middle branches. *Flowers usually showy, often pink.* Petals relatively broad and stamens long. (See p. 182.)

Ser. **ii. EUGLANDULOSI** W. Wats. *Stem rounded or obtuse angled, often pruinose, usually weak and prostrate.* Prickles either broad-based and curved or narrow-based, slender and weak. Panicle with middle branches racemose, deeply divided, rarely subcymose. *Petals moderately broad or narrow, rather small or very small,* nearly always *white. Styles often red. Stamens often short. Woodland plants.* (See p. 198.)

Ser. **i**. **HYSTRICES** Focke

(Type *R. Hystrix* Weihe)

1. Flowers small, usually 1–1·5 cm.

† Petals white, often with a greenish claw, sometimes pinkish in bud. Styles red.

332. **R. Murrayi** Sudre, 1904, *Observ. Set Brit. Rub.* 71; *R. rosaceus* ssp. *adornatus* Rogers, 1900, 80; non P. J. Muell. 1859. **Fig. 42.**

4x. *Rufescent. Stem tall, climbing,* shortly hairy, *glaucescent,* with numerous, *mostly short glands and pricklets; prickles moderate, much declining,* unequal, all except the longest gland-tipped. *Leaves rather small,* pedate; leaflets green and shortly hairy beneath, coarsely and unequally serrate undulate; *terminal leaflet* subcordate, *broad ovate, very gradually acuminate. Panicle lax, long, ± equal or subpyramidal, nodding in bud, mid branches spreading,* 3–5-flowered, *pedicels short,* felted. *Sepals* pale green, *loosely clasping.* Petals elliptic-rhomboid, retuse or erose at the apex, margin glabrous. Filaments white, rather short. Carpels glabrous. Fruit small. The uppermost leaves are a little grey felted beneath.

Endemic. E., W. Frequent; 1 (Lane near Penzance, *Ralfs.*), 6, 9, 11, 13–18, 20, 21, 23, 24, 26, 33, 34, 36–39, 49, 55, 58.

333. **R. pygmaeopsis** Focke, 1877, *Syn. Rub. Germ.* 564.

Stem low, brown to red, *slender, blunt, hairy* with few glands and acicles but numerous, very unequal, *mostly long, yellow* prickles, *all except the longest being gland-tipped. Leaves rather small,* 5-nate, prickles hooked; *leaflets* green and hairy beneath, *coarsely incise-serrate, sublobate,* undulate; *terminal leaflet* cordate, ovate or obovate, *very long-pointed. Panicle* broad, equal, long and leafy, *many flowered, with numerous,* fine, straight or falcate *prickles throughout, villous,* glandular and acicular. Sepals grey felted, prickly. Petals broad elliptic, clawed. Filaments white, rather long. Carpels glabrous. Fruit sub-ovoid, perfect.

S.E. England: frequent; 13, 14, 16–18, 20, 21, 24, 31. Belgium, N.W. Germany and the Rhineland, Austria.

334. **R. newbridgensis** Barton & Riddels. 1936, *Journ. Bot.* **74**, 204; *R. saxi-colus* var. *parisiensis* Sudre, 1906, *Batotheca Eur.* 59.

4x, 6x. *Axes deep bright red. Stem shortish, upright then arching, nearly glabrous,* glands and acicles mostly short; prickles unequal, the smaller ones gland-tipped. *Leaves 3* (4, 5)-nate, *rather small, apple green, tough,* petiole short, flat above; leaflets thinly pubescent beneath, rather doubly serrate; *terminal leaflet* emarginate, roundish to *broad oblong or obovate, short-pointed. Panicle nearly equal, with a dense rounded top, short* middle *branches* and remote, sharply ascending lower branches; *rachis rigid, wavy, angled,* hirsute with short and long glands and glandular acicles. *Sepals* green, prickly, ± *reflexed*. *Petals* elliptic-ovate *with rounded, woolly apex*, greenish white. Filaments white, long. *Fruit abundant, ovoid*. Young shoots coppery crimson.

S.E. England: locally frequent; 12–14, 16, 17, 24. France (*Seine-et-Oise*). (Cf. *R. Chenonii*, *R. apricus* and *R. rotundellus*.)

†† Petals pink, pinkish or white. Styles greenish, or tinged reddish below.

(a) Upper leaves greenish or grey felted beneath.

335. **R. ochrodermis** A. Ley, 1893, *Journ. Bot.* **31**, 15.

Western. Stem becoming ochreous and fuscous, angled, subglabrous with numerous, some *minute, stout pricklets and few, short glands and acicles*, the latter passing into the rather short, stout-based prickles. *Leaflets* 3 (5), pubescent to felted beneath, *serrulate; terminal leaflet roundish or obovate-subcuneate, short-pointed*. Panicle large, subracemose and subcorymbose above, *with very long, ascending leafy lower branches*; rachis round, subglabrous below, armed like the stem. Sepals short-pointed, reflexed. *Petals pink*, broad obovate, contracting into a slender claw, apex notched, downy. *Filaments pink-based, slightly longer than the styles*. Carpels pilose. Receptacle pilose. Fruit subglobose, deep crimson before turning black. Recalls *R. egregius*.

Endemic. E., W. Rare; 36, 40, 42, 43, 46 (near Groeslon Station, *Griffith*, 1869, as *R. Bloxamii*), 49.

336. **R. coronatus** var. **cinerascens** W. Wats. 1933 (for 1932), *London Nat.* 65.

Much branched and building up a big bush in the open, or climbing. *Stem stout and furrowed*, pale green and reddish, *glaucous*, hairy with subequal, short glands, acicles and numerous pricklets; *prickles numerous, stout-based, rather short*, slanting and curved. *Leaflets* 3–5, green and hairy to *grey-white felted* beneath, coarsely serrate; *terminal leaflet ovate, gradually acuminate*. Panicle *very large*, lax, pyramidal, *with long, spreading branches*, densely hairy and armed like the stem. Sepals prickly, loosely reflexed. *Petals ovate, pink*. *Filaments pink, longer than the red styles*. Carpels glabrous. Receptacle hirsute. Fruit small, drupels large. Recalls *R. Murrayi*.

Endemic. Rare; 13 (Horsham), 17 (Putney Heath, towards Roehampton), 38, 58 (Gallantry Bank).

337. **R. adornatus** P. J. Muell. ap. Wirtg. 1858, *Herb. Rub. Rhen.* ed. 1, no. 87; 1859, *Flora*, 231; Sudre, 1908–13, 173, t. 168.

4x. *Stem blunt-angled*, yellow to dark red, *glaucescent*, hirsute, with short and very long glands and acicles; *prickles* hairy, *subulate* from a flattened base. Leaves rather large, 5-nate, pedate; *leaflets* glabrescent above, *softly hairy* and thinly, grey felted *beneath*, angularly and unequally patent-serrate; lateral leaflets short-stalked; *terminal leaflet* subcordate, *roundish or very broadly roundish-ovate, acute or shortly acuminate*. Panicle with 2–3 simple leaves, rather long and narrowly pyramidal, the upper and middle branches patent, 1–3-flowered, deeply divided, the lower branches more spreading and panicled; *rachis hirsute*, a good many of the *prickles* strong and *recurved*. *Sepals clasping*. *Petals roundish* or broadly elliptic, subglabrous on the margin, *deep pink*, rarely pinkish. *Filaments deep pink* or white, longer than the styles. *Anthers sometimes pilose*. Carpels glabrous or pilose. Receptacle hirsute. Fruit rather large, roundish. Young shoots bronze.

England: rare; 21, 24 (Lee and Lee Clump), 35 (St Arvans), 54 (Muckton to Awthorpe). France, Belgium, Germany, *Rhineland*, Switzerland, Austria.

338. **R. saxicola** P. J. Muell. 1859, *Pollichia*, 202; non Genev. 1880; nec Rogers, 1900; *R. Koehleri* ssp. *saxicolus* (P. J. Muell.) Sudre, 1908–13, 187, t. 185.

Stem glaucescent, blunt-angled, hairy, with mostly short glands and acicles, a few acicles very long and gland-tipped; prickles unequal, moderate, slender. *Leaflets 3–5*, imbricate, softly pubescent to *white felted beneath, serrulate-dentate; terminal leaflet round*, cuspidate. *Panicle nodding, leafy, compound*, equal to *pyramidal-truncate, middle branches 3–5 (7)-flowered, spreading, the lower ones up to 10-flowered*; rachis flexuous. *Sepals* prickly, greyish green, with fine, long, leafy tips, *clasping*. *Petals* ovate, claw obsolete, downy outside, *white*, notched or pointed. *Filaments white, equalling the styles*. *Fruit* ovoid, *of numerous small drupels*, sweet.

England: local; 14 (frequent in Stubbygrove Wood, Frant), 17 (Wimbledon Common), 55 (Lea Lane, Ulverston, as *R. morganwgensis*). France (*Cauterets*), Belgium, Germany, Switzerland.

339. **R. Chenonii** Sudre, 1906, *Diagn.* 46; *R. Koehleri* ssp. *Chenonii* (Sudre) Sudre, 1908–13, 185, t. 183.

Rufescent. Very prickly. Hawkenbury. Stem angled, thinly hairy with rather short glands and pricklets, and longer glandular acicles, some very long; *prickles numerous, very broad-based, curved, the smaller ones gland-tipped*. *Leaflets* 3, 4, 5, thinly hairy and *grey felted beneath*, shallowly but *coarsely and unequally to subcompound-serrate*; basal leaflets short-stalked; *terminal leaflet roundish, obovate, cuspidate*. Panicle leafy, nearly equal with long-peduncled, spreading branches; rachis wavy, hairy with *crowded, rather small, hooked prickles*, glands and acicles. *Sepals* prickly, *partly erect*. Petals elliptic, *pinkish*, notched or pointed, glabrous. Filaments white, about equalling the greenish or red styles. Carpels pilose. Fruit roundish.

England: 14 and 16; as yet only found around Hawkenbury. N. France, Belgium.

340. **R. pilocarpus** Gremli 1870, *Beitr. Fl. Schw.* 42; *R. obtruncatus* ssp. *pilocarpus* (Gremli) Sudre, 1908–13, 177, t. 172.

Stem becoming blackish purple, hairy with short, deep purple glands and pricklets, and intermediate stout prickles, some gland-tipped; *prickles hairy*, *strong, confluent, patent or curved.* Stipules filiform. *Leaflets 3–5, rugose, tough, imbricate, thickly and softly hairy beneath, coarsely subcompound-serrate;* terminal leaflet roundish or very broadly ovate-acuminate. *Panicle large, floriferous, corymbose towards the top, with 1–2 long, compound, spreading branches below; rachis stout,* felted and hairy, armed like the stem; the terminal leaflets roundish, acute. Calyx discoid-based; sepals prickly, *petals roundish,* notched, fimbriate, pink or white; *filaments white, hardly longer than the styles. Anthers sometimes pilose. Carpels densely, and long remaining, pilose.* Fruit roundish, very good. Styles sometimes red.

England: rare; 16 (Happy Valley, Rusthall Common; south of Wrotham Heath), 19 (south end of Tiptree Heath), 20 (Tingrith). Switzerland, Bavaria, Styria.

341. **R. abietinus** (Sudre) Bouv. 1929; *R. hystrix* microg. *abietinus* Sudre, 1908–13, 181, t. 178.

Kent and Sussex. Stem blunt-angled, reddish brown, nearly glabrous; *prickles yellow, broad-based, falcate. Leaflets* 3 (5), glabrous above, pubescent and *grey felted beneath, sharply subcompound-serrate; terminal leaflet* truncate based, *roundish obovate, long-cuspidate. Panicle* leafy, *branches deeply divided,* clustered on strong panicles, otherwise cymose, 5–7-flowered, or more simple and subracemose; *rachis hirsute,* densely glandular and acicular, *prickles weak and short. Sepals* felted, shortly glandular, *patent. Petals pink,* hairy on the margin, rather *broadly elliptic. Filaments pinkish, half as long as the crimson or purple styles. Anthers pilose. Carpels pilose.* Fruit ovoid, small.

England: local; 14 (near Tunbridge Wells; lane west of Heathfield Park), 16 (Tunbridge Wells Common and around Rusthall, frequent). France (*Pyrenees,* one station; *Maine-et-Loire,* one station), Bavaria (one station).

342. **R. bavaricus** (Focke) W. Wats. 1952 (for 1951), *Lond. Nat.* Suppl., 95; *R. Koehleri* ssp. *bavaricus* Focke, 1877, *Syn. Rub. Germ.* 357; *R. hebecarpus* ssp. *bavaricus* (Focke) Sudre, 1908–13, 182, t. 180.

Stem angled, red-purple, moderately hairy with unequal glands and broad-based acicles, some long ones gland-tipped; prickles numerous, slender, some long, falcate and retroflexed. Petiole hirsute, prickles long, stipules red, filiform. *Leaflets 4, 5, densely hairy and grey-white felted beneath,* rather coarsely, unequally and sharply or crenate-serrate, jagged; *terminal leaflet short-stalked, elliptic or obovate, acuminate. Panicle dense,* subequal, leafy, racemose or subracemose above, *middle branches sometimes bunched; rachis yellow at the back, hirsute with numerous, slender, spreading, unequal prickles and longish, unequal acicles,* many of them gland-tipped. *Sepals* greenish, felted, hirsute and prickly, *patent or loosely reflexed. Petals* moderate, broad elliptic, *pinkish, hairy.* Filaments white, longer than the *yellowish styles.* Carpels pilose. *Fruit long-oblong,* abundant.

England: rare; 20 (Mardley Heath, W. H. Mills). Switzerland, Germany, Bavaria, Saxony, Austria, Tyrol, Hungary, Bakabanya.

343. **R. Mikanii** Koehl. ap. Wimm. & Grab. 1829, *Fl. Siles.* I, ii, 56; G. Braun, *Herb. Rub. Germ.* 114.

W. Kent. Stem subglabrous, glaucescent, becoming fuscous-purple, *with numerous acicles and unequal glands; prickles numerous, slender,* yellowish, mostly long, slanting. Leaves 3–5-nate, pedate, prickles hooked, stipules filiform to linear-lanceolate; *leaflets rugose, softly pilose on the veins beneath to grey felted, rather coarsely subcompound-serrate; terminal leaflet cordate, ovate or roundish rhomboid, acute. Panicle* showy, short, subcompound, broad and dense; *terminal leaflets broad, obovate-cuneate, shortly pointed; lateral leaflets very shortly stalked; rachis hirsute,* armed like the stem. Flowers *c.* 1·5 cm.; calyx felted and hairy with fine prickles and numerous long-stalked glands; *sepals ovate, slenderly tipped, ± clasping*; petals pale pink or white, obovate-cuneate, notched; *filaments about equalling the styles.* Carpels pilose. *Receptacle subglabrous. Fruit rather large, ovoid,* well developed.

England: 16 (plentiful along a valley bank off Gravelpit Lane, below Shooters Hill). Germany, *Silesia.*

(*b*) All leaves green beneath. Sepals reflexed.

344. **R. spinulifer** (*spinuliferus*) Muell. & Lef. 1859, *Pollichia,* 213; *R. Koehleri* ssp. *spinulifer* (Muell. & Lef.) Sudre, 1908–13, 185, t. 183; *R. Koehleri* sensu Rogers pro parte (set no. 127), 1900, 82; non Weihe, 1825; *R. tumulorum* Rilst. 1940, *Journ. Bot.* **78**, 164. **Fig. 43.**

4x. *A low bush, or climbing. Stem* slender, angled, reddish brown to *blackish purple in the sun,* thinly hairy with few glands and numerous, unequal acicles and pricklets; *prickles* unequal, the larger ones *subulate from a swollen base.* Leaves rather small, 3(–5)-nate; *leaflets twisted, subglabrous above, pubescent to glabrescent beneath, sharply subcompound serrate* or partly dentate; *terminal leaflet roundish to obovate-rhomboid, shortly pointed, base entire. Panicle* often with a binate and a trilobed leaf and then a few linear ones, *with very long, semi-erect lower branches,* and 1–2 (3)-flowered upper ones, forming a dense, narrow top; *rachis thinly and shortly hairy, prickles very numerous and very unequal,* straight and slightly declining or slightly bent, *some long or very long,* with fine, unequal acicles and short glands in great numbers; *pedicels armed with numerous, patent, yellowish, acicular prickles. Sepals* aciculate and glanddotted ending in long tips *strongly reflexed. Petals obovate, notched, glabrous, tapered to the claw,* spreading with sides reflexed, pink or white. Filaments white, long. Styles greenish. Carpels pilose. Fruit small, drupels rather large.

E., W. Rather frequent; 2, 14–17, 20, 21, 23, 32, 36, 49, 54. France, Belgium, Germany, Switzerland, Austria.

(*c*) All leaves green beneath. Sepals patent, semi-erect or clasping.

345. **R. horridicaulis** P. J. Muell. 1861, *Bonpl.* 284; *R. Koehleri* ssp. *saxicolus* microg. *horridicaulis* (P. J. Muell.), Sudre, 1908–13, 188, t. 185.

Stem rufescent to fuscous, angled, *hairy with very numerous*, in part *confluent*, broad-based, *recurved*, yellow to red *prickles, some very long*, and pricklets, gland-tipped acicles and shorter glands. *Leaflets* 3, 4, 5, tough, *rather small*, short-stalked and short-pointed, greyish, but not felted, and roughish with short hairs beneath, shallowly and rather patent serrate-dentate; terminal leaflet roundish cuspidate, ovate, or obovate, acuminate. *Panicle leafy, long*, rounded at the top, *the upper branches short*, patent, 1–3-flowered; *rachis hirsute, with crowded armature like the stem, some prickles being hooked. Sepals* green with white edges, green-tipped, *clasping. Petals small*, obovate-oblong, entire, pink or white. *Filaments hardly equalling the styles.* Fruit small, subglobose, drupels large.

England: very local but moderately distributed where it occurs; 14 (Rowfant, New Lodge and Marsh Green), 16 and 17 (rather frequent on the Lower Greensand south of Westerham). France, Germany, *Alsace*, Switzerland.

346. **R. apricus** Wimm. 1857, *Fl. Schlesien*, 626; *R. Koehleri* ssp. *apricus* (Wimm.) Sudre, 1908–13, 187, t. 184.

Stem brownish to deep red, *blunt*, ± *hairy* with numerous, unequal glands and acicles; *prickles slender*, rather long, *slanting. Leaflets* 3–5, short-stalked, *base entire*, glabrescent above, hairy beneath, rather *evenly serrate*, petiolules villous; *terminal leaflet elliptic or oblong-ovate, acuminate. Panicle* leafy, broad, *lax, pyramidal*, the branches oblique, 1–3 (5)-flowered, cymose, the lower ones panicled; rachis hairy, glands numerous, reddish, *prickles numerous almost acicular.* Calyx base discoid, *sepals* greenish, long, green-tipped, *erect.* Petals pinkish or greenish white, remote, elliptic, some notched. Filaments white, about equalling the red or greenish styles. Carpels glabrous. *Fruit oblong, rather small but abundant.*

England: rare; 2 (near Launcells), 4 (near Bridgerule), 24 (Tatling End, *C. Avery*), 39 (Hanchurch Hills, *E. S. Edees*). France, Belgium, Germany, Bohemia, Hungary, Austria.

var. **sparsipilus** W. Wats. 1935 (for 1934), *London Nat.* 62; *R. apricus* sensu W. Wats. 1928 (for 1927), *Rep. Bot. Exch. Cl.* **8**, 502; *R. hystrix* f. *bercheriensis* Rogers, 1900, 79.

4x. Differs in the yellow to crimson, nearly glabrous axes, the larger, white flowers, and the always red styles.

Endemic. Frequent; 13–17, 20–22, 24, 30.

347. **R. humifusus** Weihe ap. Bluff & Fingerh. 1825, *Comp. Fl. Germ.* I, 685; *R. Schleicheri* ssp. *humifusus* (Weihe) Sudre, 1908–13, 204, t. 197.

Stem slender, angled, brownish purple, *glaucescent, hairy with pallid*, mostly *rather short, acicles and glands and numerous intermediate prickles, the principal prickles numerous, slender and fragile. Petiole prickles slanting. Leaflets 3–5, long-stalked*, pubescent or pilose beneath, serrate-crenate; *terminal leaflet sub-cordate*, broad ovate, *slenderly acuminate. Panicle pyramidal*, racemose above, middle branches 1–3-flowered, pedicels long, *lower branches ascending, compound.* Sepals green, white-bordered, narrow ovate, attenuate. *Petals narrow, white,* oblong-obovate, some notched, margin downy. *Filaments white, about equalling the greenish styles.* Carpels pubescent. Fruit small. Distinguish from *R. euryanthemus.*

England: rare? 14 (in several places south of Tunbridge Wells), 16 (Shooters Hill), 18 (Langdon Hill). France (Savoie), Germany, Switzerland, Austria.

348. **R. plinthostylus** Genev. 1869, *Mem. Soc. Ac. M. et L.* **28**, 108; non Rogers, 1900, 83.

Rufescent. Stem slender, ± *glabrous* with shortish glands and pricklets; *prickles numerous, very long, yellow,* confluent, *mostly patent. Leaves rather small, bright green, pedate; leaflets 5,* glabrescent above and below, *incise-serrate; intermediate leaflets subcuneate; terminal leaflet* usually twisted, roundish ovate, *gradually long-acuminate, narrow rhomboid-ovate.* Panicle leafy, large with a narrow racemose top, the middle branches spreading, deeply divided, the lower ones sometimes very long, panicled; *rachis hairy, fiercely armed like the stem. Sepals prickly, clasping. Petals* glabrous, *pinkish, rhomboid,* veiny, erose. *Filaments white, drying pinkish, slightly longer than the reddish styles.* Carpels glabrous. *Very fertile, the fruit long-ovoid.*

England: very rare; 24 (Penn Wood, beside the footpath leading from the Church to Beamond End, and again near the exit from the wood). W. France, in the lower Loire basin, and S. France in *Tarn.*

> One cover in Herb. Genevier, labelled *R. plinthostylus,* contains three specimens from Yzernay, all of which are *R. Hystrix* not *R. plinthostylus.* Eleven specimens from three other localities are all *R. plinthostylus* and agree with the Penn bramble.

349. **R. rotundellus** Sudre, 1904, *Bull. Soc. Bot. Fr.* 23; *R. Koehleri* ssp. *rotundellus* (Sudre) Sudre, 1908–13, 188, t. 185; *R. saxicolus* sensu Genev. 1880, *Mon.* 128; non P. J. Muell. 1859.

Bucks. Staffs. Stem slender, angled, *dark violet, glaucous, glabrous,* with fine, *dark,* not very long, *acicles,* slender *pricklets and few glands;* the *longer prickles subulate.* Leaves pedate, some prickles straight; *leaflets* thinly hairy, *serrulate; terminal leaflet* subcordate, *round, acute.* Panicle rather short, lax, broad, truncate, with one broad, simple leaf, the upper branches 3–4-flowered, 1–3 other branches rather longer and panicled below; *rachis* thinly hairy, with *long, purple glands and acicles merging into acicular prickles.* Flowers 1·5–2 cm.; *sepals green* with a fine, white felted edge, prickly, *clasping; petals remote,* obovate, margin glabrous, *white with a greenish claw,* margin glabrous; *filaments white; long styles reddish below.* Carpels glabrous. Receptacle hirsute.

England: rare; 24 (Lee Clump, several bushes), 39 (Black Bank near Newcastle, *E. S. Edees*). Central France, Belgium, Switzerland, Bavaria.

Glandulosi

2. Flowers 2–2·5 (3) cm.

† Stem slightly hairy or glabrescent.

(*a*) All leaves green beneath.

350. **R. hylocharis** W. Wats. 1946, *Journ. Ecol.* **33**, 341; *R. rosaceus* var. *silvestris* R. P. Murray ex Rogers, 1894, *Journ. Bot.* **32**, 47; R. P. Murray, 1896, *Flora Somerset*, 116.

Stem reddish brown, angled *with many long glands and acicles merging into crowded, unequal prickles. Leaves large*, thin, 5-nate-pedate; *leaflets* nearly glabrous above, glabrescent beneath, *sublobate, serrate-dentate, jagged; terminal leaflet* short-stalked, cordate, ovate *attenuate or obovate acuminate.* Panicle with 1–3 simple leaves, large, broad, lax, middle branches spreading, long-stalked, 3–7-flowered, cymose; *rachis thinly hairy, prickles yellow, weak, decurrent*, patent and slanting, *passing into very long gland-tipped acicles.* Flowers 2–3 cm.; *sepals and petals sometimes 6, 7, 8; sepals* green, *attenuate and leafy-tipped, patent to erect; petals narrow, long-clawed, glabrous, rose-pink*; filaments white, long; styles greenish. Carpels glabrous.

Endemic. E., W., I. Fairly frequent in woods as well as in the open near them; 1 (Madron near Penzance, *E. F. Lees*, 1879), 2–6, 13, 16, 17, 23, 30, 39, 44, 46, 58, 59, 62. H. 40.

351. **R. scabripes** Genev. 1860, *Mem. Soc. Ac. M. et L.* 81; 1880, *Mon.* 94; *R. rosaceus* sensu Rogers, 1900, 78; non Weihe & Nees, 1825.

Mainly western. Robust. Stem becoming pruinose with numerous, short, *blackish glands, acicles* and larger pricklets; prickles numerous, very unequal, slanting. *Leaves large; leaflets 3–5, incise compound-serrate; intermediate and basal leaflets very shortly stalked; terminal leaflet very broad*, roundish or obovate, cuspidate. *Panicle long, large*, broad and leafy above with dense and compound branches, a good many others spreading and corymbose below; rachis wavy hirsute, *prickles short, unequal, confluent, some gland-tipped. Flowers starry*, 2–3 cm.; *sepals* greenish grey, white-lined, long-tipped, *from loosely reflexed and patent to clasping; petals rather broad*, elliptic-oblong, narrowed below, *rose-pink; filaments pink, much longer than the red styles.* Carpels glabrous or pilose.

E., W., I. Fairly frequent in woods in the west; 1–3, 5, 6, 8, 11, 35–39, 55. H. 3. France (*Vendée*).

(*R. apricus* var. *sparsipilus* W. Wats. on account of its large flowers may be looked for here. It is described with the main species above.)

352. **R. Koehleri** Weihe in Bluff & Fingerh. 1825, *Comp. Fl. Germ.* 681; Sudre, 1908–13, 184, t. 182; *R. Vigursii* Rilst. 1950, *Journ. Linn. Soc.* **53**, 420.

Stem turning red or brown in the sun, angled, *glaucescent*, with numerous, unequal glands and acicles, some gland-tipped and intermediate prickles; *prickles unequal, numerous, at first yellow, some very long and ± patent.* Leaves 5-nate, tough, glabrescent above, *pubescent beneath; terminal leaflet* subcordate to entire, *elliptic-obovate, acuminate*, coarsely unequal to finely *serrate* or serrate-dentate, *jagged*. *Panicle* rather long, narrow and equal *with several simple leaves* or none, the upper branches 1–3-flowered, patent, the lower ones more erect; *rachis* crooked, ± hairy, *armed like the stem*. *Sepals* glandular and prickly, long green-tipped, *loosely reflexed to patent* and on the terminal flowers often upright. *Petals elliptic, attenuate below, white or pink. Filaments far longer than* the greenish, sometimes reddish-based *styles*. Carpels glabrous or pubescent. Fruit ovoid, large.

Typically this is a vigorous plant in the open, but small forms occur. The name R. *Virgursii* has been given to dwarfish and to semi-dwarfish plants in Cornwall.

E., W., I. Rare; 2 (frequent in the south), 3 (Fancy, *Briggs*, 1870), 5, 15 (Bekesbourne Hill). H. 16 (near Clonbur). E. France, Belgium, Germany eastwards to *Silesia*, Austria, Bavaria, Switzerland, Italy.

353. **R. aculeatissimus** Kalt. 1845, *Fl. Aach. Beck.* 300; Sudre, 1908–13, t. 175 (as R. *rosaceus*); R. *rosaceus* auct. mult. non Weihe & Nees, 1822–27, *Rub. Germ.* 85; cf. W. Wats. 1949, *Watsonia*, **I**, 78.

E. Kent. Stem sharply angled, *deep red*, with few or many unequal glands, acicles and pricklets; *prickles numerous, unequal, mostly patent* from a strong base. *Leaves large, stalk short*, stipules red; *leaflets nearly glabrous above and beneath*, coarsely unequal to subcompound-*serrate-dentate* and undulate; *terminal leaflet short-stalked, broad or roundish, ovate acuminate. Panicle with 0–2, large, simple leaves, short, broad, sometimes subcorymbose with long 1–3-flowered upper branches and 1–2 semi-erect axillary branches which sometimes draw level with the top of the panicle*; rachis flexuose, thinly hairy and armed like the stem, but with finer and falcate prickles. Sepals narrow ovate, appendiculate, erect. *Petals remote, long, narrow*, notched, glabrous on the margin, *rose-pink*. Filaments long, white; styles greenish or filaments and styles coloured like the petals. Fruit large, ovoid.

England: very rare; 15 (Acrise cross roads, and Paddlesworth). N. France, Belgium, W. Germany, Bavaria.

(*b*) Upper and branch leaves, or in addition the middle stem leaves green, grey or greyish white felted beneath.

354. **R. Hystrix** Weihe in Bluff & Fingerh. *Comp. Fl. Germ.* **I**, 687; Sudre, 1908–13, 180, t. 177. **Fig. 44.**

4x. *Rufescent.* Stem low-arching, angled, glaucescent with glands, acicles and pricklets of all lengths; *prickles broad-based, very long*, straight or curved. *Leaves large, a fresh apple green*; leaflets 3–5, subimbricate, *velvety* pilose or pubescent and felted *beneath, coarsely, openly and sharply serrate, long-pointed; terminal leaflet roundish or elliptic-ovate, a little rhomboid or obovate and of the upper leaves narrow and very long-pointed. Panicle leafy, lax, pyramidal*, the lower branches remote, elongated, ascending, the *upper bracts long sub-*

lanceolate; the terminal leaflets obovate-cuneate; rachis crooked, *prickles slender, some retrorse falcate, very long.* Flowers up to 3 cm.; sepals felted, ovate, long-pointed, erect; petals 5–7, pink, elliptic, ovate or obovate, notched, glabrous or slightly pilose; filaments long, *pink; styles yellowish or reddish based.* Carpels glabrous. Fruit roundish, ovoid.

E., W., I. Frequent in moist woods; 3, 4, 7, 13, 15–17, 19, 21–24, 35, 36, 38–41, 48, 57, 67. H. 1 (near Dingle Church), 35. Rare on the Continent; France, Belgium, W. Germany, Bavaria.

355. **R. semiglaber** (Rogers) W. Wats. 1946, *Journ. Ecol.* **33**, 341; *R. Marshallii* var. *semiglaber* Rogers, 1897, *Rep. Bot. Exch. Cl.* **1**, 479; 1900, 84.

Western. Rufescent. Stem angled, with short glands and acicles, some long and gland-tipped; *prickles confluent, patent or slanting.* Leaves pedate, *smallish*, leaflets 3–5, glabrous above, *pubescent and grey felted beneath, shallowly serrate-dentate, jagged; terminal leaflet cordate, ovate or obovate-oblong or roundish, broadly acuminate. Panicle* small, nearly equal, entirely racemose or compound, having several racemose branches below; rachis thinly hairy, pedicels felted, all *very densely armed.* Flowers *c.* 2·5 cm.; *sepals clasping; petals rather broad*, rhomboid-ovate, notched, margin glabrous, *rose-pink*; filaments white, long; styles yellowish green. Carpels glabrous. *Fruit large, ovoid, of numerous drupels, crimson*, then black.

Endemic. E., W., I. Rather frequent in the west; 4, 33–35, 40, 48–51 (not 36, 42). H. 2 (Muckross).

356. **R. infestus** Weihe in Boenn. 1824, *Prod. Fl. Monast.* 153; non Focke, Sudre and other authors, who have transferred the name to *R. taeniarum* Lindeb; *R. setulosus* Lef. & Muell. 1859, *Pollichia*, 199; Sudre 1908–13, t. 115 (as *R. Colemannii*). **Fig. 45.**

Stem erect then arching, becoming brownish red, *with* glands, acicles and *often very many, strong, curved prickles touching and coalescing. Leaves* 5-nate, *deep green; leaflets* glabrous above, *pubescent to greenish grey felted beneath, coarsely serrate*; basal leaflets rather shortly stalked; *terminal leaflet* roundish or *broadly ovate, acuminate* from a subcordate or subtruncate base. *Panicle short, few-flowered, occupying most of the flowering branch*, with 1, 2 or more simple leaves and broad, linear, green bracts, *branches shortish, semi-erect*; rachis thinly hairy, the *prickles crowded, strong, straight to hooked, of various sizes*, pedicels *long, felted.* Flowers 2–3 cm., *incurved; sepals prickly, greenish, ovate, acute, concave-patent to loosely erect; petals 5, 6, 7, roundish or broad oblong with a cuneate base, white; filaments* white at first, *reddening*, investing and concealing the yellowish or greenish styles. *Carpels glabrous.* Receptacle pilose. *Fruit large, ovoid-oblong, sweet. Flowering early June.*

It was this species that Bloxam distributed, correctly, as *R. infestus* Weihe, from near Coventry railway station. One of his specimens now in Herb. Babington consists of two panicles of *R. infestus* and a stem-piece with two leaves of *R. villicaulis*. Babington rejected Bloxam's identification, and formed the opinion that the correct name was *R. Colemannii* Bloxam, of which a bush was then growing in the Cambridge Botanic Garden. He proceeded to draw up a description partly from the mixed Coventry specimen and partly from the true *R. Colemannii* from Packington, Leicester, and published it in his *British Rubi* 1869, as *R. Colemannii* Bloxam, quoting as synonyms *R. Colemannii* Bloxam in Kirby's *Fl. Leic.* and *R. infestus* Bloxam, MS. Sudre's t. 115 as *R. Colemannii* is correct for the Coventry plant, both as to stem-piece and panicle; his description, he says, is made from the Coventry plant, and copying from Rogers adds that that was Bloxam's type (of *Colemannii* Bloxam). The plate is a good representation of *R. infestus* Weihe, but the description does not agree with the plate, nor was the Coventry plant the type of Bloxam's *R. Colemannii*. Roger's description, from which Sudre seems to have borrowed a few particulars, includes *R. crudelis* W. Wats. England: local; 16, 17 (frequent), 21, 22, 24, 38. France (*Oise, Aisne*), Denmark, N.W. Germany.

†† Stem moderately or densely hairy.

(*a*) Leaflets coarsely serrate, the principal teeth often patent. Petals pinkish or pink.

357. **R. adenolobus** W. Wats. 1935 (for 1934), *London Nat.* 61; *R. Koehleri* var. *cognatus* sensu Rogers, 1900, 83, nequaquam *R. cognatus* N. E. Brown, 1892. **Fig. 46.**

4x. *Vigorous. Stem blunt with flat sides, reddish brown to fuscous, glaucescent with rather few, longish glands, numerous long, pale acicles and long, slender prickles, straight or bent. Leaves large*, pale green, 3–5-nate, pedate, stipules narrow linear, *petiole with some geniculate prickles; leaflets strigose above*, softly hairy to *grey felted beneath*, coarsely, somewhat doubly, serrate, to rather finely, compound dentate-sinuate on the upper leaves; *terminal leaflet* elliptic oblong or obovate, subcuneate, or broad ovate, *long-acuminate, base nearly entire. Panicle with several simple leaves, which are glandular above and grey pubescent beneath, often very large*, with the middle and upper branches semi-erect cymose, with intricate, longish pedicels and several spreading, numerous flowered, panicled branches below; *rachis* dusky hairy, armed like the stem, *some prickles geniculate. Sepals ovate-lanceolate, attenuate*, loosely reflexed to semi-erect; *petals narrow* obovate, pilose on the margin, *pink, pinkish, or lilac* (rarely white); filaments white, equalling or twice as long as the greenish styles. Carpels pilose or pubescent. Fruit subglobose or slightly oblong. Some stalked glands and acicles may be found on the margins of all leaves but especially on the basal lobe.

E., W., S. Rather common; 4, 8, 11, 13, 15–24, 30, 34–36, 38, 39, 41, 49, 52, 62, 63, 66, 67, 69, 91, 103, 110. France (Seine-et-Oise).

358. **R. dasyphyllus** (Rogers) Druce, 1908, *Brit. Pl. List,* 22; *R. Koehleri* ssp. *dasyphyllus* Rogers, 1899, *Journ. Bot.* **37**, 197.

4x. *Axes red. Common. Stem* low-arching, angled, *with long glands, acicles and numerous, slender prickles.* Leaves 3, 4, 5-nate, prickles hooked; *leaflets thick, tough, glabrous above, softly hairy and slightly felted beneath,* serrate and conspicuously *jagged,* the base fringed with acicles and stalked glands; terminal leaflet broad elliptic-obovate, cuspidate. *Panicle long and narrow,* sub-racemose above, the pedicels patent *with several, short, remote, erect branches in the axils below;* rachis nearly straight, rigid. Flowers *c.* 2 cm.; *sepals patent with tips ascending or erect,* ultimately sharply reflexed; *petals* 5–8, pilose, *narrow oblong-obovate, deep pink; filaments pink; styles salmon.* Carpels pilose, Young fruit pale green, then coral red to black, oblong.

E., W., S., I. Very common and recorded for nearly all parts of Great Britain, but not for Cornwall, Isle of Wight, the fen counties 28, 29, 31, nor Lincoln. Known in N. Ireland (H. 37–40). France (*Seine-et-Oise*) and Denmark.

359. **R. Marshallii** Focke & Rogers 1895, *Journ. Bot.* **33**, 103.

*5x. *Stem low, reddish brown,* angled, *hirsute,* glands few but stout *pricklets and intermediate prickles numerous, some gland-tipped; prickles numerous, hairy, subulate, long patent.* Leaves rather small; *leaflets* 3–5, thick, green and softly hairy or also felted beneath, *coarsely* and *deeply to slightly compound-serrate;* terminal leaflet cordate, ovate or oblong-obovate, acuminate. *Panicle not leafy, very long, narrowed, the branches short,* spreading, 3–7-flowered, cymose; *rachis* felted and *hirsute with numerous, long, patent, red prickles.* Flowers *c.* 2 cm.; sepals patent, then reflexed; *petals* ovate, notched, fimbriate, *pink;* filaments white or pink, not much longer than the greenish or bright red styles, which overtop the connivent stamens after the petals have dropped. Carpels hairy. *Fruit short, subglobose, of few large drupelets, dark brown* before ripening. A large form in Squerryes Park, Westerham, v.-c. 16, is **5x.** *All but the longest prickles are gland-tipped.* The species is perhaps derived from a cross between *R. phaeocarpus* and *R. vestitus.*

Endemic. E., I. 12–17, 21–24, 30, 32–34, 36, 57. H. 38.

360. **R. Billotii** P. J. Muell. 1861, *Bonpl.* 283; *R. adornatus* microg. *Billotii* (P. J. Muell.) Sudre, 1908–13, 174, t. 169.

Rufescent. Surrey and Kent. Stem angled, *intricately hairy* with unequal glands and acicles, some gland-tipped; *prickles pale, subulate.* Petiole short, prickles mostly patent. *Leaflets* 3–5, short-stalked, nearly glabrous above, *silkily hairy* or also felted *beneath,* finely dentate-serrate or coarsely and un-equally serrate and *jagged;* terminal leaflet deeply emarginate or entire, elliptic-obovate, *sharply and finely cuspidate,* the midrib prickles patent. *Panicle not leafy, very long and narrowed to the top, the branches 3–5-flowered, patent; rachis* crooked, *densely villous with very long, slender, declining and falcate prickles* and very long glands and gland-tipped acicles. *Sepals* yellowish, grey felted, large, ovate-lanceolate, *attenuate with long, green tips, aculeate, erect*; petals narrow elliptic, pink, fringed; filaments white or tinged pink, slightly longer than the greenish styles. Carpels glabrous or pilose. Fruit roundish. *Flowering early July.*

England: local; 4 (Sticklepath), 16 (Mereworth Woods; Oldbury Hill, Ightham; Crockham Hill Common West; Shooters Hill), 17 (Woods around Farleigh and Selsdon; Newlands Corner). France (*Vosges*, one station), Belgium (one station).

361. **R. Hartmanii** (Gandog. ex Sudre) W. Wats. 1946, *Journ. Ecol.* **33**, 342; *R. fusco-ater* ssp. *Hartmanii* (Gandog. ex Sudre) Sudre, 1908–13, 173, t. 167; *R. horridus* Hartm. 1832, *Handb. Skand. Fl.* **2**, 139; non K. F. Schultz, 1819, *Fl. Dan. Suppl.* t. 77.

4x, 5x. *Kent. Sussex. Stem* blunt, *fuscous, hirsute*, with few short glands and pricklets; *prickles* unequal, the intermediate ones often gland-tipped, *the large ones numerous, stout-based, many curved. Leaves* 3–5-nate, pedate; *prickles hooked*; stipules filiform; *leaflets imbricate, thick, softly hairy and grey felted beneath, coarsely* incise *serrate*, undulate; *terminal leaflet* subcordate to emarginate, *roundish or ovate, acuminate.* Panicle rather narrow and equal, or broader below with perhaps two widely spreading, 5–7-flowered, axillary branches; rachis felted and hirsute, armed like the stem. *Flowers c. 2 cm.*; *sepals prickly*, leafy-cuspidate, patent; *petals* rather *broad ovate* or obovate, crumpled, fimbriate, pinkish; *filaments white, equalling the greenish styles. Carpels bearded.* Fruit ovoid-oblong, abundant. Recalls *R. fuscus.*

England: locally plentiful; 14 and 16 (around Tunbridge Wells, Hawkenbury and Eridge). S.E. Sweden and *Bornholm.*

362. **R. Lapeyrousianus** (Sudre) W. Wats. 1956, *Watsonia*, **3**, 290; *R. Koehleri* ssp. *Lapeyrousianus* Sudre, 1900, *Rub. Pyr.* 79; *R. hebecarpus* ssp. *Lapeyrousianus* (Sudre) Sudre, 1908–13, 183, t. 180.

Surrey. Robust. Tall. Stem sharp-angled, the sides concave, with a few long and short hairs, glaucescent and fuscous, with fine, unequal glands and acicles; *prickles numerous, yellowish,* unequal, *some very long* declining and falcate. *Leaves large,* 5-nate, prickles slanting and retrorse-falcate; *leaflets* shortstalked, nearly glabrous above, softly pubescent and felted beneath, coarsely crenate, patent serrate, *all very long-pointed*; terminal leaflet notched at the base, ovate-elliptic. *Panicle very long, narrow*, subpyramidal, *the upper branches* 1–3-flowered, *deeply divided or fasciculate*, sharply ascending, *pedicels long*; rachis felted, closely and shortly hairy, with short, and very long, glands and subulate prickles. Sepals grey glandular and aculeate, patent; *petals* elliptic, pilose, *pink*; filaments white, about equalling the greenish styles. Carpels pilose. *Receptacle hirsute, the hairs protruding below the head of carpels.* Fruit large, roundish ovoid, abundant.

Glandulosi

England: rare; 17 (St Ann's Hill, Chertsey, *J. G. Baker*, 1867; *C. Avery*, 1950; Blackdown Common, Chobham Ridges, *C. Avery*). France (*Ariège*, two stations).

(b) Leaflets sharply incise subcompound-serrate. Petals white.

363. **R. asperidens** (Sudre) Bouv. 1923; *R. Koehleri* ssp. *asperidens* (Sudre) Sudre, 1908–13, 186, t. 184.

S.E. England. Axes and arms fuscous-purple. Stem low-arching, angled, glaucescent, with *slender acicles* and glands; *prickles red-based, very slender,* patent, slanting and falcate. *Leaves bright green,* 5-nate-subpedate, petiole-prickles slanting or slightly bent; *leaflets very softly hairy beneath, ciliate, sharply incise subcompound-serrate;* basal margins fringed with glands and acicles; *terminal leaflet* short-stalked, subcordate, elliptic or rhomboid-ovate, *slenderly falcate-acuminate. Panicle* short, lax, broad, *branches patent or but slightly ascending,* 1–5-flowered, long-peduncled, the *pedicels* long and *armed with numerous, fine prickles;* rachis hairy, prickles numerous, falcate. Flowers 2–2·5 cm. *Sepals green or greenish with a white edge,* narrow ovate, leafy-tipped, prickly, patent. *Petals* 5–8, *narrow obovate, tapered below, remote,* white, glabrous. Filaments white, long. Styles greenish. Carpels glabrous. Fruit ovoid, moderate.

England: locally plentiful; 8, 13–17 (from St Leonards Forest, near Upper Grouse Farm, to Paddlesworth, E. Kent, and around Tunbridge Wells, Westerham, Addington and West Wickham in W. Kent). France, Belgium, Germany, eastwards to *Saxony, Bavaria,* Switzerland.

364. **R. emarginatus** P. J. Muell. 1858, *Flora,* 164; *R. Lejeunei* ssp. *emarginatus* (P. J. Muell.) Sudre, 1908–13, 178, t. 174; *R. carneiflorus* P. J. Muell. 1859, *Pollichia,* 169; *R. napaeus* Focke in Asch. & Grab. 1902, *Syn. Mitteleur. Fl.* **6,** 543; *R. oblongatus* P. J. Muell. 1859, *Pollichia,* 184.

Stem purple, slender, angled, hairy with ± glandular and unequal *acicles, all slender and some passing into gland-tipped intermediate prickles; prickles slender, unequal, long,* slanting to slightly bent. *Leaves* rather large, 5-nate, pedate, *yellowish green; leaflets thin, all short-stalked,* strigose above, *densely pectinate-pilose on the veins beneath to ± grey felted on the upper leaves,* sharply unequal to subcompound crenate-serrate, *jagged, the mucros long; terminal leaflet ± deeply cordate,* roundish to *elliptic-obovate,* rather suddenly acuminate. *Panicle rather short, slightly tapered* to the truncate top, *branches* 3–1-flowered, *all patent;* pedicels and rachis hirsute, armed with numerous, *very slender, straight and curved prickles,* stalked glands and very unequal *acicles, some as long as the prickles and gland-tipped. Sepals* very aciculate, long-tipped, *patent-erect. Petals narrow* elliptic-rhomboid, mostly *notched* and nearly glabrous, *pink.* Filaments long, pinkish. Styles reddish-based. Carpels glabrous. *Receptacle very pilose, especially below the lowest carpels. Fruit roundish, carpels rather large.*

England: very rare; 17 (one patch in a damp spot situated between Little-heath Wood and Gee Wood, near Selsdon, 29 July 1950). Belgium, Germany (around Weissenburg), *Alsace (Baden), S.W. Bavaria* (Waging), Switzerland (near Lugano on L. Lugano and near Winterthur, Canton Zürich), N. Italy (near Luino on L. Maggiore).

(*c*) Leaflets ± shallowly and evenly serrate or serrate-dentate.

365. **R. fusco-ater** Weihe ap. Bluff & Fingerh. 1825, *Comp. Fl. Germ.* **1**, 681; Sudre, 1908–13, 172, t. 166.

4x. *Stem* blunt, *deep purple-black, glaucous,* with unequal glands and numerous pricklets; *prickles numerous, unequal, subulate.* Leaflets 3–5, deep green above, yellowish green, hairy and felted beneath; *terminal leaflet* cordate, *round, acute. Panicle short, very broad, with spreading, cymose* middle, and oblique, panicled lower branches, *pedicels long*; rachis densely hairy, prickles from straight to hooked. Flowers 1·5–2·5 cm.; *sepals* broad ovate, shortly pointed, *patent to loosely clasping; petals broad* fimbriate, *deep rosy pink; filaments also deep rosy pink, rather short, remaining erect and retaining their colour after the petals fall*; styles deep pink or greenish. Fruit roundish, ovoid, rather large. Prefers open situations.

E., W. Fairly well distributed, but nowhere abundant; 4 (W. Lynn, *W. H. Mills*, 1948), 13–15, 17, 21, 23, 24, 30, 36, 37, 39, 49, 57, 62. France, W. Germany, Switzerland, Austria.

366. **R. absconditus** Lef. & Muell. 1859, *Pollichia*, 167; *R. fusco-ater* ssp. *Hartmanii* microg. *absconditus* (Lef. & Muell.), Sudre, 1908–13, 173, t. 167.

Robust. Rufescent. Stem angled with rather few, short glands and frequent acicles, some long and gland-tipped; *prickles numerous, yellow,* unequal, *long, slender, much declining. Leaflets large,* 3, 4, 5, *yellowish green,* rough above, *softly hairy to felted beneath; terminal leaflet roundish or sometimes almost reniform,* deeply cordate, abruptly acuminate. Panicle large, broad and lax, or narrower and subpyramidal, ± leafy, with 1–3-flowered branches above and 2–3 long-peduncled panicled branches below; *rachis wavy, very hairy, with crowded yellow prickles, some hooked*; pedicels long, armed with subulate prickles. *Flowers* 2·5–3·5 *cm.*; *sepals large, patent; petals* ovate or obovate, pointed, *pink,* margin glabrous; *filaments pinkish,* long; styles yellowish or reddish. Carpels pilose. Receptacle pilose. Fruit large, ovoid.

England: rather uncommon; 14 (Eridge Rocks), 16, 17 (Booker Common), 20, 24, 36 (Wormbridge Common), 39. France (*Oise*), Bavaria.

367. **R. oegocladus** (*oigocladus*) Muell. & Lef. 1859, *Pollichia*, 134; *R. fusco-ater* ssp. *oegocladus* (Muell. & Lef.) Sudre, 1908–13, 173, t. 167.

Stem villous, with short glands and numerous, short acicles, unequal, *cushion-based, glandular acicles and intermediate prickles; prickles yellowish, subulate, ± patent. Leaves yellowish green,* petiole prickles straight to very retrorse-falcate; leaflets softly hairy and greyish, rarely felted beneath; *terminal leaflet cordate, roundish, cuspidate. Panicle lax and open with widely spreading branches; rachis densely villous, prickles subulate,* the terminal leaflets *closely strigose above,* the lateral leaflets subsessile. Flowers 1·5–2 (2·5) cm.; *sepals greenish, leafy-tipped,* felted, shortly glandular and with acicles, *loosely reflexed,* sometimes becoming erect; *petals 5–8,* pilose, roundish to elliptic-ovate, *pink, or pink and white* (particoloured); filaments pink or white, styles slightly greenish or reddish. Carpels glabrous. Receptacle hirsute.

England: rare; 15 (Bigbury; Covert Wood), 16 (Shooters Hill frequent), 21, 22 (Wytham), 24, 35 (St Arvans, *H. J. Riddelsdell*, 1925). France, W. Germany and Switzerland, one station each.

368. **R. aristisepalus** (Sudre) W. Wats. 1946, *Journ. Ecol.* **33**, 342; *R. Schleicheri* ssp. *inaequabilis* var. *aristisepalus* (*aristisefalus* errore typographico) Sudre, 1908–13, 205; *R. velatus* sensu Rogers, 1900, 92; non Lef. ex Genev. 1872.

Western. Stem blunt, furrowed, with short, unequal, deep red glands and *numerous, longish acicles* and pricklets; *prickles short, unequal, very strongly de-clining or recurved.* Stipules narrow-linear. *Leaflets 5, very hairy* to felted beneath; *terminal leaflet very broad, cordate, ovate, acuminate, crenate-serrate. Panicle nodding, long, with ovate to linear-lanceolate bracts,* branches 1–4-flowered, ± *deeply divided; rachis very villous, prickles* fine, slanting to hooked, glands mostly sunken, together with *much longer acicles;* pedicels and sepals finely aculeate. *Flowers c. 2 cm.;* sepals ovate-lanceolate, finely leafy-tipped, patent to clasping; *petals moderately broad, obovate, tapered below,* notched, pale pink or white; filaments white, *hardly equalling the* yellowish or red *styles. Carpels densely pubescent.* Fruit small, subglobose. Recalls *R. fuscus.*

Endemic. E., W. Local; 35 (The Narth), 36, 37 (Bromsgrove, Lickey), 39, 40, 42, 44.

369. **R. hypochlorus** (Sudre) Bouv. 1923; *R. obtruncatus* ssp. *mutabilis* microg. *hypochlorus* Sudre, 1908–13, 176, t. 172.

Stem yellowish brown with *few glands and acicles but numerous, long, inter-mediate prickles; prickles yellowish, very slender, small-based,* unequal, slightly slanting and falcate. *Leaflets 3–5, yellowish green,* velvety and grey felted beneath, finely and unequally serrate; *terminal leaflet elliptic or slightly obovate, with a nearly entire, narrow base.* Panicle hardly narrowed, with several trilobed or ovate leaves, the upper branches 1–3-flowered, oblique, the lower ones remote, panicled, ascending; rachis villous, armed like the stem. *Flowers c. 1·7–2·3 cm.; sepals aculeate, patent to clasping; petals rhomboid-ovate, remote, pink,* apex glabrous, entire; *filaments* white, *equalling* the greenish *styles.* Carpels pilose. Fruit roundish.

E., I. Rare, heaths and hedges; 16 (Crockham Hill), 18 (Galleywood Common). H. 16 (Oughterard, *Marshall*, 1895, as 'Koehleri, off type'). N.W. France (Valois, Sarthe, Maine-et-Loire).

370. **R. hebeticarpus** (*hebecarpos*) P. J. Muell. 1861, *Bonpl.* 282; Sudre, 1908–13, 181, t. 179.

Devon. Stem blunt with long and short *intricate hair*, few glands and numerous, unequal pricklets and tubercular acicles, some gland-tipped; *prickles red*, unequal, the longest ones *subulate from a narrow base. Leaflets 3–5, firm*, glabrescent above, *grey-white felted* and shortly hairy beneath, *serrate-dentate*; terminal leaflet elliptic-obovate, acuminate. Panicle dense, hardly narrowed above, with 1–3-flowered patent, and two or three spreading, panicled branches; rachis clothed and armed like the stem. Flowers *c.* 2·5 cm.; *sepals patent; petals narrow elliptic-obovate, tapered to the base*, notched and pilose, pink; *filaments white, equalling the reddish styles. Carpels densely hairy. Fruit of numerous drupels*, ovoid, abundant.

England: seen only at the roadside at Lee Moor, S. Devon, 1939. Vosges, Eastern Bavaria.

371. **R. spinulatus** Boul. 1868, *Ronces vosg.* 101; *R. Koehleri* microg. *spinulatus* (Boul.) Sudre, 1908–13, 185, t. 183.

N.W. Kent. Axes red brown. Stem blunt with *numerous, long glands, acicles, pricklets and slender, unequal prickles.* Leaflets 3–5, imbricate, glabrous above, softly hairy to grey felted beneath, unequal crenate-serrate, the principal teeth prominent; *terminal leaflet* subcordate, *roundish, finely acuminate.* Panicle rather long, equal, broad, with several simple leaves which are glandular above and felted beneath, the middle branches long-peduncled, spreading, the lower ones short, erect, all few-flowered; *rachis* hirsute, aciculate below, *prickles long, very slender.* Flowers 2–2·5 cm.; *sepals* ovate, appendiculate, *patent to erect; petals narrow, white*; filaments white, longer than the greenish styles. Carpels glabrous. Fruit moderate, roundish-ovoid, perfect.

England: confined to a few woods in N.W. Kent. France, Switzerland, Styria.

Ser. ii. EUGLANDULOSI W. Wats.

(Type *R. Bellardii* Weihe)

1. Largest prickles moderately strong, with a broad, compressed base, curved.

372. **R. Schleicheri** Weihe in Boenn. 1824, *Prod. Fl. Monast.* 152; Sudre, 1908–13, 198, t. 194.

Stem trailing, *red, glaucescent, villous*, with numerous *yellowish to brown* glands, acicles and prickles. Petiole convex above. *Leaflets* 3 (4, 5), *rather small*, hairy beneath, *the midrib glandular*, unequal to *subcompound incise-serrate; terminal leaflet short-stalked*, ovate, *obovate-rhomboid or elliptic, slenderly acuminate*, narrowed to a sometimes broad, truncate, notched base. *Panicle nodding, racemose above*, middle branches bunched, bracts long linear; *rachis*

crooked, hirsute and felted, *prickles hooked, numerous on the pedicels. Flowers
1–1·5 cm.*; sepals green-tipped, aculeate, patent to half erect; *petals narrow,
obovate-spatulate,* some notched, *pinkish or white,* if white, at first greenish in
bud, glabrous; filaments white, longer than the greenish styles. Carpels
thinly pubescent. *Fruit small, of about 10 drupels,* oblong, of an agreeable
taste. Flowering about 9 June.

England: rare, woods; 15, 17, 23, 36, 58. France, Belgium, Netherlands,
Switzerland, Germany, *Bavaria,* Moravia, Austria.

var. **grandidentatus** Barber, 1911, *Fl. Oberlausitz.*

Differs in the bright red-based prickles, petiole finely furrowed, terminal
leaflet base rounded, softly hairy beneath, with no glands on the midrib,
the *sharp, coarse, deep teeth,* and the *carpels glabrous.* Makes a larger bush with
the stem only thinly hairy.

England: rare; 14 (Stubbygrove Wood, Frant, and the vicinity), 16 (Bay-
ham Road, Tunbridge Wells). Germany (*Oberlausitz*).

var. **eriocladus** Sudre, 1909, *Rub. tarn.* 52.

*N.v.v. Terminal leaflet broad-based, truncate or cordate-ovate-acuminate,
the long point often oblique, twisted and folded, margin lobate or compound, sharply
and coarsely incise-serrate, densely, softly hairy and sometimes greenish-grey felted
beneath.* Panicle rachis hirsute; flowers *c.* 1·5 cm., petals white, villous.
Carpels covered with dense, long hairs.

England: 62 (Wass Bank, Rogers, 1890; *W. H. Mills,* 1953; Helmsley
Moor and neighbourhood, in plenty, *W. H. Mills,* 1946). France (*Tarn*),
Germany, *Bavaria.*

var. **longisetus** Sudre & Schmid. ex Schmid. 1, Dec. 1911, *Ronces Leman*;
Sudre, 1908–13 (Oct. 1912), 199.

Stem angled, furrowed, thinly hairy, glabrescent, *glaucous; prickles nearly
straight.* Leaflets unequally serrate. *Panicle* pyramidal, *dense at the apex, the
upper leaves grey felted beneath; rachis fuscous,* thinly felted and pilose, *prickles
numerous, nearly straight, purplish glands numerous with a few, very long, gland-
tipped acicles.* Sepals ending in a long linear appendage, erect on the fruit.

England: 16 (Tunbridge Wells Common, in plenty on a bank above the
London Road). E. France, Belgium, Germany, *Bavaria* and *Silesia.*

373. **R. viridis** Kalt. 1845, *Fl. Aach. Beck.* 284. **Fig. 47.**

Rufescent, glabrescent. Stem sharp-angled, furrowed, with few unequal glands,
some subsessile, and acicles and pricklets, *prickles unequal, strong-based, fragile-
pointed. Leaves a pleasant apple-green,* pedate, *petiole subglabrous, channelled
throughout, prickles hooked, stipules filiform; leaflets 5,* imbricate, *glabrous
above, silkily pubescent beneath, compound-serrate* with the principal teeth
larger; terminal leaflet cordate, roundish-ovate, acuminate. *Panicle leafy,* lax,
branches half erect, 3–5-flowered, corymbose or subcymose, *thinly hairy to
glabrescent, prickles slender, long, some straight, others strongly curved.* Flowers

1·5–2 cm.; *sepals* greenish, long-tipped, *reflexed; petals white, narrow, glabrous;* filaments white, long; styles greenish. Carpels glabrous. Fruit large, ovoid.

England: rather local; woods, wet heaths and banks; 12, 17, 22, 24. S. France, Belgium, W. Germany, Switzerland, Austria, Hungary.

374. **R. dissectifolius** (Sudre) W. Wats. 1946, *Journ. Ecol.* **33**, 308; *R. Schleicheri* ssp. or microg. *dissectifolius* Sudre, 1906, *Diagn.* 47; 1908–13, 200, t. 194.

Essex. Stem low, short, glabrescent, *reddish purple; prickles straighter than in* R. *Schleicheri. Leaflets* 3 (4, 5), glabrescent on both sides, *coarsely, deeply* and unequally to subcompound-*serrate*, prickles hooked; *terminal leaflet rhomboid-ovate or -obovate*, drawn out into a broad point. *Panicle* often pyramidal *with dense, short-pedicelled flowers, the upper branches often in pairs; rachis* crooked, shortly hairy, striate; *the terminal leaflets coarsely dissected. Flowers* c. 1·5 cm.; *sepals* green, white-edged, with a blunt, leafy appendage, *clasping; petals white*, elliptic, glabrous; filaments slightly longer than the green styles. *Carpels glabrous.* Fruit subovoid, good. Discovered by *Mr G. C. Brown.*

England: 18, 19 (in several localities), 20 (Ringshall, *W. H. Mills*). France, Belgium, Bavaria, Hungary.

375. **R. graciliflorens** (Sudre) W. Wats. 1956, *Watsonia*, **3**, 290; *R. tereticaulis* microg. *graciliflorens* Sudre, 1901, *Fl. Pyr.* 175; *R. Schleicheri* ssp. *inaequabilis* microg. *graciliflorens* Sudre, 1908–13, 204, t. 197.

W. Kent. Stem thinly pilose, not pruinose, weakly armed. *Leaflets* 3 (5)-nate, rather large, thinly hairy beneath, *evenly and finely serrate, not glandular on the midrib; terminal leaflet cordate, roundish to ovate, long acuminate. Panicle very short and narrow, the 1–2-flowered upper branches short and spreading*, the lower axillary branches, when there are any, remote, short, ± erect. *Flowers minute; calyx erect; filaments much longer than the greenish styles.* Carpels glabrous. *Fruit small, of about 20 drupels. Leaves and habit of a diminutive* R. hirtus.

England: very rare; 16 (Pembury Walk, plentiful; *Evans, Simpson* and *Watson*, 1934). South of France, Bavaria, Switzerland.

2. Largest prickles weak, base narrow, hardly compressed.

† Glands purple-stalked when growing in the open.

(*a*) Petals white.

376. **R. hirtus** Waldst. & Kit. 1805, *Pl. rar. Hung.* **2**, 150; Sudre 1908–13, 221, t. 204.

4x. *Robust. Axes and leaves very hairy.* Stem ± pruinose, with unequal glands and some acicles; *prickles unequal, long, subulate, fragile. Leaflets* 3–5, green and *softly hairy beneath*, unequally serrate; *terminal leaflet cordate, roundish, ovate. Panicle* hanging, leafy, *large, broad, lax, pyramidal, middle and lower branches spreading widely, prickles almost acicular.* Flower buds with a discoid, often brownish base, *flowers 2–2·5 cm.*; sepals ovate, erect; petals oblong-obovate, glabrous; filaments white, slightly longer than the greenish, rarely pinkish styles. *Carpels pilose. Receptacle slender, cylindrical. Fruit rather small*, subglobose, *drupels small*, sweet and aromatic. *Flowering early June.* Woods, often in deep shade.

E., I. Frequent in S.E. England; 8, 14–17, 19–21 (Horsendon Hill, Middlesex, *E. Lees*, 1848, as *R. Lejeunei*), 24, 32, 62 (Thirsk and Oldstead). H. 39. (Not France), Belgium, Germany, Switzerland, eastwards to Russia, the Caucasus and Asia Minor.

var. **gymnocarpus** (Boul. & Pierrat) Sudre, 1908–13, 222.

Stem less hairy. Leaf smaller, unequally *serrulate*, closely pubescent beneath. *Panicle dense*, compound, leafy, thinly hairy. Stamens hardly longer than the styles. *Carpels glabrous.*

England: 18 (Danbury Common, *C. E. Britton*). France (Vosges), Germany, Bavaria, Silesia, Hungary.

377. **R. Bellardii** Weihe ap. Bluff & Fingerh. 1825, *Comp. Fl. Germ.* 1, 688; Syme, 1864, *Eng. Bot.* 3, 191, t. 454; Sudre, 1908–13, 206, t. 198.

4x, 5x. *Robust. Stem reddish brown*, pruinose, *round, slightly hairy*, with crowded, rather long glands and acicles; prickles moderately long, weak, slanting. *Leaves 3-nate, large; leaflets* short-stalked, thinly hairy on both sides, *evenly serrate or serrulate*, rarely, when very luxuriant, jagged; *terminal leaflet broad elliptic or elliptic-obovate, cuspidate, base ± entire; the lateral leaflets similar. Panicle broad, rather short* with *1–3 simple leaves, glandular on both sides*, sub-corymbose at the top, *mid-branches spreading, flowers long-stalked*; rachis densely and shortly hairy, armed like the stem. *Flowers up to 2·5 cm.*; sepals *triangular-ovate-attenuate, convex-reflexed then clasping; petals oblanceolate, apex fimbriate*; filaments white (drying pinkish), barely equalling the greenish styles. Carpels glabrous. Receptacle pilose. *Fruit subovoid, soon turning blood red. Flowering from early June.*

E., W. Damp oakwoods. Apparently rather rare; 5, 11, 26, 29, 32, 39, 48, 55, 62. Central Europe from Wales to Königsberg, and from South Sweden and Denmark to N. Italy and Bohemia.

378. **R. pallidisetus** Sudre, 1904, *Obs. Set of Brit. Rub.* 15; *R. divexiramus* sensu Rogers, 1900, 86; non P. J. Muell. 1867.

Western. Axes deep purple-red, glaucescent. Stem roundish, becoming nearly glabrous, with *rather short glands and pricklets* and a few *longer, gland-tipped acicles*; prickles few, subequal, slender. *Petiole long. Leaflets 5*, softly hairy beneath, *sharply subcompound-serrate; terminal leaflet subcordate, broad, elliptic, cuspidate. Panicle long*, with dense 1–3-flowered upper, and distant, racemose lower branches, rachis shortly hairy, with short, slanting prickles and *unequal glands including a few that are very long-stalked.* Flowers *c.* 2 cm.; sepals green, white-edged, long-tipped, *erect; petals 5, 6, 7, narrow oblong-obovate*, glabrous on the apex, *pale pink*; filaments white, longer than the *greenish or bright red styles. Carpels pilose.* Fruit roundish ovoid, good.

E., W. Rather rare, usually in the open; 3–5, 34–36, 40, 49, 52. Germany, Alsace, Hungary (*Bakabanya*).

379. **R. Guentheri** Weihe in Bluff & Fingerh. 1825, *Comp. Fl. Germ.* 679; W. Wats. 1949, *Watsonia*, I, 82; *R. hirtus* ssp. *Guentheri* (Weihe) Sudre, 1908–13, 224, t. 205.

S.E. England. Distinguished from R. *hirtus*, which it somewhat resembles, by the deep green, thickish *leaflets*, which are pilose and *grey-white felted beneath*; by the terminal leaflet being elliptic-ovate-acuminate and rather coarsely toothed; and by the smaller flowers about 1·5 cm., with petals notched and pilose on the apex. *Stamens barely as long as the red styles.* Fruit small, composed of rather few, large drupels. Rachis often zigzag. Sepals *convexo*-reflexed before becoming erect.

England: rare; 14, 15 (Denstead Wood, west of Canterbury), 19 (Danbury Common), 21 (Woods near Ruislip). France, Germany, Switzerland, Austria.

380. **R. perplexus** P. J. Muell. ex Wirtg. 1862, *Herb. Rub. rhen.* ed. I, no. 155; *R. scaber* Kalt. 1845, *Fl. Aach. Beck.* 289; non Weihe, 1825; *R. Kaltenbachii* sensu Focke, 1877, *Syn. Rub. Germ.* 375; non Metsch, 1856; *R. hirtus* ssp. *Kaltenbachii* (Metsch) Sudre, 1908–13, 228, t. 205; non Rogers, 1900.

4x. *Stem* slender, arching and *trailing*, angled, slightly furrowed, *pruinose, with scattered, short hairs, numerous, rather short, unequal glands and acicles,* some gland-tipped; *prickles very fine,* slanting. *Leaves large, thin;* leaflets 3, 4 (5), glabrescent on both sides, coarsely unequal, patent-serrate; *terminal leaflet cordate, ovate to obovate, very long and gradually pointed. Panicle small,* short and racemose, *or large,* lax and compound, *very broad-topped with several patent, panicled or corymbose, often clustered, middle,* and longer axillary *branches; pedicels long and slender,* pubescent-felted; *rachis zigzag.* Flowers *c.* 2 cm.; *sepals triangular-ovate, attenuate, with a leafy appendage, erect;* petals narrow, apex pointed, glabrous; *filaments white, equalling the red styles. Carpels* glabrous. Fruit small.

England: rare; 13 (Midhurst Close Walks, near the old well), 14, 17 (Ranmore), 24. Belgium, Germany (*Rhineland*), Switzerland, Austria.

381. **R. rubiginosus** P. J. Muell. 1858, *Flora,* **42,** 166; non Braeucker, 1882; *R. hirtus* ssp. *Kaltenbachii* microg. *rubiginosus* (P. J. Muell.) Sudre, 1908–13, 229, t. 206; non *R. hirtus* ssp. *rubiginosus* (P. J. Muell.) Rogers, 1900.

Stem angled, *glabrous, pruinose, purplish, often with yellowish, very numerous unequal, slender glands, acicles, and prickles.* Leaves rather large, *mostly 3-nate,* rarely a completely 5-nate-pedate leaf; *leaflets strigose above, glabrescent beneath*; terminal leaflet moderately broad ovate to obovate, *long-acuminate* from a subcordate base, unequally crenate-serrate to slightly compound-serrate. *Flowering branch sharp angled, furrowed, flexuose, glabrous below to slightly felted above; panicle* sometimes bent at the top, *floriferous*, with broad, lax, middle branches, spreading, divided deeply, and with 1–2 axillary, ascending, panicled, numerous flowered branches, pedicels long, felted, the whole *armed with firm, acicular, purple prickles* and numerous, deep purple acicles and stalked glands. Panicle terminal leaflets narrow elliptic-rhomboid. *Sepals deep green with a white felted edge*, narrow, prolonged into a *long, green, linear tip*, aciculate and densely glandular, *reflexed and flat to patent and erect*; petals 5, 6, small, *rather broad-elliptic*, glabrous, white or faintly pinkish; filaments white, equalling the purplish red styles. Carpels glabrous.

England: very rare; woods above Coldrum, W. Kent, 1949. S. France, Belgium, Germany, *Alsace, Bavaria*, Hungary.

382. **R. nigricatus** Muell. & Lef. 1859, *Pollichia*, 204; *R. hirtus* ssp. *nigricatus* (Lef. & Muell.) Sudre, 1908–13, 226, t. 205.

N.v.v. Ireland. Stem angled, purple, glaucescent, ± intricately villous, with numerous, rather short glands and acicles, some gland tipped; prickles numerous, subulate, slanting. *Leaves rather small, 3 (4, 5)-nate*, petiole prickles straight or slightly falcate; *leaflets with base entire, glabrescent beneath, serrulate*; terminal leaflets broad elliptic, acuminate. Panicle rather broad, pyramidal-truncate with a few, simple leaves, the upper branches 2–3-flowered, spreading, the lower branches with numerous flowers, semi-erect, peduncles short, pedicels long and felted; rachis pilose, *prickles acicular, glands and acicles blackish purple*. Sepals grey felted, densely glandular, clasping. Petals ovate-lanceolate or narrow oblong. Filaments white, slightly longer or shorter than the red-based styles. Carpels glabrous or pubescent. *Fruit abundant*, aromatic.

Ireland: abundant near Wexford; *Marshall*, 1896, as '*R. rubiginosus* P. J. Muell., *fide* Rogers and Focke'. France, Belgium, Germany, Switzerland, Austria.

(b) Petals pink.

383. **R. purpuratus** Sudre, 1900, *Rub. Pyr.* 82; 1908–13, 191, t. 188.

Kent and Bucks. Stem angled, *purplish, glaucescent, glabrescent*, with numerous unequal glands, acicles and some pricklets; prickles numerous, rather short, unequal, declining. *Leaves* 3–5-nate, pedate, *rather small*; leaflets hairy beneath, unequally incise-serrate; *terminal leaflet* broad *ovate or obovate-oblong*, short pointed. *Panicle small*, pyramidal, *purple with glands*, with usually one simple leaf, narrow racemose above, with patent middle, and 1–3, distant, racemose, oblique, few-flowered branches; lateral leaflets subsessile; rachis thinly hairy and felted above. *Flowers* 1·5–2 *cm.*, calyx base discoid, *sepals green with a contrasting white border, erect; petals* 5, 6, 7, narrow obovate, pointed, some apiculate, margin glabrous, remote, *rosy pink*, pinkish or pure white; filaments white, longer or shorter than the *reddish or red styles*. Carpels glabrous. Fruit small, subglobose. *Flowering from the first week of June.*

England: rare; 16 (Shooters Hill, in fair quantity), 24 (Fulmer). France, Belgium, Germany, *Saxony*, Silesia, Austria, *Styria*, Hungary.

384. R. Durotrigum R. P. Murray, 1892, *Journ. Bot.* **30**, 15.

Dorset. Robust. Stem red to purple, angled, glabrescent, glands and acicles numerous, passing into prickles. *Leaves large*, (3) 5-nate; leaflets subglabrous above, thinly hairy beneath, *coarsely incise, ± compound-serrate; terminal leaflet short-stalked, cordate, roundish-ovate, acuminate.* Panicle with several simple leaves, some lobed, narrowed above, with upper and lower branches alike, *few-flowered*; rachis crooked angled, hairy, *prickles very weak, subulate. Flowers c. 3 cm.*; *sepals large, greenish*, hirsute, glandular, ovate, attenuate, *leafy-pointed, erect; petals long (14 mm.), narrow, tapered below, pink or pinkish*; filaments white, barely equalling the reddish-based styles. Carpels pilose. *Flowering July.*

England: only known in east Dorset, where it is well distributed.

The Anglesey bramble referred to by Rogers was *R. semiglaber*. The Fittleworth, W. Sussex bramble assigned here by Rogers was *R. rosaceus.* Switzerland, Germany, *Bavaria and Silesia.*

2. Glands yellowish-stalked, rarely becoming reddish.

(*a*) Stem glabrous or nearly so.

385. R. leptadenes (Sudre) W. Wats. 1946, *Journ. Ecol.* **33**, 342; *R. serpens* ssp. *leptadenes* Sudre, 1908–13, 219, t. 203; *R. echinatus* P. J. Muell. 1858, *Flora*, **41**, 171; non Lindl. 1829. **Fig. 48.**

4x. Stem prostrate, pruinose, with numerous, very long, cushion-based, gland-tipped acicles and bright red to black-headed *glands; prickles* numerous, unequal, *very slender. Leaves large, deep green*, pedate; leaflets 3, 4, 5, nearly glabrous above, ± hairy beneath, subcompound patent-crenate-serrate; *terminal leaflet cordate, ovate, gradually acuminate or broad elliptic-cuspidate.* Panicle long with several simple leaves, the *upper branches deeply divided, 1–3-flowered*, the lower ones racemose; *pedicels* felted, long and slender; *furnished with very numerous, very long, glandular acicles*; shortly hairy at the top, glabrous below. Flowers 1·5–2 cm.; *sepals deep green, white-bordered, erect*; petals elliptic-rhomboid, some notched, margin glabrous, pinkish or white; filaments white, longer than the greenish styles. Carpels subglabrous. *The leaves are fringed all round with short-stalked glands and acicles.*

England: rare; 24 (Penn Wood towards Beamond End; Angling Spring Wood, Great Missenden). France, Belgium (*Malmedy, W. Watson*), Germany, Switzerland, Hungary.

386. **R. Lintonii** Focke ex Bab. 1887, *Journ. Bot.* **25**, 331; *R. granulatus* ssp. *Lintonii* (Bab.) Sudre, 1908–13, 140, t. 135.

Stem at first erect, angled, *pruinose*, with a moderate number of *rather long*, unequal, *pallid glands and acicles; prickles* not numerous, unequal, *very slender and long*, slanting and falcate, *mostly seated on the angles. Leaflets* 3, 4, 5, *entirebased*, *softly hairy and grey felted beneath, sharply subcompound-serrate; terminal leaflet roundish* to *obovate-subcuneate, finely acuminate*, intermediate and basal leaflets very shortly stalked. *Panicle branches all ± sharply erect*, racemose above, the pedicels slender, sometimes more compound and broader below; *rachis* flexuose *with long glands and acicles, prickles very slender. Flowers* c. 1·5–2 cm.; *sepals* hirsute and prickly, ovate, *slenderly tipped, erect; petals* elliptic-obovate, *white*, pilose on the apex; filaments white, about equalling the greenish styles. *Fruit very good.*

Endemic. Wood margins and hedges, rare; (not 6, *euryanthemus*), 20 (Whippendell Wood, Watford, and Chorley Wood Common), 28 (Sprowston), (not 35, *acutifrons*), 39 (Whitmore). Frequent H. 35 (near Rockingstone, Ballyness Bay, *Y. Heslop-Harrison*).

(*b*) Stem densely hairy.
i. Flowers moderately large, c. 2 cm. Styles greenish or yellowish.

387. **R. elegans** P. J. Muell. 1858; *Flora*, **41**, 170; *R. serpens* ssp. *angustifrons* Sudre, 1908–13, 217. **Fig. 49.**

4x. *Robust and hairy. Rufescent.* Stem roundish to angled, pruinose, with numerous glands, acicles and pricklets up to half as long as the prickles; *prickles slender, subequal. Leaves yellowish green*, petiole prickles curved; *leaflets* 3–5, ± pilose beneath, unequally *incise* or subcompound, *sometimes coarsely crenate-patent-serrate, the principal teeth often larger and jagged; terminal leaflet with base ± entire, elliptic or obovate, very long, often falcate-pointed. Panicle* nodding, long and broad pyramidal, leafy, the upper leaves glandular, *branches long-peduncled, spreading*; rachis hairy, armed like the stem, the *terminal leaflets very long-pointed.* Flowers c. 2 cm.; *sepals* triangular-ovate-attenuate, *clasping; petals* 5–7, narrow obovate, apiculate, glabrous, *pinkish* in bud, sometimes flushed red outside; *filaments white, much longer or rarely shorter than the yellowish styles. Receptacle hirsute, especially below the lowest carpels.* Fruit subglobose.

England: local; woods in semi-shade; 7, 13, 17–20, 24, 30, 35, 36, 38. France, Belgium, Germany, Switzerland, Hungary.

388. **R. incultus** Muell. & Wirtg. ex Wirtg. 1862, *Herb. Rub. rhen.* ed. 1, no. 153; *R. rivularis* ssp. *incultus* (Wirtg.) Sudre, 1908–13, 209, t. 200.

W. Sussex. All arms straw-coloured. Stem angled, *a good deal hairy*, still more aciculate and glandular; prickles fine and short, slanting and falcate. Leaves rather large, 3, 4, 5-nate, pedate, prickles declining or subfalcate;

leaflets short-stalked, rather narrow, entire-based, shortly, and *at first softly hairy* and green beneath, *coarsely, unequally and angularly serrate, a little jagged; terminal leaflet rather narrow, elliptic-obovate, acuminate.* Panicle dense, narrowed towards the abrupt, leafy top; rachis hirsute, densely and finely armed like the stem, *the upper leaves glandular on the back and margin; the terminal leaflets obovate-cuneate. Flowers c. 1·8 cm.*; sepals narrow, triangular-ovate-attenuate, densely long-glandular and aciculate, ± *erect; petals white, fimbriate, narrow elliptic*; filaments white and about equalling the yellowish styles. Carpels glabrous.

England: rare; 13 (Fittleworth to Bedham, W. H. Mills, 1953). Central Europe; central France, Belgium, Rhineland, Pomerania, Silesia, Bavaria, Austria, Hungary. Thinly distributed.

ii. Flowers small, 1–1·5 cm.

389. **R. hylonomus** Lef. & Muell. 1859, *Pollichia*, 224; *R. serpens* ssp. *chlorostachys* microg. *hylonomus* (Lef. & Muell.) Sudre, 1908–13, 219, t. 203. **Fig. 50.**

Wintergreen. Rufescent. Stem blunt, with *crowded, long, slender acicles and glands* and patent to falcate, unequal *prickles. Leaves yellowish green*; leaflets 3, 4, 5, hairy to glabrescent above and below, *rather coarsely, unequally patent-serrate; terminal leaflet* short-stalked, emarginate, elliptic-ovate, *very long acuminate.* Panicle long, narrow, leafy and compound, with short, remote, axillary branches, or reduced to a raceme; rachis densely hairy. *Calyx base discoid, sepals green, white-edged, triangular-lanceolate, acuminate, long-pointed* very strongly reflexed after flowering, then slowly becoming erect; *petals remote, oblanceolate, acute, glabrous on the margin, pinkish* or white; *filaments slightly or very red; styles sometimes purple.* Carpels pilose. Fruit roundish.

England: frequent in dark woods and on sunny commons; 11, 13–17, 20, 22–24, 36. France, Germany, Switzerland.

390. **R. analogus** Lef. & Muell. 1859, *Pollichia*, 232.

Middlesex. Stem turning brownish, angled, with *numerous unequal reddish glands and acicles; prickles numerous, sharply declining and recurved. Leaves dark green*, prickles hooked; *leaflets 3, scantily hairy beneath*, shallowly crenate-serrate with short mucros; *terminal leaflet shortly stalked, obovate, shortly pointed.* Panicle with *small, narrow leaves to the top*, subpyramidal, lax, branches few-flowered; *rachis dark green*, hairy, armed like the stem, *terminal leaflets oblong-obovate-acute.* Sepals green with narrow, leafy, glandular appendages, prickly, erect. *Petals very small, oblanceolate, glabrous, white with a green base. Filaments white, hardly longer than the greenish styles.* Carpels glabrous.

England: very rare; 21 (Bayhurst Wood, east side). France, Germany, Austria, Hungary.

391. **R. lusaticus** Rostock in Wagner, 1886, *Fl. Loeb. Berg. Progr.*; *R. hirtus* ssp. *serpens* var. *lusaticus* (Rostock) H. Hofmann in Wunsche, *Pfl. Sachs.*; *R. rivularis* ssp. *lusaticus* (Rostock) Sudre, 1908–13, 209, t. 200.

W. Kent. Stem blunt-angled, *densely hairy, glaucescent,* with numerous glands and acicles, some very long; *prickles golden yellow,* unequal, *slanting.* Stipules linear. *Leaflets 3, fresh green* above, thinly greyish hairy to *felted beneath,* with fine, pale, patent prickles and acicles on the midrib, *finely serrate-crenate; terminal leaflet elliptic or obovate, subcuneate,* acuminate. Panicle short, narrow, racemose, felted and hairy, prickles weak, mostly patent, terminal leaflet subrhomboid. *Flowers 0·75–1 cm.*; *sepals* very white within, greyish green felted without, prickly and glandular, *erect; petals white, narrow,* glabrous on the margin; filaments white, *longer than the reddish-based styles.* Carpels glabrous. Fruit small, of few rather large drupels. Habit of *R. tereticaulis.*

England: very rare; 16 (Pembury Walk). Central France, Switzerland, Germany, *Alsace, Bavaria, Saxony, Silesia.*

DRAWINGS

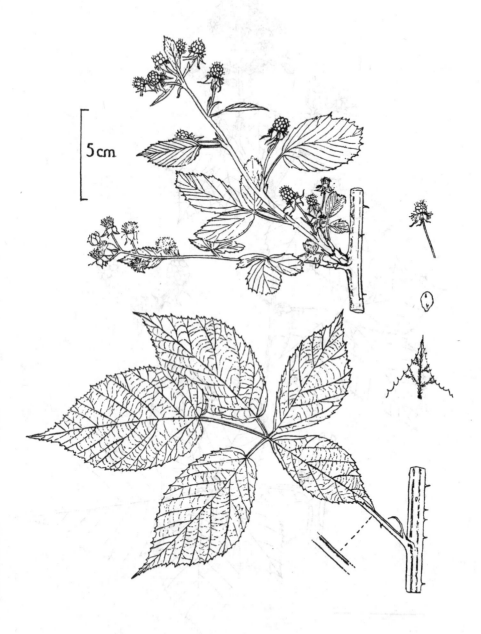

Fig. 1. R. nessensis

5 cm.

Fig. 2. R. affinis

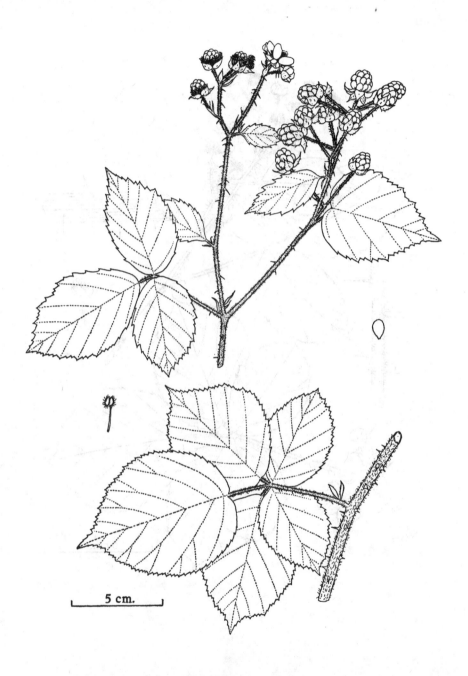

Fig. 3. R. Balfourianus

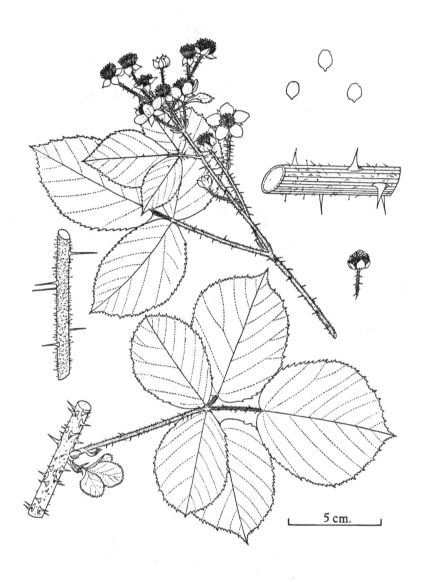

Fig. 4. R. tuberculatus

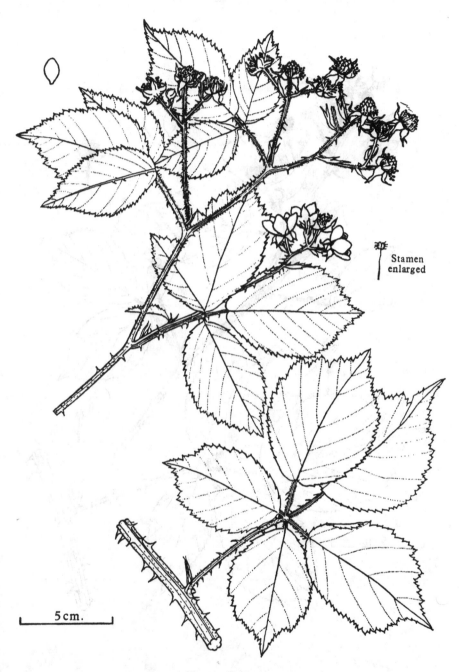

Stamen
enlarged

5 cm.

Fig. 5. R. gratus

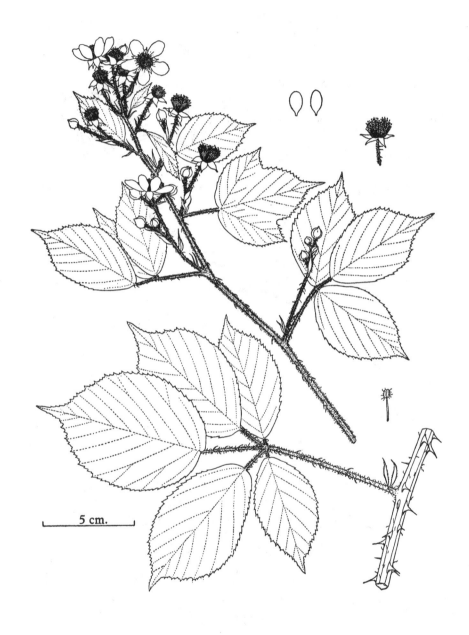

5 cm.

Fig. 6. R. confertiflorus

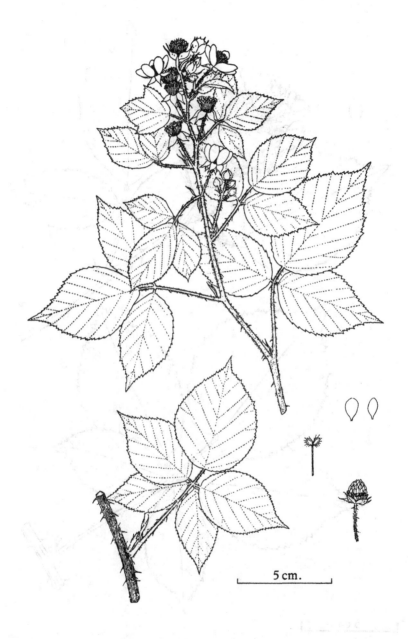

Fig. 7. R. sciocharis

5 cm.

Fig. 8. R. carpinifolius

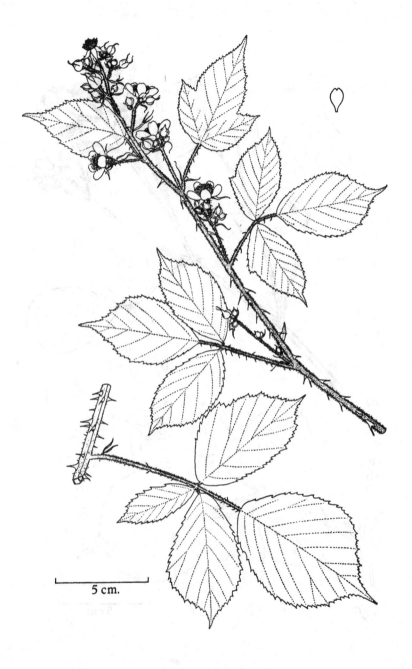

5 cm.

Fig. 9 R. Questieri

Fig. 10. R. oxyanchus

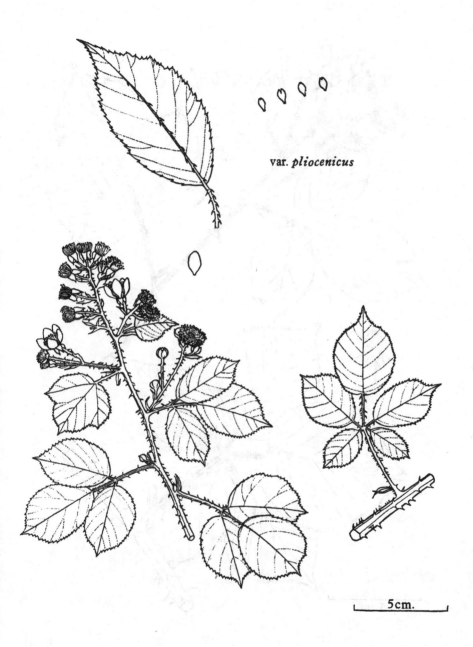

var. *pliocenicus*

5cm.

Fig. 11. R. egregius *and var.* pliocenicus

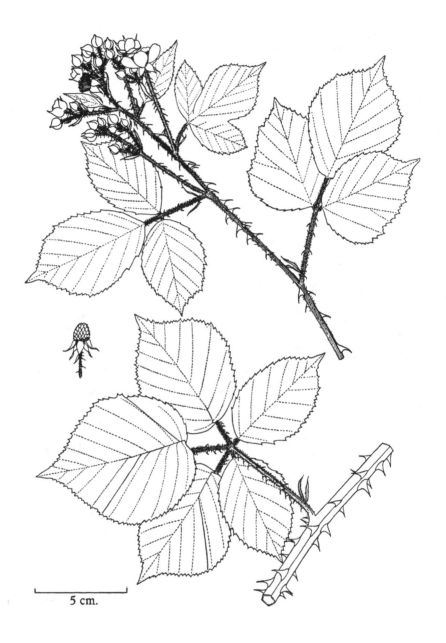

5 cm.

Fig. 12. R. leucandrus

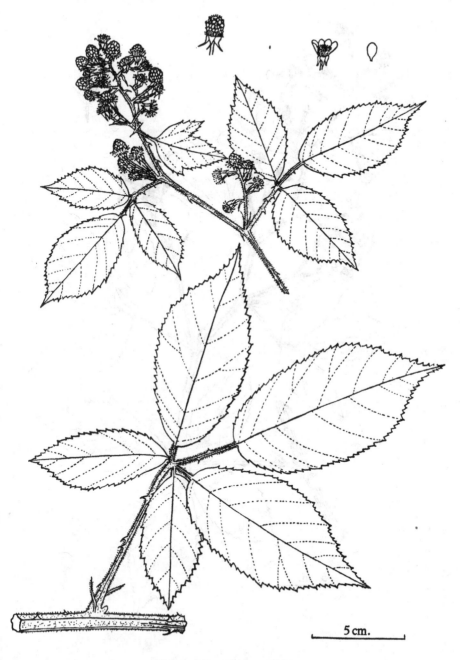

Fig. 13. R. subinermoides

5 cm.

Fig. 14. R. silvaticus

var. *parvifolius*

5 cm.

Fig. 15. R. pyramidalis *and var*. parvifolius

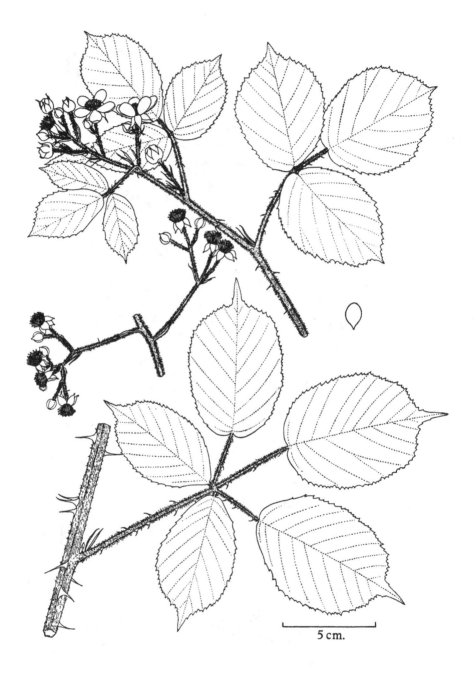

Fig. 16. R. londinensis

5 cm.

Fig. 17. R. villicaulis

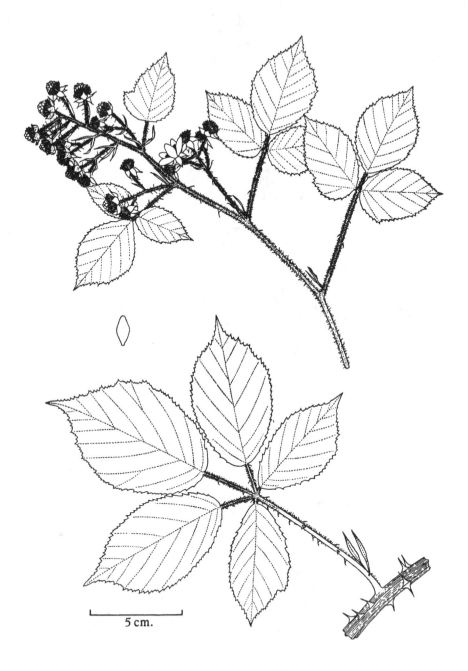

5 cm.

Fig. 18. R. rhombifolius

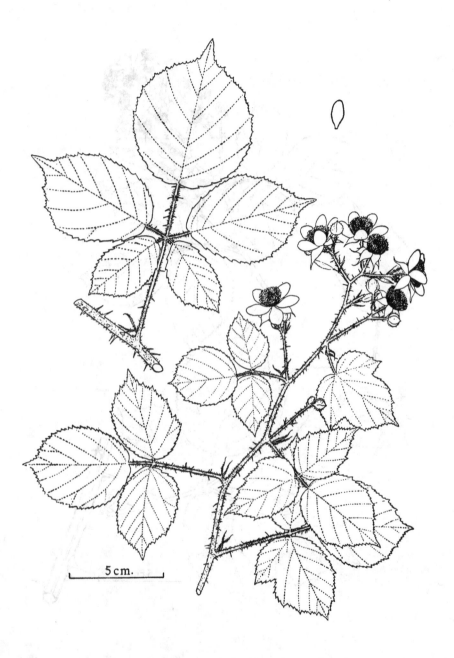

5 cm.

Fig. 19. R. alterniflorus

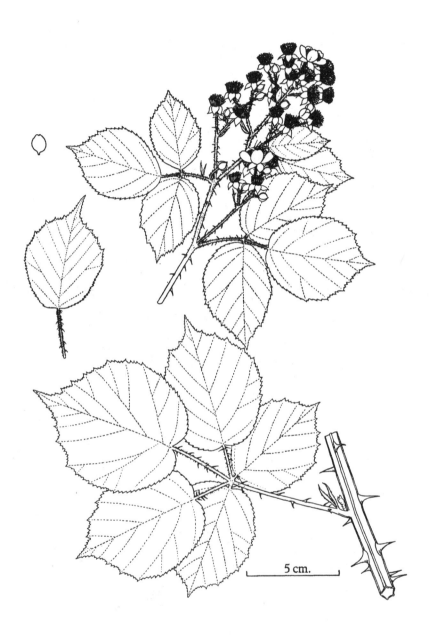

Fig. 20. R. cardiophyllus *and var.* fallax

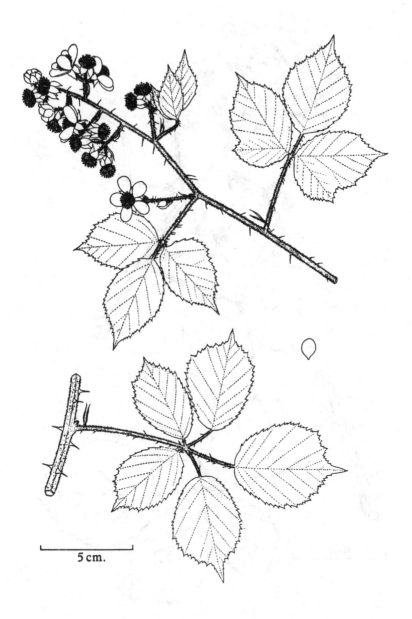

Fig. 21. R. Winteri

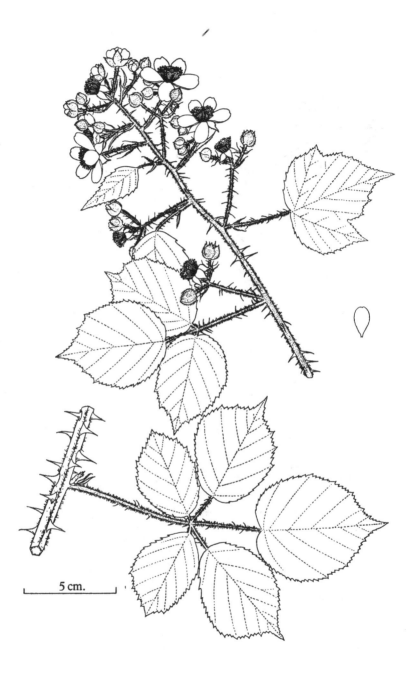

5 cm.

Fig. 22. R. cuspidifer

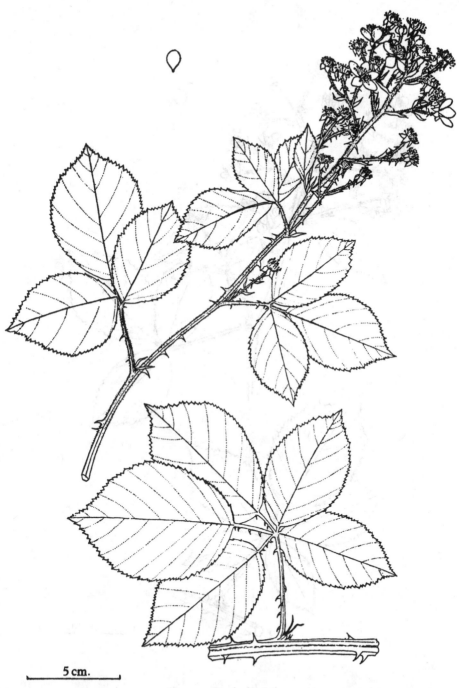

5 cm.

Fig. 23. R. hylophilus

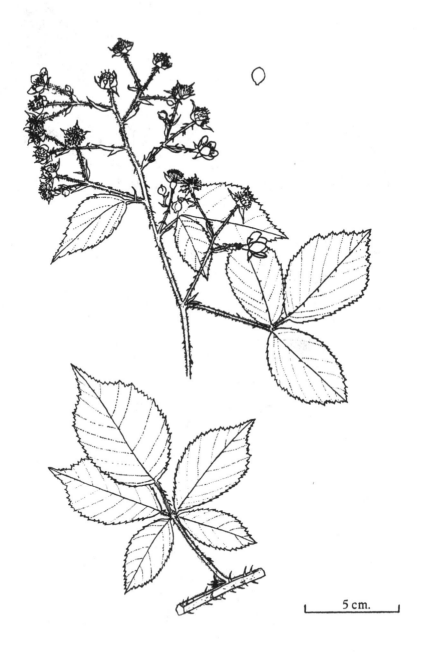

Fig. 24. R. Sprengelii

Fig. 25. R. hebeticaulis

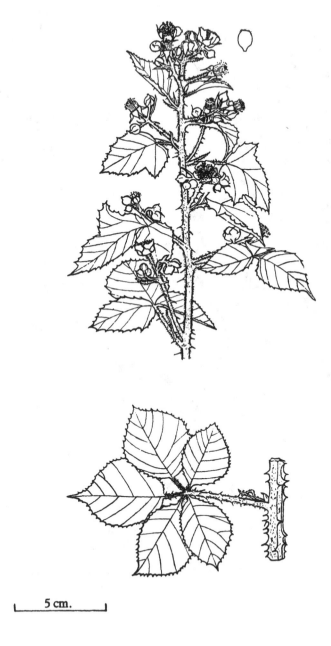

Fig. 26. R. leucostachys

5 cm.

Fig. 27. R. Boraeanus

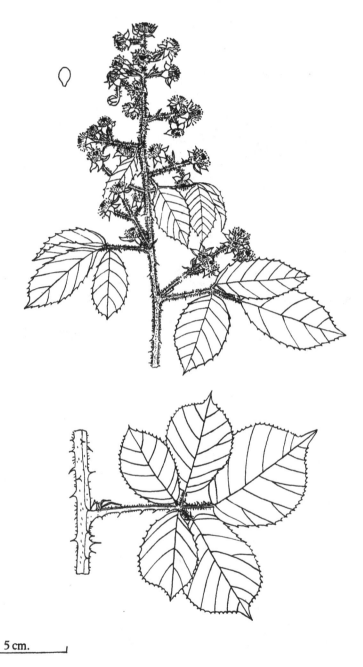

5 cm.

Fig. 28. R. Drejeri

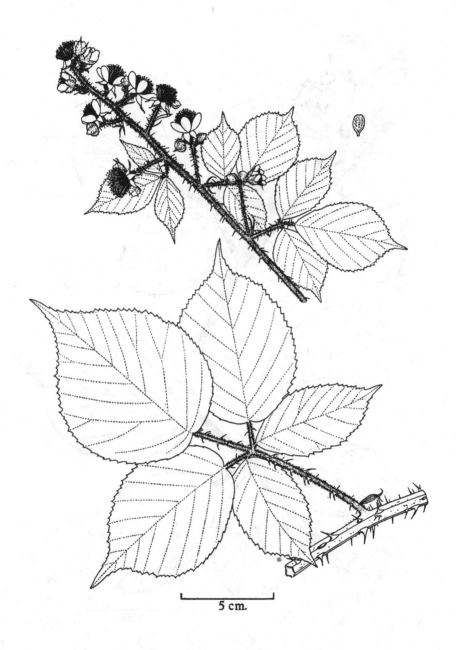

Fig. 29. R. Leyanus

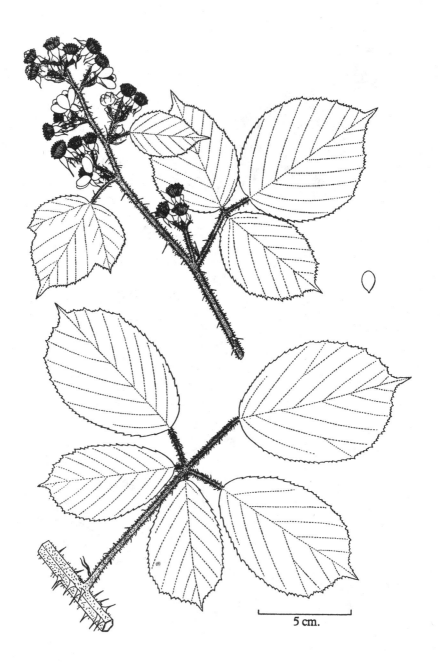

Fig. 30. R. sectiramus

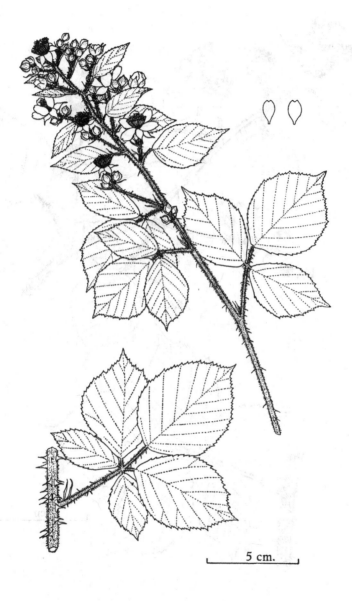

Fig. 31. R. Genevieri

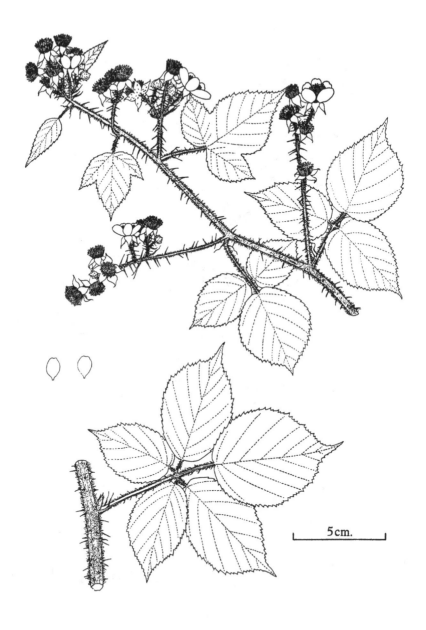

Fig. 32. R. micans

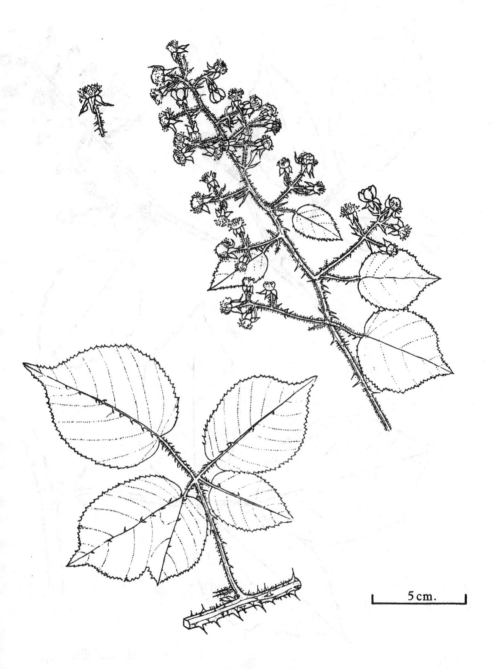

Fig. 33. R. trichodes

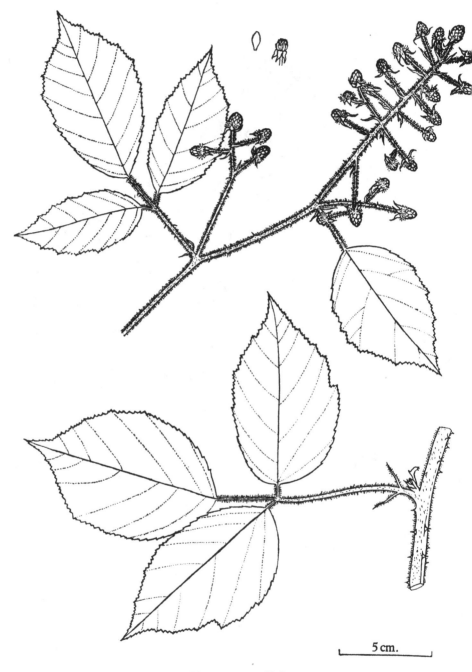

Fig. 34. R. pallidus

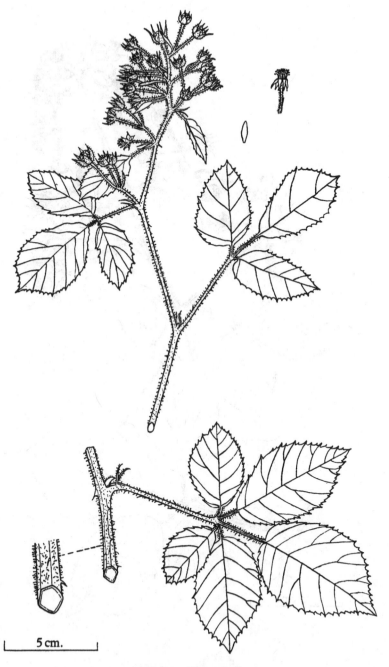

Fig. 35. R. scaber

5 cm.

Fig. 36. R. rufescens

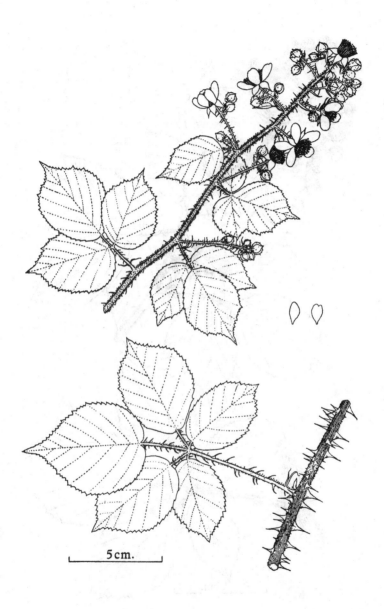

5 cm.

Fig. 37. R. heterobelus

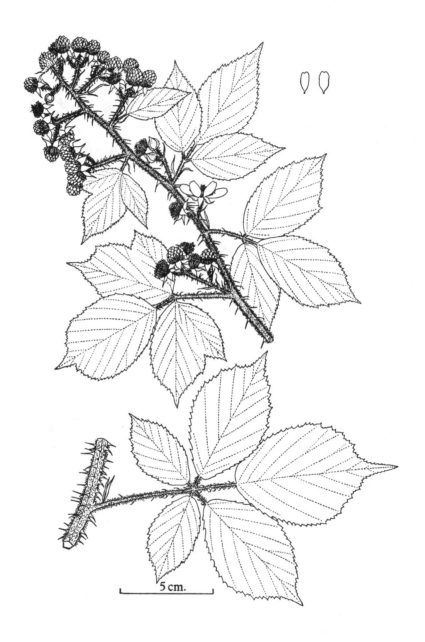

Fig. 38. R. retrodentatus

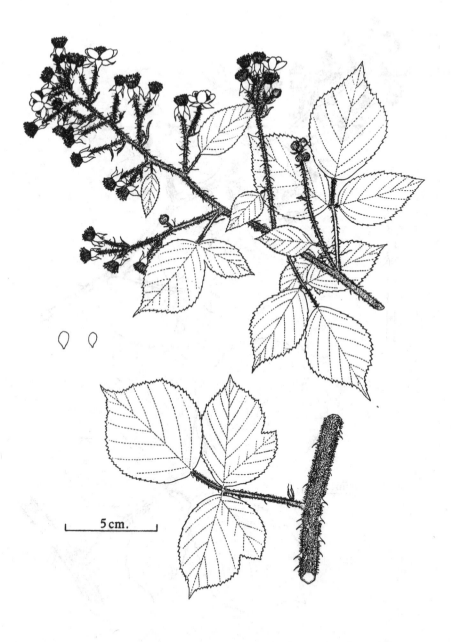

Fig. 39. R. rotundifolius

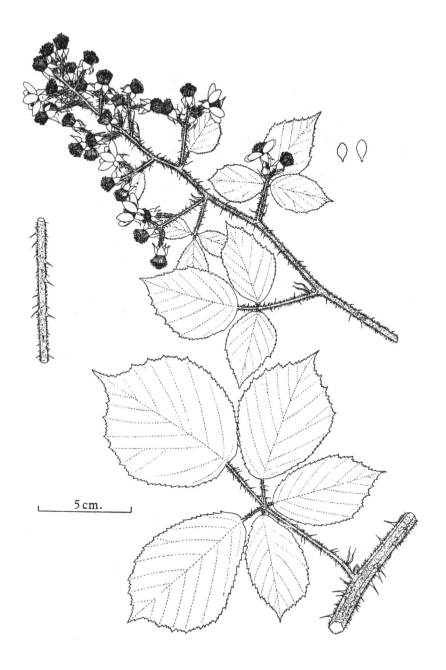

5 cm.

Fig. 40. R. Lejeunei

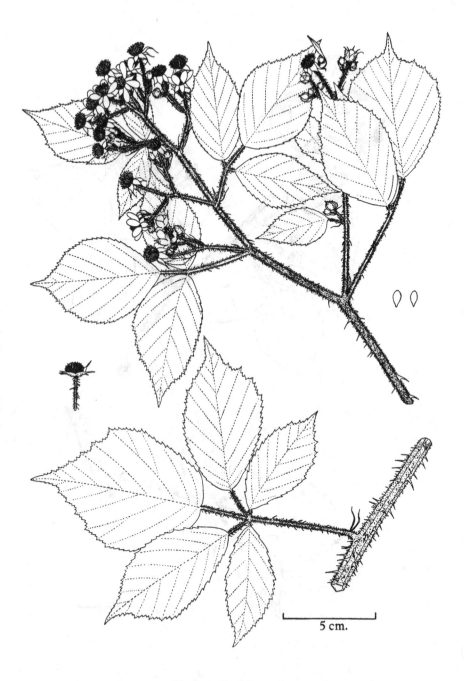

5 cm.

Fig. 41. R. Rilstonei

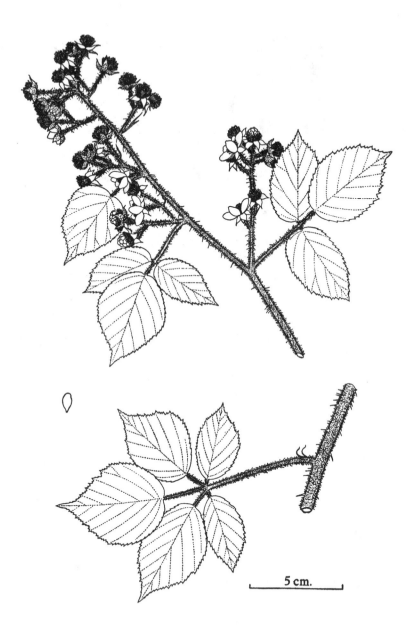

Fig. 42. R. Murrayi

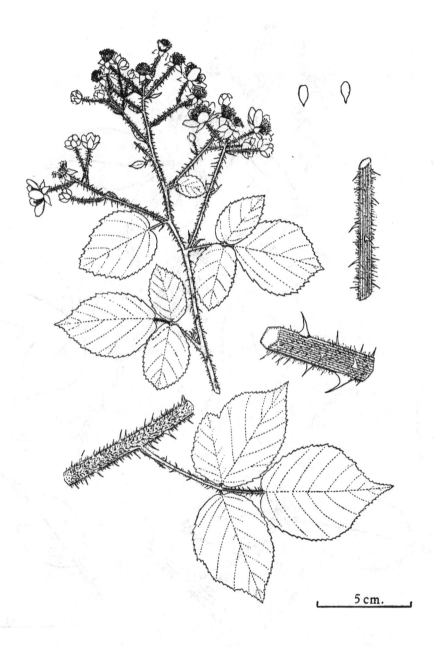

Fig. 43. R. spinulifer

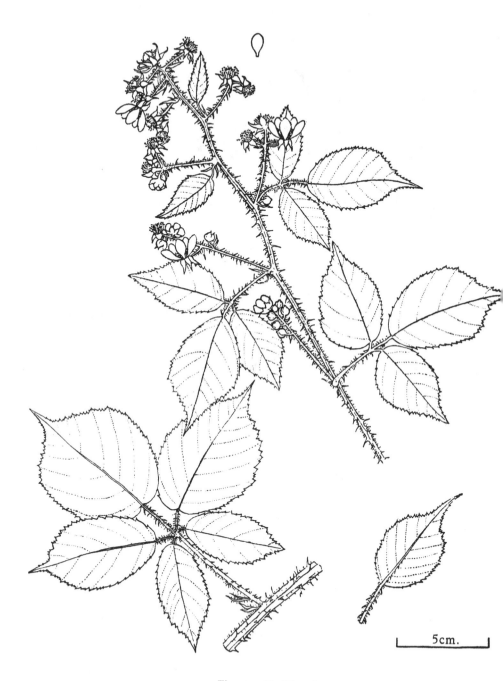

Fig. 44. R. Hystrix

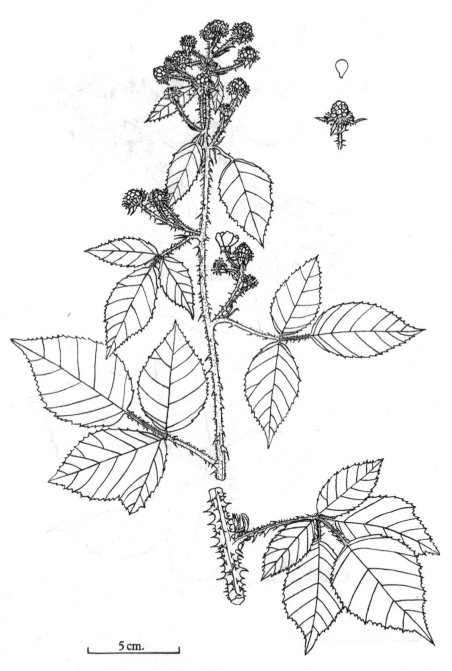

5 cm.

Fig. 45. R. infestus

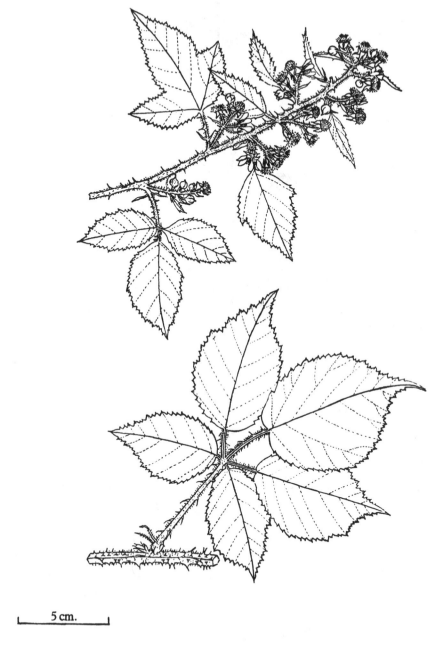

5 cm.

Fig. 46. R. adenolobus

Fig. 47. R. viridis

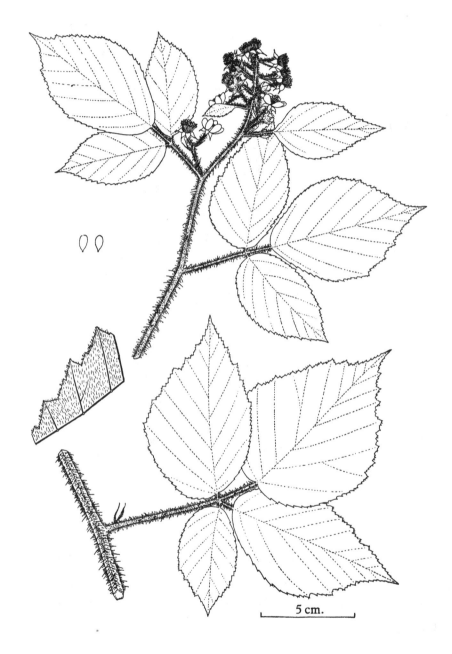

Fig. 48. R. leptadenes

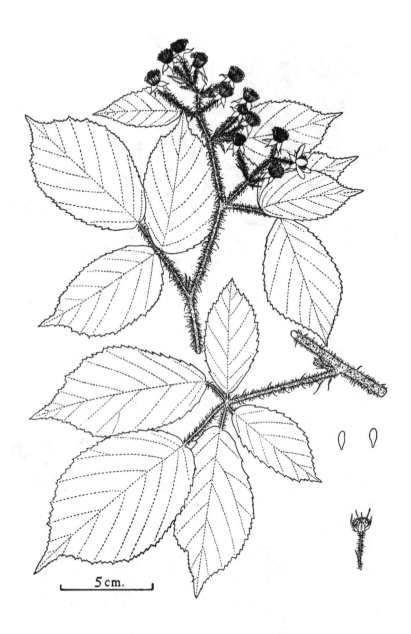

5 cm.

Fig. 49. R. elegans

259

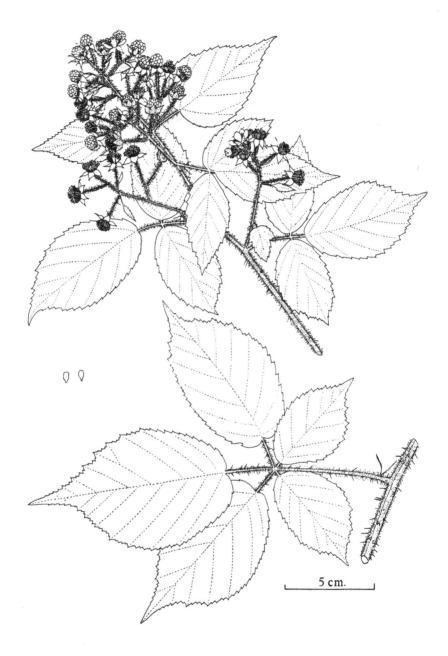

Fig. 50. R. hylonomus

GLOSSARY

(It must be pointed out that although the inflorescence of *Rubus* is called a 'panicle', the flower terminating the main axis is the first to flower, even when the inflorescence is simple and is called a 'raceme'. These two terms are therefore not stricly correct: the inflorescence is always definite. If the order of flowering is compared with that of the 'strings' of a currant or a gooseberry bush it will be seen what is meant.)

acicle, a short or long, needle-like, sometimes gland-tipped, structure, having the consistency of a fine prickle.

aculeolate, furnished with pricklets.

acuminate, gradually narrowed to a long point.

allotetraploid, a plant of four chromosome sets, two sets of which agree together but differ from the other two sets. (*See* autotetraploid.)

autotetraploid, a plant of four chromosome sets, all of which sets agree together. (This definition and that for 'allotetraploid' are only approximate. Apart from the difficulty which is met with in comparing the shapes of the chromosomes in *Rubus*, there appears to be a tendency in allotetraploids for the differences to be lessened; and in autotetraploids for differences to arise.)

apiculate, furnished on the apex with a minute, abrupt, sharp point.

appendiculate, ending in an elongated expanded tip.

apospory, seed-formation which proceeds from a purely vegetative cell, not from a gamete or a fusion of gametes.

biseriate, in two rows.

clone, a group of independent organisms derived vegetatively (i.e. without a sexual process) from a common ancestor.

crenate, having teeth with a rounded apex and separated by a sharp V-shaped incision.

cuneate, wedge-shaped.

cuspidate, having an abrupt, narrow point.

cymose, said of a branch of the panicle which has two more or less opposite branches which each end in a flower and are again more or less oppositely branched.

dentate, having teeth pointing outwards formed by two concave sides and separated from the next tooth by a U-shaped depression.

digitate, said of a leaf, the stalks of the five leaflets of which spring direct from the apex of the leafstalk.

17-3

dioecious, having stamens and styles separated on different plants.

distichous, set in two rows on opposite sides of an axis.

ecesis, germination of a seed and growth of the seedling to a mature plant in a new habitat.

egg-nucleus, the female reproductive unit, which in *Rubus* tetraploids after having 'reduced' may, in the presence of a male nucleus, 'double' and develop spontaneously.

emarginate, notched.

endemic, not found abroad.

equal, said of a panicle which is equal, or nearly equal, in breadth throughout.

fasciculate, bunched.

felted, faced with dense, minute, tufted hairs, which lend, when numerous, a grey or white appearance.

filiform, slender as a thread.

fimbriate, fringed.

fuscous, dark chocolate-coloured.

gamete, a male or female sexual cell.

generative nuclei, naked male nuclei, two in number, which pass out of the pollen tube into the embryo sac, one uniting with the central nucleus there to form the endosperm, the other with the egg nucleus to form the seed.

genotype, the complement of genes which in association determine the form and behaviour of the phenotype so far as the environment allows.

glabrous, hairless, smooth.

glabrescent, hairy at first, becoming smooth.

glaucous, generally used as equivalent to pruinose.

glaucescent, becoming glaucous with age.

hirsute, having dense, straight hairs, making the surface roundish.

jagged, said of the teeth of a leaf which jut out, with the point turned rather backwards.

lineolate, having short, raised streaks.

microgene, a low order of species; the term and conception are now obsolete.

migration, dispersal of a seed by any agency to a new habitat.

mucronate, provided with an abrupt, short, stiff, prickle-like tip, looking like an excurrent midrib.

panicled, said of a branch of an inflorescence having the divisions compound and scattered, not opposite.

pectinate, an arrangement of simple hairs spreading like the teeth of a comb, from the veins on the underside of a leaf, and appressed to the surface.

pedate, said of a leaf when the lower leaflets spring from the stalks of the leaflets next above.

pedicel, flower stalk.

peduncle, the lower unbranched portion of a compound panicle branch.

phenotype, the breeding plant body in its environment.

pilose, furnished with long, spreading, more or less straight, simple hairs.

pruinose, covered with a dull bloom which can be rubbed off.

pubescent, faced with short, fine, simple hairs, lying rather along ribs.

racemose (strictly, 'pseudo-racemose'), said of a panicle which has the flower stalks simple.

reniform, cordate, obtuse, broader than long.

retrorse, turned sharply downwards.

retuse, having a hollowed apex.

sigmoid, shaped like a nearly straight S, reversed, as when seen reflected in a mirror.

seriate, in a row.

serrate, having sharp, forward-pointing teeth, separated by sharp V-shaped incisions.

sinuate, having long, shallow, saucer-shaped depressions here and there between the teeth.

strigose, having simple hairs lying appressed and pointing all in the same direction.

subulate, awl-shaped with a somewhat expanded base.

thyrsoid, said of a dense, elongate, tapered panicle.

villose, villous, having closely placed, wavy hairs, making a soft surface.

PRINCIPAL WORKS CONSULTED

ADE, A. In Vollman's *Flora von Bayern*, 1914. (Exhaustive. The descriptions are after Sudre.)

ARESCHOUG, F. W. C. *Some Observations on the genus Rubus.* Lund, 1886–87.

BABINGTON, C. C. *Synopsis of the British Rubi.* London, 1846–48.

BABINGTON, C. C. *The British Rubi.* London, 1869.

BABINGTON, C. C. *Manual of British Botany*, ed. 8. London, 1881.

BERTRAM, W. *Flora von Braunschweig.* 1908. (Excellent.)

BLOXAM, A. 'Rubi', in Kirby, *Flora of Leicestershire.* 1850.

BOULAY, N. 'Rubi', in Rouy et Camus, *Flore de France*, 6. 1900.

BOULAY, N. 'Rubi', in Coste, *Flore de la France*, 2. Paris, 1903.

CHARLET, A., MAGNEL, L. et MARECHAL, A. 'Contributions à l'étude de la dispersion des *Rubus* en Belgique', in *Bulletin de la Société Royale de Botanique de Belgique*, 1918–1920. (Supplementing Sudre's list of 1910.)

FOCKE, W. O. *Synopsis Ruborum Germaniae.* Bremen, 1877.

FOCKE, W. O. 'Rubi', in Koch, *Synopsis der Deutschen und Schweizer Flora*, 2. Leipzig, 1892.

FOCKE, W. O. 'Rubi', in Buchenau, *Flora der nordwestdeutschen Tiefebene.* Leipzig, 1894.

FOCKE, W. O. 'Rubi', in Buchenau, *Kritische Nachtrage zur Flora der nordwestdeutschen Tiefebene.* Leipzig, 1904.

FOCKE, W. O. 'Rubi', in Ascherson und Graebner, *Synopsis der mitteleuropäischen Flora*, 6, i. Leipzig, 1902–3.

FOCKE, W. O. *Rubi Europaei.* 1914.

(The work of 1902–3 is the best, but all are beyond praise.)

FOERSTER, A. *Flora Excursoria Aachen.* 1878. (Important.)

FORD, E. B. *Mendelism and Evolution.* 1940.

FRIDERICHSEN, K. 'Rubi', in Raunkiaer, *Dansk Ekskursions-Flora.* Copenhagen, 1922.

FRITSCH, K. *Exkursionsflora für Österreich.* 1922.

GARCKE, F. A. *Flora von Deutschland*, ed. 22. 1922.

GENEVIER, G. *Essai monographique sur les Rubus du bassin de la Loire.* Angers, 1869.

GENEVIER, G. *Monographie des Rubus du bassin de la Loire.* Paris, 1880.

(Both deal with the Rubi of the Loire basin. Important; but to be used with caution in conjunction with his herbarium.)

GREMLI, A. *Beiträge zur Flora der Schweiz.* 1870. (Important.)

GREMLI, A. *Excursionsflora für die Schweiz.* Aarau, 1881. (Important.)

GUSTAFSSON, A. *Genesis of the European Blackberry Flora*, in *Lunds Universitets Årsskrift*, 39. 1943.

Rubi of Great Britain and Ireland

HESLOP-HARRISON, Y. 'Chromosome numbers in the British *Rubus* flora', in *New Phytologist*. 1953.

HOFMANN, H. In Wünsche und Schorler, *Pflanzen Sachsens*. 1919.

KALTENBACH, J. H. *Flora des Aachener Beckens.*—Nachtrag. Aachen, 1845. (Important.)

KELLER, R. 'Rubi', in *Übersicht über die Schweiz*. 1919. (The descriptions are after Sudre.)

KRAUSE, E. H. L. In Prahl's *Flora Schleswig-Holstein*. 1890.

LANGE, J. *Haandbog den danske Flora*. Copenhagen, 1886–88. (Incorporating Friderichsen and Gelert's work. First-rate.)

LEJEUNE, A. L. S. In Lejeune et Courtois, *Compendium Florae Belgicae*, II. 1831.

LEES, ED. In *Phytologist*, 3 and 4, 1848–53.

LEES, ED. In Steele's *Handbook of Field Botany*. Dublin, 1847.

LIDFORSS, B. For list of papers see Gustafsson, 1943.

MAAS, G. 'Rubi', in Ascherson und Graebner, *Flora Nordostdeutschen Flachlandes*. 1898.

MUELLER, P. J. In *Flora*, 1858; *Pollichia*, 1859 and *Bonplandia*, 1861. (Excellent and important for British Rubi.)

REICHENBACH, L. *Flora Germanica excursoria*. 1830–32.

ROGERS, W. M. 'Essay', in *Journal of Botany*. 1892–93.

ROGERS, W. M. *Handbook of British Rubi*. 1900.
(These pioneer treatises on British Rubi will always be precious to British botanists.)

SCHMEIL, O. und FITSCHEN, J. *Flora von Deutschland*, ed. 26. 1920.

SPRIBILLE, F. R. In Schube's *Flora Schlesien*. 1904.

SUDRE, H. *Rubi Europae*. 1908–13.

SUDRE, H. *Batotheca Europaea*. 1903–17.

SUDRE, H. 'Les *Rubus* de Belgique', in *Bulletin de la Société royale de botanique de Belgique*. 1928–29. (Corrects some errors in British names, creates others; generally repeats Focke's errors.)
(The life-size drawings in the first work are excellent. The last work includes a list, localities and a key.)

VAARAMA, A. Cytological studies on some Finnish species and hybrids of the genus *Rubus* L. in *Journ. Sci. Agr. Soc. Finland*, II. 1939.

VARIOUS CONTRIBUTORS. *Icones Florae Danicae*. 1761.

WATKINS, A. E. *Heredity and Evolution*. 1935.

WATSON, W. C. R. '*Rubus* list', in *Journal of Ecology*, 33. 1946.

WEIHE, K. E. und NEES V. ESENBECH, C. G. *Rubi Germanici*. 1822–27. (Life-size figures. The foundation work for the study of European Rubi.)

INDEX

Page numbers in bold type are those on which the description is given

GROUPS HIGHER THAN SPECIES

(Subgenera are in bold capitals, sections and subsections in capitals, and series and subseries in ordinary type.)

INDEX TO SPECIES

*Names of accepted species, subspecies and varieties are printed in
ordinary type and synonyms in italics*

Index

Index

Printed in the United States
by Bookmasters

Printed in the United States
By Bookmasters